CONTROL THEORY IN BIOMEDICAL ENGINEERING

CONTROL THEORY IN BIOMEDICAL ENGINEERING

Applications in Physiology and Medical Robotics

Edited by

OLFA BOUBAKER
University of Carthage, National Institute of Applied Sciences and Technology, Tunis, Tunisia

Academic Press is an imprint of Elsevier
125 London Wall, London EC2Y 5AS, United Kingdom
525 B Street, Suite 1650, San Diego, CA 92101, United States
50 Hampshire Street, 5th Floor, Cambridge, MA 02139, United States
The Boulevard, Langford Lane, Kidlington, Oxford OX5 1GB, United Kingdom

Notices
Knowledge and best practice in this field are constantly changing. As new research and experience broaden our understanding, changes in research methods, professional practices, or medical treatment may become necessary.

Practitioners and researchers must always rely on their own experience and knowledge in evaluating and using any information, methods, compounds, or experiments described herein. In using such information or methods they should be mindful of their own safety and the safety of others, including parties for whom they have a professional responsibility.

To the fullest extent of the law, neither the Publisher nor the authors, contributors, or editors, assume any liability for any injury and/or damage to persons or property as a matter of products liability, negligence or otherwise, or from any use or operation of any methods, products, instructions, or ideas contained in the material herein.

Library of Congress Cataloging-in-Publication Data
A catalog record for this book is available from the Library of Congress

British Library Cataloguing-in-Publication Data
A catalogue record for this book is available from the British Library

ISBN: 978-0-12-821350-6

For information on all Academic Press publications
visit our website at https://www.elsevier.com/books-and-journals

Publisher: Mara Conner
Acquisitions Editor: Sonnini R. Yura
Editorial Project Manager: Mariana Henriques
Production Project Manager: Kamesh Ramajogi
Cover Designer: Vicky Pearson Esser

Typeset by SPi Global, India

Contents

Contributors

Tanvir Ahmed
Bio-Robotics Lab, University of Wisconsin-Milwaukee, Milwaukee, WI, United States

Honglin An
Department of Biomedical Engineering, Faculty of Engineering, National University of Singapore, Singapore, Singapore

Mohammad Reza Askari
Department of Chemical and Biological Engineering, Illinois Institute of Technology, Chicago, IL, United States

Md. Assad-Uz-Zaman
Bio-Robotics Lab, University of Wisconsin-Milwaukee, Milwaukee, WI, United States

Alireza Bahramian
Biomedical Engineering Department, Amirkabir University of Technology, Tehran, Iran

Joanna Balbus
Department of Pure and Applied Mathematics, Wroclaw University of Science and Technology, Wroclaw, Poland

Mohamed Benrejeb
Laboratory of Research in Automation (LA.R.A), National School of Engineers of Tunis, Tunis El Manar University, Tunis, Tunisia

Olfa Boubaker
University of Carthage, National Institute of Applied Sciences and Technology, Tunis, Tunisia

Brahim Brahmi
Musculoskeletal Biomechanics Research Lab, McGill University, Montreal, QC, Canada

Rachel Brandt
Department of Biomedical Engineering, Illinois Institute of Technology, Chicago, IL, United States

Catherine Jiayi Cai
Department of Biomedical Engineering, Faculty of Engineering, National University of Singapore, Singapore, Singapore

Xinchen Cai
Department of Biomedical Engineering, Faculty of Engineering, National University of Singapore, Singapore, Singapore

Ines Chihi
Laboratory of Energy Applications and Renewable Energy Efficiency (LAPER), Tunis El Manar University, Tunis, Tunisia

Ali Cinar
Department of Chemical and Biological Engineering; Department of Biomedical Engineering, Illinois Institute of Technology, Chicago, IL, United States

Ong Kwok Chin Douglas
Department of Biomedical Engineering, Faculty of Engineering, National University of Singapore, Singapore, Singapore

Iman Hajizadeh
Department of Chemical and Biological Engineering, Illinois Institute of Technology, Chicago, IL, United States

Nicole Hobbs
Department of Biomedical Engineering, Illinois Institute of Technology, Chicago, IL, United States

Juan J. Huaroto
Department of Engineering, Faculty of Science and Philosophy, Universidad Peruana Cayetano Heredia; Department of Mechanical Engineering, Universidad Nacional de Ingenieria, Lima, Peru

Md Rasedul Islam
Richard J. Resch School of Engineering, University of Wisconsin-Green Bay, Green Bay, WI, United States

Sajad Jafari
Biomedical Engineering Department, Amirkabir University of Technology, Tehran, Iran

Ernest N. Kamavuako
Department of Informatics, Center for Robotics Research, King's College London, London, United Kingdom

Raouf Ketata
National Institute of Applied Sciences and Technology (INSAT), Tunis, Tunisia

Krystian Kubica
Faculty of Fundamental Problems of Technology, Department of Biomedical Engineering, Wroclaw University of Science and Technology, Wroclaw, Poland

Seenivasan Lalithkumar
Department of Biomedical Engineering, Faculty of Engineering, National University of Singapore, Singapore, Singapore

Abir Lassoued
University of Carthage, National Institute of Applied Sciences and Technology, Tunis, Tunisia

Hela Lassoued
National Institute of Applied Sciences and Technology (INSAT), Tunis, Tunisia

Chwee Ming Lim
Department of Biomedical Engineering, Faculty of Engineering, National University of Singapore, Singapore, Singapore

Phoebe Lim
Department of Biomedical Engineering, Faculty of Engineering, National University of Singapore, Singapore, Singapore

Mohammad Habibur Rahman
Bio-Robotics Lab, University of Wisconsin-Milwaukee, Milwaukee, WI, United States

Krishna Ramachandra
Department of Biomedical Engineering, Faculty of Engineering, National University of Singapore, Singapore, Singapore

Mudassir Rashid
Department of Chemical and Biological Engineering, Illinois Institute of Technology, Chicago, IL, United States

Hongliang Ren
Department of Biomedical Engineering, Faculty of Engineering, National University of Singapore, Singapore, Singapore

Fathalla A. Rihan
Department of Mathematical Sciences, College of Science, United Arab Emirates University, Al Ain, United Arab Emirates

Nouran F. Rihan
Faculty of Pharmacy, Clinical Program, Cairo University, Cairo, Egypt

Mert Sevil
Department of Biomedical Engineering, Illinois Institute of Technology, Chicago, IL, United States

Etsel Suárez
Department of Engineering, Faculty of Science and Philosophy, Universidad Peruana Cayetano Heredia; Department of Mechanical Engineering, Universidad Nacional de Ingenieria, Lima, Peru

Leoni Goh Yi Ting
Department of Biomedical Engineering, Faculty of Engineering, National University of Singapore, Singapore, Singapore

Farzad Towhidkhah
Biomedical Engineering Department, Amirkabir University of Technology, Tehran, Iran

Zion Tszho Tse
University of Georgia, Athens, GA, United States

Emir A. Vela
Department of Energy and Mechanical Engineering, Universidad de Ingenieria y Tecnologia-UTEC, Barranco, Peru

Preface

As I write the preface for this book, *Control Theory in Biomedical Engineering*, COVID-19 continues to spread, with more than 4 000 000 cases around the world and more than 270 000 deaths as declared the World Health Organization (WHO). In this crucial moment, when developed countries like the United State, Spain, the United Kingdom, Italy, France, Germany and Canada are facing difficult choices in their healthcare systems, regardless of their sophisticated biomedical systems, advanced robotic systems, and innovative technologies in resuscitation services. Developing countries with very modest medical materials but high-level medical skills, like Tunisia, are trying to control the crucial situation relying only on mathematical models of the pandemic. Restrictive confinement measures are being implemented as a way to "control" the spread of the virus. In all ways, "control" remains the target solution.

In China, however, where COVID-19 began as an epidemic before making its way around the world in a matter of months, no new domestic cases were reported. According to recent press reports, several sophisticated medical robots are being deployed there in an effort to combat the spread of the deadly virus. Robotics is being used to sanitize hospitals, some of which use ultraviolet light to clean, to reduce workers' exposure to the virus as much as possible. Besides disinfection and street patrols, the robots are also deployed to deliver food and support nurses in communication with patients in quarantine to reduce human-to-human contact. Dancing robots lead patients in exercises and entertain bored quarantined patients. Robots are used to patrol public spaces to identify people who may be running a fever. Robots completely replaced humans in a coronavirus care hospital in Wuhan, China, where a humanoid robot worked 24/7 measuring heart rates and blood oxygen levels via smart bracelets and rings worn by patients.

Undoubtedly, mathematical modeling, control theory, and medical robotics are fundamental sciences for healthcare systems. Modeling-based control is of huge importance as it can be used to understand feedback paths in physiological systems, establish medical diagnoses, understand the interrelationship among physiological variables in the human body, and predict the dynamic behavior of some diseases and epidemics. They can contribute to regulate human physiology via commercial artificial organs and assistive technologies. Furthermore, control theory is probably of great significance in the pervasiveness of medical robots in surgery, exploration, diagnosis, and therapy. Control theory is absolutely at the heart of medical sciences and the future of medicine.

On one hand, the human body has a natural and autonomous control process responsible for maintaining human life called homoeostasis. Homeostasis is the property of a physiological system to regulate its internal environment to a given set point in the presence of specific stimulus-producing changes in that variable. The control process in the human body is ensured through the coordination of the control center and the natural sensors and effectors. The roster of vital parameters concerned by homeostasis is long and includes blood pressure, blood oxygen, heart rate, blood calcium, blood glucose, and so on. In this framework, there are several well-known examples of mathematical model-based control. The most eminent ones are the endocrine system, the immune system, the neurological system, the cardiac and pulmonary system, and the locomotor system. These models are widely used in the medical field and have shown great advances in prediction, diagnosis, and therapy. However, further progress is still expected for several domains, especially to prevent, detect, and control several chronic diseases (e.g., chronic respiratory diseases, cardiovascular diseases, diabetes mellitus, thyroid disease, cancers, HIV/AIDS, etc.) and to predict the evolution of and control devastating epidemics (severe acute respiratory syndrome (SARS), yellow fever, cholera, Ebola, hepatitis B, H1N1/09 virus, COVID-19, etc.).

On the other hand, control theory deeply impacts the everyday life of a large part of the human population like disabled and elderly people using assistive and rehabilitation robots for improving their quality of life and increasing their personal independence. For this large community, the body process requires external control laws in order to regulate its natural functions via artificial organs, including artificial arms, legs, pancreas, heart, lungs, liver, and so on, and assistive technologies, including wheelchairs, prostheses, walkers, exoskeletons, and companion robots.

Medical robotics has undergone a profound revolution in the past three decades. Nowadays, they are used to perform delicate surgical procedures in specific disciplines including orthopedics, urology, cardiac surgery, neurosurgery, ophthalmology, pediatric surgery, general surgery, and so on. Robots help in drug administration and disinfection duties and perform a growing number of other health tasks. Compared to manual machines in healthcare, medical robotic systems offer a wide range of advantages. They can eliminate human fatigue and contamination risk as well as improve dexterity, precision, and many other capabilities of doctors and surgeons. They reduce patient trauma, offer faster recovery, and limit scarring. The most widely used surgical robot worldwide is the da Vinci surgical system, which is used in 67 countries and used to perform more than 7.2 million surgical procedures.

This book is a timely and comprehensive guide for graduate students and researchers in both control engineering and biomedical engineering fields. It is also useful for medical students and practitioners who want to enhance their understanding of physiological processes and medical robotics. Written by eminent scientists in the field, this book is a unique reference illustrating the many facets of control theory in biomedical engineering through concrete examples.

It contains 12 chapters organized into two parts. Part I introduces applications of control theory in physiology, and Part II describes the related applications in medical robotics.

Chapter 1 presents key definitions and principles of modeling and control physiological systems. A comprehensive literature survey is compiled to provide an overall picture of this application area and reveals its impact in our human life. Challenges and future trends are also presented.

In Chapter 2, a simplified mathematical model of cholesterol homeostasis is introduced to describe the rate of changes in the cholesterol level in the blood and liver. The proposed model helps to understand causes of cholesterol homeostasis disorders, the different ways to lower its concentration, and possible investigations in its therapy.

Chapter 3 presents an artificial pancreas system for the treatment of type 1 diabetes. The proposed system is a fully automated insulin delivery system working with no meal and physical activity announcements. It uses an adaptive-learning model with a predictive control algorithm designed to compute a safe and optimal insulin amount.

Chapter 4 presents a mathematical model of tumor-immune interactions under chemotherapy treatment. The model is described by a system with delay differential equations and governed by an optimal control law. It shows the impact of the optimal treatment after few days of therapy.

In Chapter 5, a cardiac arrhythmia classification method based on the fuzzy logic controller and the genetic algorithm is introduced. The proposed approach is used to classify the MIT-BIH arrhythmia database recordings into five cardiac cases.

Chapter 6 reviews the handwriting models proposed in literature. The relationship between hand or/and forearm electromyography (EMG) signals and writing parameters is described. Advantages and drawbacks are summarized.

Chapter 7 highlights the impact of medical robotics to improve human abilities. Different classification approaches of these robots, including surgical robots, rehabilitation, and assistive devices, is presented via a historical viewpoint and a comprehensive survey of the literature.

In Chapter 8, the state of the art of wearable devices for upper limb amputees is presented. New directions are also exposed to go further in continuous development.

In Chapter 9, a review of existing upper limb exoskeletons was done to find key challenges that need to be solved for improved functionality.

Chapter 10 focuses on the human locomotion system and its complex dynamics. Double-pendulum based models are exploited to design controllers for walking on a treadmill. The step length, time, velocity, and position on the treadmill are considered as walking parameters. For the designed controllers, time series of those parameters are compared to real experimental data of human walking.

In Chapter 11, an inexpensive endoscope design for nasopharyngeal cancer investigation is introduced. This cancer is the most widespread one in Asia. The prototype is a small portable device with the main aim of being affordable and home-based. A comparison between the current nasopharyngoscope and the proposed prototype is evaluated via metrics and needs. A list of acceptance criteria is also specified to ensure safety regulations and effectiveness.

Chapter 12 proposes a new design of an emerging class of "invertebrate-like" continuum robots for minimal invasive surgery. It is a combined actuation and tunable stiffness mechanism obtained by using 3D-printed auxetic structures as well as origami and kirigami structures. It uses simple and cost-effective structures that can also be highly miniaturized.

In conclusion, control is a crucial theory in medicine and this book treats all the previous themes together in a very educational way. I owe an enormous debt of gratitude to all those who helped me through the years that have led up to this achievement. I want to thank EVERYONE who ever said anything positive to me or taught me something. I heard it all, and it meant something.

To Professor *Mekki Ksouri* from the National Engineering School of Tunis and Private Higher School of Engineering and Technology, Tunisia, thank you for being a leader in whom I have confidence, honor, and respect. I will always be happy to have been your PhD student, part of your research team, and having had the chance to represent you. Your training and guidance in control theory were of undeniable value.

Thank you to everyone who strives to grow and help others grow with generosity. All my thoughts go here to Professor *Hichem Kallel* from the National Institute of Applied Sciences and Technology, Tunisia, who initiated my training in robotics and helped me to progress in this field. To Professor *Quanmin Zhu* from the University of the West of England, Bristol, UK, thank you for the undeniable encouragement and generosity, which were essential for my last five books. I greatly appreciate your guidance, support, and friendship, and perhaps most of all, your examples.

The following researchers are particularly acknowledged for the honor they bestowed upon me by contributing artwork and for their professional efforts to provide comments and reviews: Professor *Hongliang Ren* from the National University of Singapore, Singapore; Emeritus Professor *Mohamed Benrejeb* from the National Engineering School of Tunis, Tunisia; and Professor *Ali Cinar* from the Illinois Institute of Technology, USA.

I am also immensely grateful to everyone on the Elsevier editorial team who helped me so much. Special thanks to the ever patient Acquisitions Editor *Sonnini Yura*, from Brazil, who was behind the idea of editing this book, *Mariana Henriques*, from Brazil, the amazing Editorial Project Manager for her help, guidance and for being a great cover designer, and Kamesh Ramajogi, the production project manager from India, for his patience and skills.

Thanks to my former doctoral student, Dr. *Abir Lassoued*, Assistant Professor at the National School of Engineers of Tunis, Tunisia, for all didactic efforts to design many diagrams and figures joining control theory to physiology.

Finally, thanks to my daughter, *Sarra Boubaker*, External Doctor at the Faculty of Medicine of Sousse, Tunisia, for her continuous encouragement, but mostly for fruitful discussions and comments on the accuracy and proof of medical information.

Olfa Boubaker
University of Carthage, National Institute of Applied Sciences and Technology, Tunis, Tunisia

PART I

Applications in physiology

CHAPTER 1

Modeling and control in physiology

Abir Lassoued, Olfa Boubaker
University of Carthage, National Institute of Applied Sciences and Technology, Tunis, Tunisia

1 Introduction

According to World Health Organization (WHO) statistics for 2019, the major causes of death in the world are cardiovascular diseases, cancer and infectious diseases (World Health Organization, 2019). To have a better understanding of these physiological processes, it is imperative to describe these complex systems by mathematical models. Indeed, many challenging problems are still pending for which these models can be exploited in several contexts, namely to help doctors to establish a diagnosis, to understand the interrelationship among physiological variables and to predict the dynamic behavior of some diseases.

Modeling physiological systems has a long history interconnected with the history of medicine. In this field, the research work of Denis Noble in 1960 on cardiac cell mode (Noble, 1960) can be considered as the cornerstone for the development of the current research works related to modeling in biomedical engineering. Various research works related to mathematical modeling in physiology have been proposed since then, leading to an extensive bibliography revealing the prosperity of this field. These include a handbook (Bronzino, 1999), several valuable books (Ottesen and Danielsen, 2000; Beuter et al., 2003; Ottesen et al., 2004; Takeuchi et al., 2007; Cherruault, 2012; Banks, 2013; Hacısalihzade, 2013; Carson and Cobelli, 2014; Reisman et al., 2018; Cobelli and Carson, 2008; Devasahayam, 2012; Sanft and Walter, 2020), a number of survey papers (Brown, 1980; Bekey and Beneken, 1978; Perelson, 2002; Makroglou et al., 2006; Boutayeb and Chetouani, 2006; Anderson and Quaranta, 2008; Leng and MacGregor, 2008; Bertram, 2011; Glynn et al., 2014; Chowell et al., 2016; Zavala et al., 2019), and software tools (Hester et al., 2011). Mathematical models in biology (Roberts, 1976; Brown and Rothery, 1993; Allman and Rhodes, 2004; Edelstein-Keshet, 2005)

Control Theory in Biomedical Engineering
https://doi.org/10.1016/B978-0-12-821350-6.00001-9

can be categorized into several classes: linear and nonlinear models, continuous or discrete models, deterministic or stochastic models, parametric or nonparametric models, and lumped or distributed parameter models.

The human body contains a natural and autonomous control process that can maintain human life (Cherruault, 2012; Banks, 2013; Hacısalihzade, 2013). However, in many cases, certain failures in the body process require external control laws in order to regulate its natural behavior (Morari and Gentilini, 2001; Iii et al., 2011). The external controllers can be based on different control algorithms such as predictive, adaptive and optimal control laws principles (Swan, 1981; Hajizadeh et al., 2018; Turksoy and Cinar, 2014). These different types of controllers can be also applied in commercial artificial organs and assistive technologies in order to simulate natural human functions (Turksoy and Cinar, 2014; Smith et al., 2018).

This chapter provides comprehensive information on mathematical modeling and control-based modeling in physiology in order to explain the complexity of the human body process. In addition, the chapter examines challenges in this wealth domain.

The chapter is structured as follows. Section 2 presents a comprehensive review of mathematical modeling in physiology. Section 3 investigates various control principles by pointing out their applications in physiology. Finally, Section 4 provides the possible challenges for modeling and controlling systems in medicine. It also describes commercial artificial organs and assistive technologies that guarantee easier life for patients.

2 Mathematical modeling in physiology

A physiological model is a mathematical representation that approximates the behavior of an actual physiological system. Physiological models can serve mainly for the following purposes: (1) *to understand* the physiological system, (2) *to predict* its dynamics, and (3) *to control* the system under desirable conditions. However, the main problem in physiological systems is to find the appropriate mathematical models to describe real problems, especially disease dynamics. In fact, mathematical modeling represents a valuable tool for understanding the human body processes.

2.1 Modeling methodology

Fig. 1 describes the modeling process generally adopted for designing mathematical models for physiological systems (Kuttler, 2009). There are various modeling approaches and their number is still increasing. Some modeling

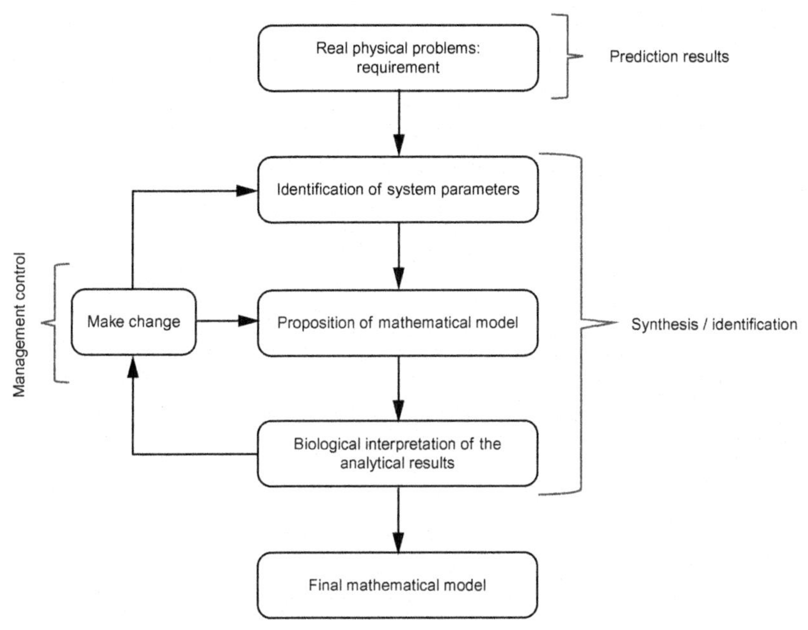

Fig. 1 The modeling process for real physiological problems. *(The figure has been adopted from Kuttler, C., 2009. Mathematical Models in Biology.)*

methods were based on linear systems due to their simplicity analysis. However, nonlinearities are ubiquitous in physiology. Peculiar critical aspects of physiology should be described and analyzed by nonlinear systems. The nonlinear dynamic of physiological systems is a daunting challenge owing to its considerable complexity. In the following, we present the main modeling approaches and some classes of mathematical models in physiology. We also include some examples of applications of mathematical models in physiology.

2.2 Modeling approaches

There are different modeling approaches in the literature for physiological systems (Ottesen and Danielsen, 2000; Ottesen et al., 2004; Takeuchi et al., 2007; Cobelli and Carson, 2008; Carson and Cobelli, 2014). Of them, the compartmental approach is the most familiar. In this subsection, we describe the most recognizable approaches.

2.2.1 Compartmental modeling approach

The compartmental modeling approach is the most familiar and one of the oldest approaches used to describe interactions between physiological variables. For more details about this approach, see the survey paper by Brown (1980), *Compartmental Modeling and Tracer Kinetics* (Anderson, 2013), *Compartmental Analysis in Biology and Medicine* (Jacquez, 1972), and chapter 7 in *Introduction to Biomedical Engineering* (Enderle and Bronzino, 2012). In this framework, several applications in different areas of physiology were represented by the compartmental approach, such as in diabetes dynamics (Chiarella and Shannon, 1986), the respiratory system (Similowski and Bates, 1991), tumor resistance to chemotherapy (Alvarez-Arenas et al., 2019), metabolic systems (Cobelli and Foster, 1998; Staub et al., 2003), pharmacokinetics (Garcia-Sevilla et al., 2012b), and so on. A software tool allowing implementation of linear compartmental models is also available and is described in Garcia-Sevilla et al. (2012a).

Compartmental modeling is very attractive to users because it formalizes physical intuition in a simple and reasonable way. According to this method, the governing law is conservation of mass. Compartmental models are lumped parameter models, in that the events in the system are described by a finite number of changing variables (Cobelli and Foster, 1998). Each compartment characterizes both the physical-chemical proprieties and its environment and the corresponding mathematical model is a collection of ordinary differential equations (ODEs). Each equation defines the time rate of change of amount material in a particular compartment. Thus, the basic equations of compartment model with n compartments are defined as (Brown, 1980):

$$\frac{dx_i}{dt} = f_{i0} + \sum_{\substack{i=1 \\ j\neq i}}^{n} \left(f_{ij} - f_{ji}\right) - f_{0i}\,; x_i(0) = x_{0i}\,; i = 1, 2, \ldots, n.$$

where x_i is the amount of material in compartment i, x_{i0} is its corresponding initial value and f_{ji} is the mass flow rate of compartment j from compartment i. Fig. 2 illustrates the compartment structure. The index 0 denotes the environment of the physiological system.

2.2.2 Equivalent modeling approach

The main goal of the equivalent approach is to describe physiological systems by using equivalent physical systems such as electrical or mechanical systems. The equivalent models are used in physiology to simplify their dynamic analysis. Each variable in a physical system has its corresponding variable in an analog physical system. In the literature, some basic

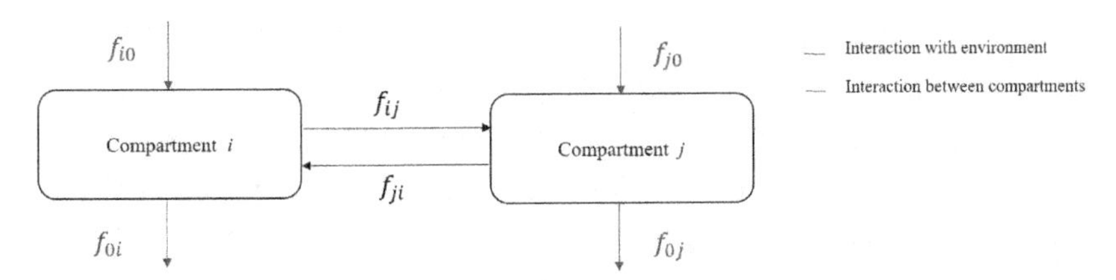

Fig. 2 Basic compartment structure. *(The figure has been modified from Brown, R.F., 1980. Compartmental system analysis: state of the art. IEEE Trans. Biomed. Eng.)*

physiological measurements have their analog physical variables (Aidley, 1998). Table 1 describes some of these.

In neurophysiology and cell physiology the use of equivalent circuits is well known, where the classic examples are circuits simulating the generation of membrane potential (Aidley, 1998), the cable properties of nerve and muscle fibers, and action potential and neuromuscular transmission. Using analogue circuits to model and simulate physiological systems is also considered as a modern approach for teaching physiology to postgraduate medical students. The aim is to promote qualitative as well as quantitative analogue thinking about physiological processes (Rupnik et al., 2001; Ribarič and Kordaš, 2011; Sever et al., 2014).

In the last decades, particular physiological systems were modeled by equivalent electronic circuits such a pulmonary ventilation (Ghafarian et al., 2016) and blood circulation (Ismail et al., 2018). In Ismail et al. (2018), for example, to understand the cardiovascular system, the thermodynamics of vessels was represented by an electronic circuit as described in Fig. 3. In this circuit, the voltage represents the blood pressure when the current represents the blood flow. C and R are the compliance and the resistance of the systemic arterial tree, respectively. L is the impedance of proximal aorta.

For the use of mechanical equivalent models, several systems are also proposed. For example, the reader can found a pneumatic model given for the respiratory system (Shi et al., 2016) and a mass-spring-damper inverted pendulum model for a bipedal-compliant walking system (Joe and Oh, 2019). The related model is given in Fig. 4.

2.2.3 Data-driven modeling approach

Data-driven modeling (DDM) approach is an empirical approach that does not involve mathematical equations derived from physical processes but instead involves analysis of time series data. Examples include linear

Table 1 Physical, mechanical and electrical analogues.

	Mechanical analogues			Electrical analogues		
Physiological measurements	**Name**	**Notation**	**Symbol**	**Name**	**Notation**	**Symbol**
Pressure	Force	F	–	Voltage	V	–
Volume	Displacement	x	–	Charge	q	–
Flow	Velocity	$v=\frac{dx}{dt}$	–	Current	$I=\frac{dq}{dt}$	–
Viscous drag	Viscous resistance	$B=\frac{F}{v}$		Resistance	$R=\frac{V}{I}$	
Compliance	Compliance	$C=\frac{x}{F}$		Capacitance	$C=\frac{q}{V}$	

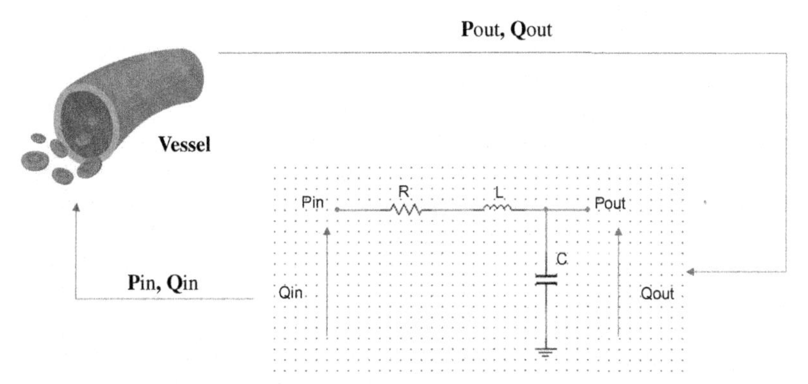

Fig. 3 Equivalent electronic circuit of blood vessels system.

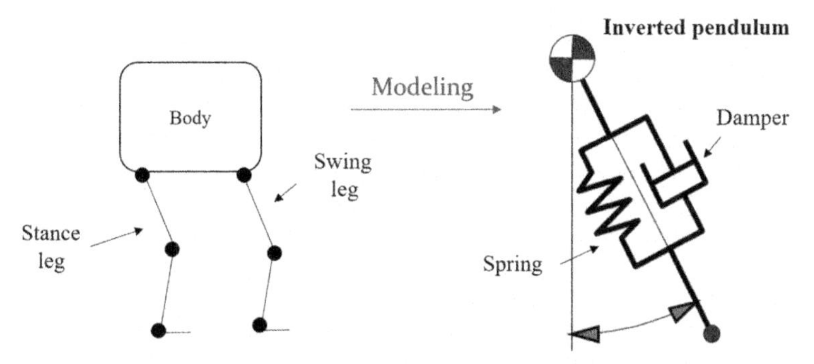

Fig. 4 Equivalent mechanical model of the human bipedal walking process. *(The figure has been adopted from Joe, H.M., Oh, J.H., 2019. A robust balance-control framework for the terrain-blind bipedal walking of a humanoid robot on unknown and uneven terrain. Sensors 19 (19), 4194.)*

regression models and popular biologically inspired computational models including neural networks and genetic algorithms. DDM is based on analyzing the data about input, internal and output variables of a system without explicit knowledge of the physical behavior as described by Fig. 5. The availability of sufficient patient historical data has paved the way for the introduction of machine learning and its application for intelligent and improved systems for diabetes management (Dutta et al., 2018; Paoletti et al., 2019), immunology (Fong et al., 2018), brain diseases (Karim et al. 2018; Oxtoby et al., 2018), and so on.

2.3 Classification of mathematical models

Physiological systems are dynamical systems and then generally described mathematically by differential equations. A model in physiology is usually

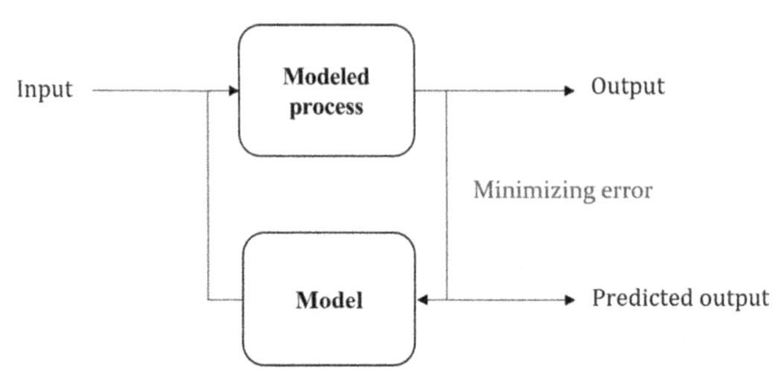

Fig. 5 Data-driven modeling approach principle. *(The figure has been adopted from Solomatine, D., See, L.M., Abrahart, R.J., 2008. Data-driven modeling: concepts, approaches and experiences. In: Practical Hydroinformatics. Springer Berlin Heidelberg, Berlin, Heidelberg, pp. 17–30.)*

constructed using a basic and natural law intimately tied to many other interdisciplinary areas, such as biophysics, and biochemistry, and involves electrical and mechanical analogues. Different classes of models exist and can be described as follows: lumped models vs. distributed parameter models, continuous-time vs. discrete-rime models, deterministic vs. stochastic models, parametric vs. nonparametric models and so on. Lumped models are generally described by ODEs, whereas distributed parameter systems are described by partial differential equations (PDEs). For example, readers can find different models of diabetes and blood flow dynamics in the following survey papers (Makroglou et al., 2006; Cobelli et al., 2009; Kokalari et al., 2013) including lumped and distributed parameter models. A review on recent progress on modeling and characterizing early epidemic growth patterns from infectious disease including the simplest model SIR is found in Chowell et al. (2016).

Discrete models are characterized by a countable number of states where both input and output signals are discrete signals. These types of systems are easily implementable in physiological applications since experimental real results are already discrete. In literature, several physiological systems are described by discrete models such as ECG signals (Huang et al., 2019). Discrete models such as logical, finite state machine or Boolean networks have also an important impact in modeling process for physiological systems.

A stochastic model represents a set of random variables that can be continuous or discrete (Kulkarni, 2016). In fact, a stochastic measure contains uncertainty. Thus, the results of stochastic systems are impossible to predict even when the dynamics and the initial states are given. These types of models are very useful in physiological systems, where stochasticity and random variables have an important impact in real applications. In recent research, environmental and demographic stochasticity are the two main types of stochasticity. Several physiological systems in literature are described by stochastic models such as the plasma membrane system (Sato et al., 2019) and gastric emptying system (Yokrattanasak et al., 2016).

A parametric model is a finite-dimensional model of statistical models. Specifically, a parametric model is a family of probability distributions that has a finite number of parameters. A model is "nonparametric" if all the parameters are in infinite-dimensional parameter spaces (Bickel and Doksum, 2001). The reader can find many examples of physiological systems written in parametric and nonparametric models in Marmarelis (2004).

2.4 Structural identifiability

Mathematical models in physiology are generally described by a set of consistent differential equations and a set of physiological parameters to be estimated in an accurate way. However, systems in physiology are naturally characterized by poor observability. Observability is a modeling property that describes the possibility of inferring the internal state of a system from observations of its output. Indeed, the possibility for the clinician to observe practically and quantify the relevant phenomena occurring in the body through clinical tests is very limited due to complexity of dynamics of such systems due to the high number of interacting and unmeasured variables. Systems in physiology are also characterized by poor controllability due to the limited capacity of such systems to drive the state of the system by acting on some control variables. All these factors may severely hinder the practical identifiability of these models, i.e., the possibility to estimate the set of parameters from clinical data. Identifiability is a structural property of a model introduced in Bellman and Åström (1970) that defines the amount of useful information that can be generated from the output (clinical data for physiological systems). Hence, the importance of designing clinical protocols that allow for estimating the model parameters in the quickest and more reliable way. In fact, structural identifiability becomes a particular case of observability if the parameters are considered as constant state variables. Identifiability has been largely studied by researches and the reader can find several survey papers in this framework (Pia Saccomani et al., 2003; Boubaker and Fourati, 2004; Chis et al., 2011; Kabanikhin et al., 2016; Villaverde and Barreiro, 2016; Bezzo and Galvanin, 2018; Villaverde, 2019) where many aspects of identifiability are studied, like the role of initial conditions (Pia Saccomani et al., 2003) and optimal design of clinical tests (Bezzo and Galvanin, 2018). Different application papers of structural identifiability have been published (Xia and Moog, 2003; Raue et al., 2009; Miao et al., 2011; Eberle and Ament, 2012; Tuncer et al., 2016; Pironet et al., 2019). A software tool to test global identifiability of biological and physiological system is described in Bellu et al. (2007).

2.5 Practical identifiability

Practical identifiability concerns parameter and state estimation processes via online or offline techniques. Since the 1970s, great efforts have been made to describe physiological systems in explicit mathematical models for which several online or offline techniques for parameter estimation are developed and the entire procedure of estimation, from model formulation to computer selection, is re-examined (Rideout and Beneken, 1975). Nowadays, parameter estimation techniques are of ever-increasing interest in the fields of medicine and biology for which few books and book chapters (Marmarelis and Marmarelis, 1978; Khoo, 1999; Westwick and Kearney, 2003; Heldt et al., 2013; Ho, 2019), survey papers (De Nicolao et al., 1997; Giannakis and Serpedin, 2001), and application papers (Tong, 1976; Misgeld et al., 2016) are available. A software package is also available that solves structural/practical identifiability problems (Galvanin et al., 2013), as described in Fig. 6. This framework first conducts a thorough analysis to identify and classify the nonidentifiable parameters and provides a guideline for solving them. If no feasible solution can be found, the framework instead initializes the filtering technique prior to yield a unique solution.

2.6 Application examples

There are several well-known examples of mathematical models in physiological systems, such as the endocrine system, immune system and cardiovascular system. In this subsection, we describe the simplest models of these renowned systems and give an extensive bibliography.

2.6.1 The endocrine system models

The endocrine system is the set of glands in the body that produce hormones directly into the circulatory system in order to regulate physiological and behavioral activities (Neave, 2008). Hormones are used to communicate between organs and tissues. The endocrine process is primordial for the

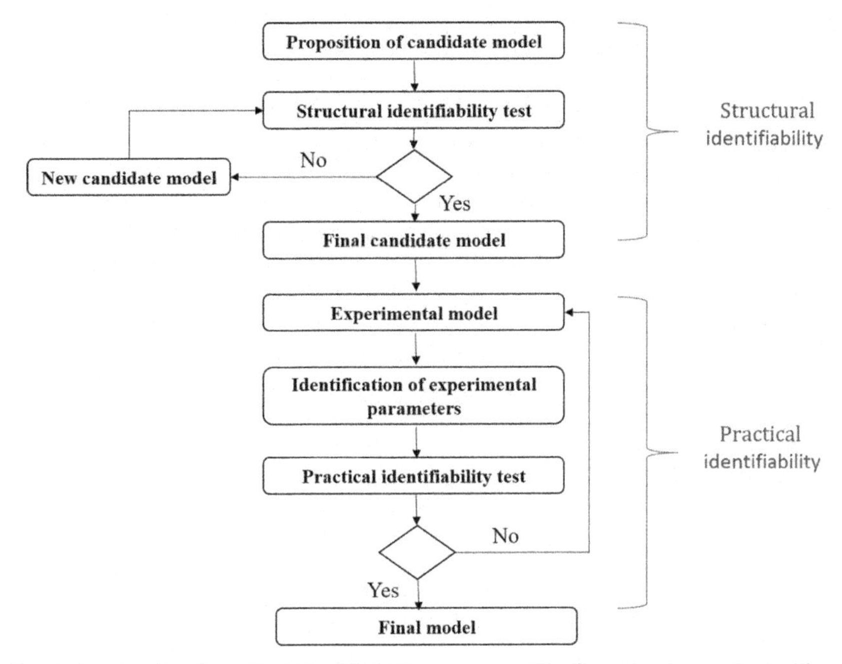

Fig. 6 Structural and practical identifiability approach. *(The figure has been adopted from Galvanin, F., et al., 2013. A general model-based design of experiments approach to achieve practical identifiability of pharmacokinetic and pharmacodynamic models. J. Pharmacokinet. Pharmacodyn. 40 (4), 451–467.)*

healthy functioning of the body. In fact, hormone hyperfunction and hormone hypofunction have different impacts on many diseases. Many glands exist within the human body. We provide only three examples and the related expanded mathematical models in the literature in Table 2.

The most predominant endocrine disease in the world is diabetes. In the literature, several mathematical models have been constructed in order to analyze the glucose–insulin regulation system (Kansal, 2004). The basic model, called the minimal model, is characterized by minimal number of parameters (Bergman et al., 1979). This system is described by the following equation:

$$\frac{dG}{dt} = aG - b,$$

where G is the glucose concentration and (a, b) are the system parameters.

2.6.2 The tumor-immune system model

The immune system is an organization of cells and molecules. When an organism or tumor cell appears in the body, the principal role of the immune

Table 2 Mathematical models of some endocrine systems.

Gland	Related mathematical model	Reference
Pancreas	Glucose-insulin regulation system	Bergman et al. (1979), Mari (2002, Kansal (2004), Boutayeb and Chetouani (2006), Makroglou et al. (2006), Gani et al. (2009), Balakrishnan et al. (2011), Ajmera et al. (2013), Palumbo et al. (2013)
Hypothalamo-pituitary-gonadal (HPG) axis	Reproductive function	Clément (2016)
Thyroid	Thyroid disease	Liu et al. (1994)

system is to identify and then to eliminate all foreign bodies (Gałach, 2003). In fact, the immune response begins when tumor cells are detected as undesirable cells in the body. Then, the lymphocyte cells are stimulated in order to coordinate the counterattack. In the literature, the interaction between tumor cells and lymphocyte cells in the immune system looks likes the interaction between species in the predator-prey model (Bell, 1973; Ben Saad et al., 2019). One of the most basic mathematical models describing tumor-immune system interactions is the Bell model proposed in Bell (1973). This model is based on the classic Lotka-Volterra model (Volterra, 1928). Other simple models can be found in research papers (Mayer et al., 1995; Sachs et al., 2001) and in survey papers (Adam and Bellomo, 2012; Weerasinghe et al., 2019). Thus, the immune response can be described by the following simple mathematical model (Sotolongo-Costa et al., 2003):

$$\begin{cases} \dfrac{dX}{dt} = aX - bXY, \\ \dfrac{dY}{dt} = dXY - fY - kX + u. \end{cases}$$

where X and Y are the number of tumor and lymphocyte cells, respectively, and (a, b, d, f, k, u) are system parameters.

2.6.3 The cardiovascular system

The cardiovascular system is composed of the heart and vessels. Its main function is to pump the blood in the body in order to supply all tissues and organs with oxygen and other nutrients (Quarteroni et al., 2009). In fact, cardiovascular disease is one of the major causes of death worldwide (World Health Organization, 2019). Thus, to prevent disease, it is primordial to model cardiovascular systems. Mathematical modeling of the cardiovascular system has a long history (Quarteroni, 2001). The first mathematical model of the cardiovascular system was proposed in Grodins (1959). This system is described by the following equation:

$$RC\frac{dV}{dt} + V = CP,$$

where V is the diastolic volume, P is the venous filling pressure, C is the compliance of the relaxed ventricle and R is the total viscosity.

In literature, the cardiovascular system is considered a complex and critical system. Thus, many researchers have approached the modeling process of this system via different viewpoints (Leaning et al., 1983; Shim et al., 2004; Liang and Liu, 2005; Abdolrazaghi et al., 2010; Shi et al., 2011; Ambrosi et al., 2012; Bessonov et al., 2016; Quarteroni et al., 2017; Bora et al., 2019). A computational environment for human cardiovascular system modeling and simulation is described in Larrabide et al. (2012).

2.7 Chaos in physiology

As reported by Rössler, physiology is the mother of chaos (Rossler and Rossler, 1994). Several researches demonstrate that chaos is a common feature in complex physiological systems (Mackey and Glass, 1977; Glass et al., 1988; Goldberger et al., 1990; West and Zweifel, 1992; Elbert et al., 1994; Wagner, 1996; Glass, 2001; Aon et al., 2011; Li, 2015; Nazarimehr et al., 2017). A chaotic system is defined as a nonlinear system with unpredictable dynamic behaviors and extreme sensitivity to initial conditions (Lassoued and Boubaker, 2016; Boubaker and Jafari, 2018). In physiology, chaos can be a sign of health or disease (Goldberger and West, 1987). In fact, some healthy functions of physiological systems are characterized by chaotic dynamic behaviors, namely the human brain behaviors (Baxt, 1994). However, other physiological systems are characterized by ordered and regular dynamics. In this case, chaotic dynamic behaviors prove the existence of peculiar pathologies, namely those for cardiovascular measures (blood pressure, heart rate, etc.) (Wagner, 1998). Several physiological systems like the cardiovascular system are capable of five kinds of behavior: (1) equilibrium (fixed point), (2) periodicity (limit cycle), (3) quasi-periodicity (limit torus), (4) deterministic chaos (strange attractor) and (5) random behavior (no attractor) as described in Fig. 7. Systems adopt one

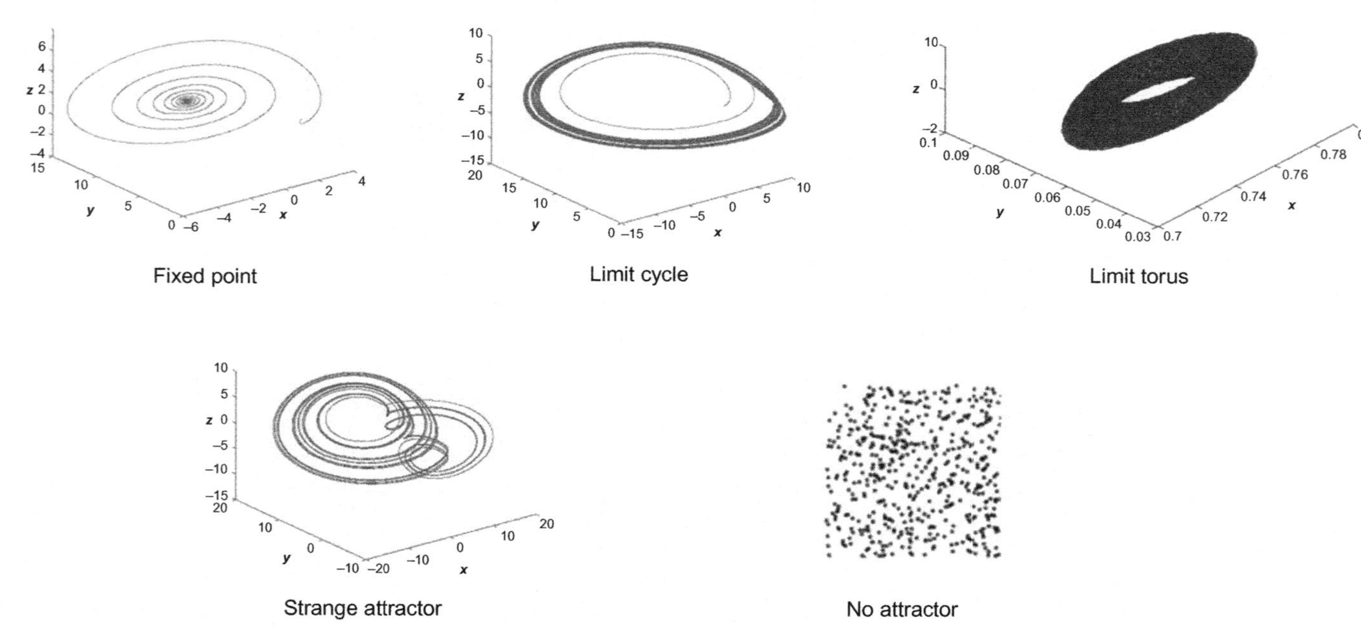

Fig. 7 Five possible dynamic behaviors exhibit by physiological systems.

or more these behaviors depending on the function they have evolved to perform (Sharma, 2009).

Recently, several research works have attempted to model and to analyze chaotic physiological systems. The glucose-insulin regulatory system is the most popular of these dynamics (Ginoux et al., 2018; Shabestari et al., 2018; Rajagopal et al., 2019). In (Rajagopal et al., 2019), for example, the corresponding mathematical model is described by three differential equations where the related state variables are insulin concentration, blood glucose concentration and the population density of β-cells. Chaotic dynamic behaviors for the late system are observed when various anomalies are detected in the glucose-insulin regulatory system such as hypoglycemia, hyperinsulinemia and when the body cells resist accepting insulin (type 2 diabetes). Fig. 8 describes some of these particular chaotic attractors.

Moreover, many researchers claim the importance of chaotic behaviors in the human brain (Freeman, 1992; Schiff et al., 1994; Sarbadhikari and Chakrabarty, 2001; Aram et al., 2017; Rostami et al., 2019) especially in migraine headache (Bayani et al., 2018), attention deficit disorder (Baghdadi et al., 2015) and epilepsy (Panahi et al., 2017, 2019). The mathematical model used to analyze epileptic seizures confirms that the normal behavior of the human brain is chaotic. However, the abnormal epileptic behavior of the human brain is periodic. On the other hand, the mathematical model of this system is a nonlinear neural network representing different interconnected parts in the brain (Schiff et al., 1994).

One of the most interesting physiologic problems is cancer. Indeed, cancer is considered as the major cause of death worldwide (World Health Organization, 2019). Cancer can affect different physiological systems.

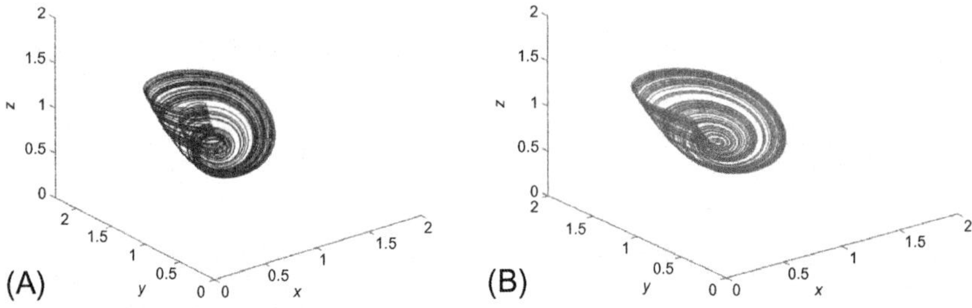

Fig. 8 Projections of the chaotic attractor related to the glucose-insulin regulatory system onto the spaces (x, y, z): (A) Hypoglycemia, (B) Hyperinsulinemia, where $x(t)$ is the insulin concentration, $y(t)$ is the blood glucose concentration and $z(t)$ is the population density of β-cells.

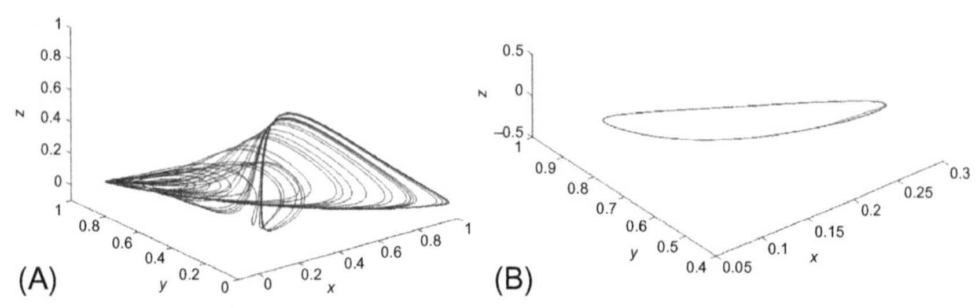

Fig. 9 Projections of the attractor related to the chaotic-cancer system onto the spaces (x, y, z): (A) chaotic attractor, (B) periodic attractor, where $x(t)$ is the logistic growth of cancer cells population, $y(t)$ is the competition between tumor cells and healthy and $z(t)$ is the effector cells.

Thus, the complexity and diversity of cancer diseases have attracted the interest of the scientific community in order to analyze and understand the dynamic behaviors of cancer cells. Therefore, many mathematical models were proposed to define tumor evolution with ordinary differential equations. Some models exhibit chaotic attractors and regular attractors (fixed point, limit cycle and periodic orbits) (Valle et al., 2018). One of these models is the three-dimensional chaotic-cancer system proposed by Itik and Banks (2010). The three state variables are the logistic growth of cancer cells population, the competition between tumor cells and healthy cells, and the effector cells. Both periodic and chaotic behaviors are dependent on the strength of the immune response against cancer and describe the evolution of tumor cells as shown in Fig. 9. A patient is healthy when the effector cells are equal to zero, or more precisely when the chaotic-cancer system converges to an equilibrium point.

3 Control in physiology

The human body contains a natural and autonomous control process that can maintain human life. This natural control process is designed by homoeostasis. In this section, we present the principals and different examples of human homeostasis.

3.1 The homeostasis principal

In physiology, control refers to the process of stabilizing a physiological variable to a specified set point, either by reversing perturbations via negative feedback closed loops or via anticipatory open loops.

In the human body, the control process is designed by homeostasis. The term homeostasis was invented by the celebrated French physiologist Walter

Bradford Cannon (Cannon, 1929). Cannon combined two words from Ancient Greek ὅμος (hómos, "similar") + ιστημι (histēmi, "standing still")/*stasis* (from στάσις) into a Modern Latin form (Davies, 2016). Cannon wrote, "The constant conditions which are maintained in the body might be termed equilibria. That word, however, has come to have exact meaning as applied to relatively simple physico-chemical states, in closed systems, where known forces are balanced. The coordinated physiological processes which maintain most of the steady states in the organism are so complex and so peculiar to living beings—involving, as they may, the brain and nerves, the heart, lungs, kidneys and spleen, all working cooperatively—that I have suggested a special designation for these states, homeostasis" (Davies, 2016).

The homeostasis principle is then the property of a physiological system to regulate its internal environment to a given set point in presence of a specific stimulus producing changes in that variable. The objective is to maintain stable and relatively constant physiological behavior. The control process in the human body is ensured through the coordination of the control center and the natural sensors and effectors as shown in Fig. 10. The control center is composed of the nervous system and the endocrine system.

3.2 Homeostasis examples

Fig. 11 introduces several examples of homeostasis, including energy and fluid balances. A review on physiological energy homeostasis can be found in Chapelot and Charlot (2019).

In the human body, the most typical example of homeostasis is the temperature control process described in Fig. 12. In this example, the temperature should be kept close to 37°C. When a stimulus occurs, the equilibrium is restored either by sweating or by reducing blood circulation to the skin. Thus, any change that either raises or lowers the set point temperature is

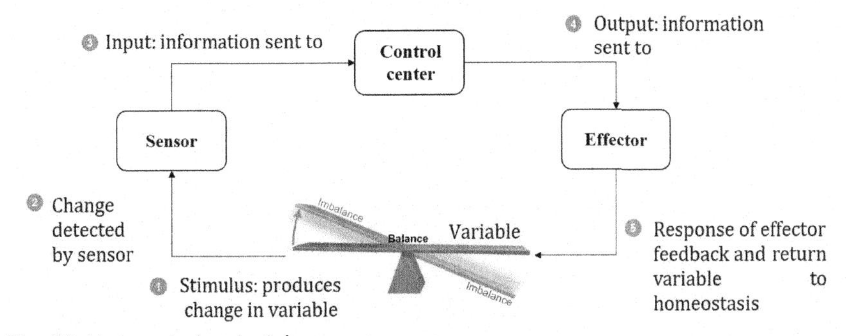

Fig. 10 Homeostasis principle.

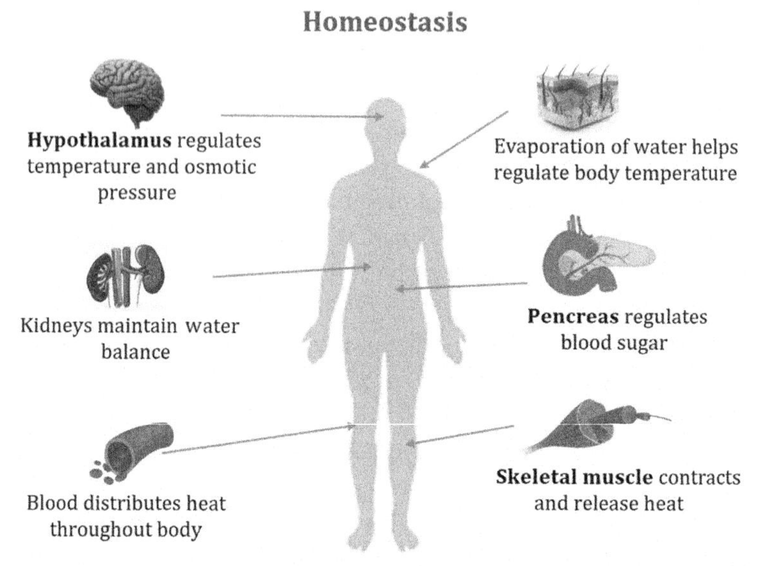

Fig. 11 Homeostasis examples.

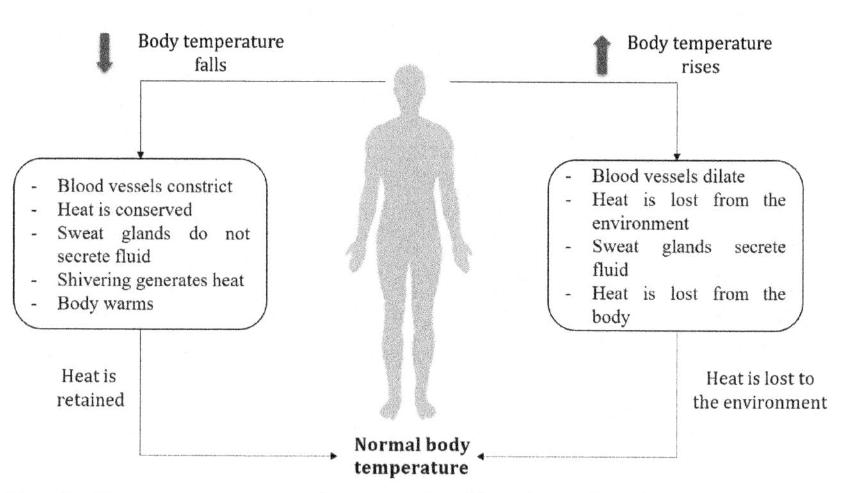

Fig. 12 Human temperature homeostasis. *(The figure has been adopted from Abozenadah, H., et al., 2018. Allied Health Chemistry, Foundations of General, Organic, and Biological Chemistry.)*

automatically having negative feedback. More details about the regulation process of temperature in the human body can be found in Benzinger (1969). Fig. 13 explains the control of body temperature by the hypothalamus causing constriction or dilation of skin capillaries and sweat production.

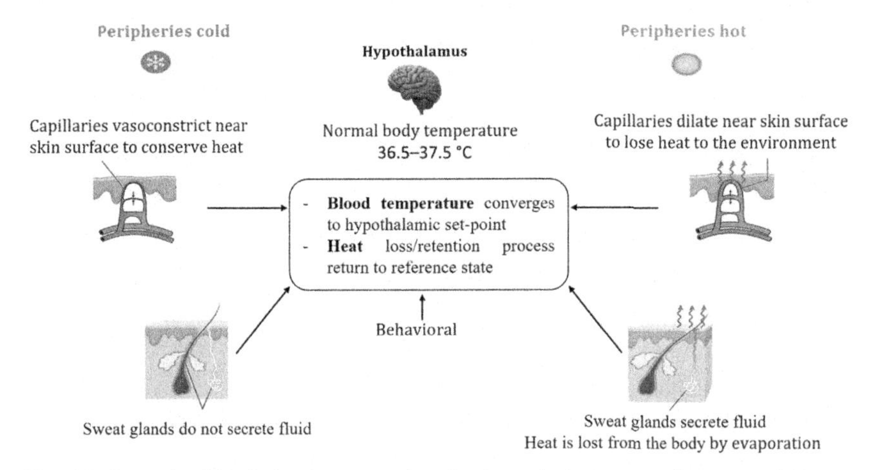

Fig. 13 Control of body temperature by the hypothalamus causing constriction or dilation of skin capillaries and sweat production.

Let's note here that the roster of vital parameters concerned by homeostasis is long. It includes blood pressure, heart rate, blood calcium and blood glucose.

In order to maintain homeostasis in the cardiovascular system, blood pressure and heart rate must be converging to reference states. In fact, different external stimuli, such as stress, affects the equilibrium of blood pressure. Blood pressure can so increase above or toward normal. Fig. 14 describes this control process. Moreover, when blood pressure increases

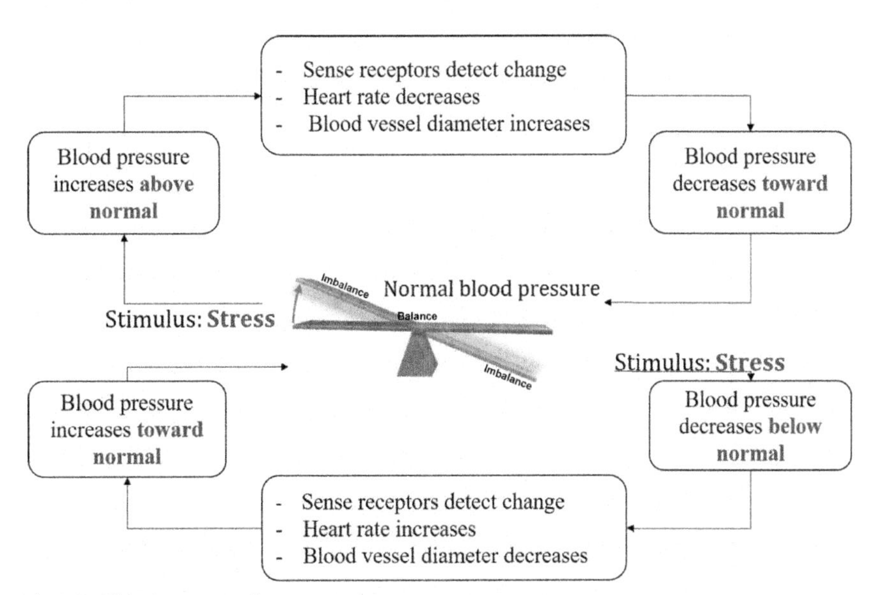

Fig. 14 Blood pressure homeostasis.

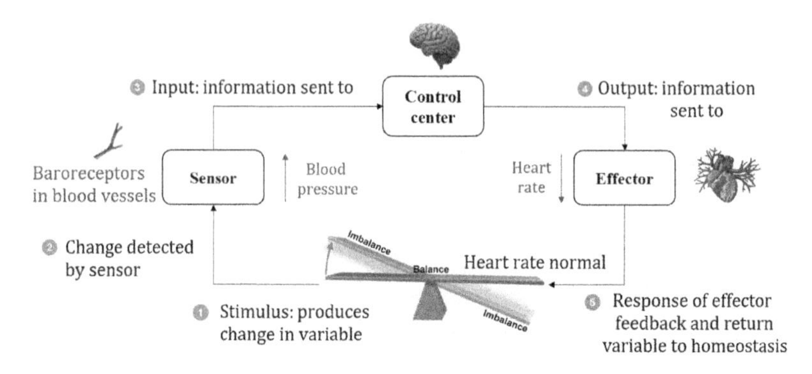

Fig. 15 Heart rate homeostasis.

heart rate increases, which affects heart rate homeostasis equilibrium. Fig. 15 describes this control process.

Calcium homeostasis regulates calcium flow to and from the bones. A normal blood calcium level is about 10 mg/dL. This level is critical for normal body functions. When the calcium level changes (hypocalcemia or hypercalcemia) many problems with blood coagulation, muscle contraction, nerve functioning and bone strength can be detected. Fig. 16 describes the control process of blood calcium homeostasis.

Glucose homeostasis is the balance of insulin and glucagon in the blood in order to maintain human health. In fact, glucose is considered as a principal source of energy and brain tissues do not synthesize it. In fact, different

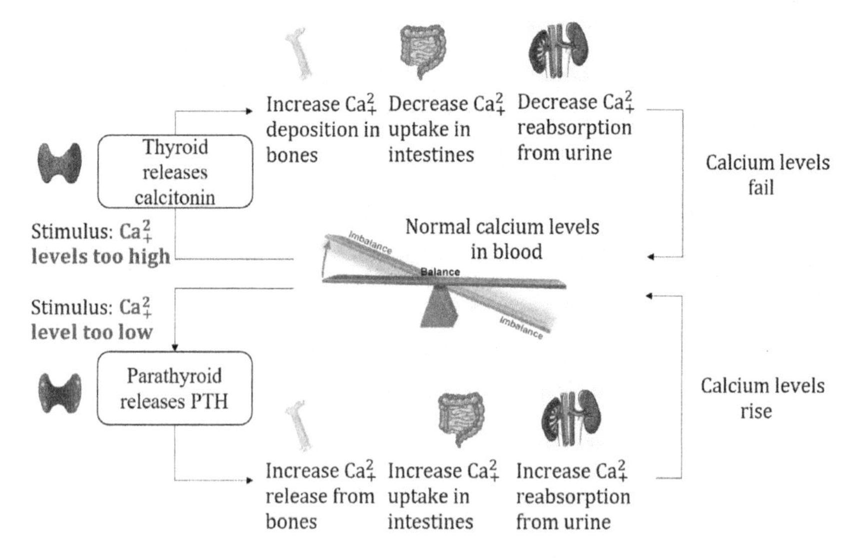

Fig. 16 Blood calcium homeostasis.

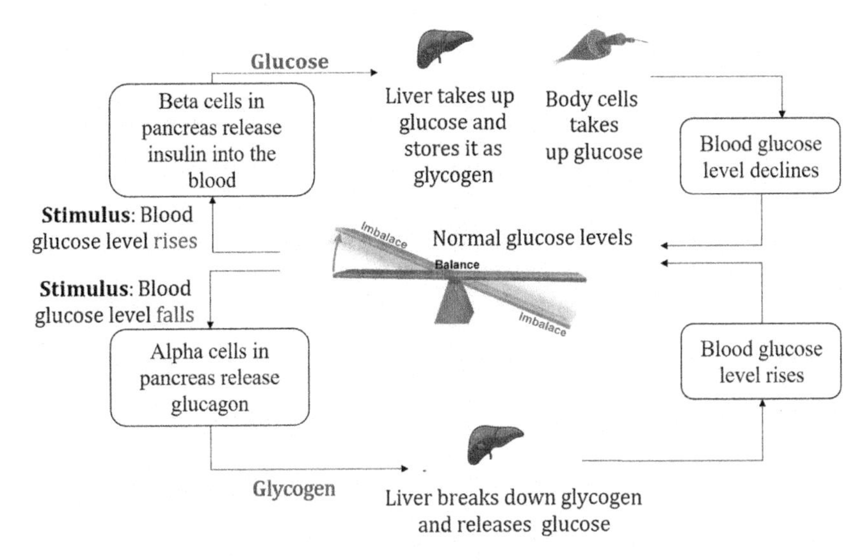

Fig. 17 Blood glucose homeostasis. *(The figure has been adopted from Abozenadah, H., et al., 2018. Allied Health Chemistry, Foundations of General, Organic, and Biological Chemistry.)*

external stimuli, such as eating or physical activity, affect the equilibrium of glucose blood level. The blood glucose level can so increase or decrease and affect the body's equilibrium. In order to maintain blood glucose homeostasis, the control process includes a closed feedback loop involving the pancreatic islet cells, the liver, the brain and the muscle, as described in Fig. 17.

3.3 Control strategies in homeostasis

In the literature, three basic control strategies guaranteeing homeostasis are reported (Houk, 1988): (1) negative feedback control, (2) feed-forward control and (3) adaptive control. The control process of the human body via negative feedback is described by the block diagram shown in Fig. 18. In this strategy, natural sensors detect stimulus as external or internal perturbations related to a given physiological variable, compare their levels to a given set point and then send the error information in the form of electrical signals to the brain. Then, the brain transmits the control signal to one or more natural effectors (muscles and glands), which respond to the instruction of the brain as a corrective action in order to maintain the physiological variable in a stable steady state.

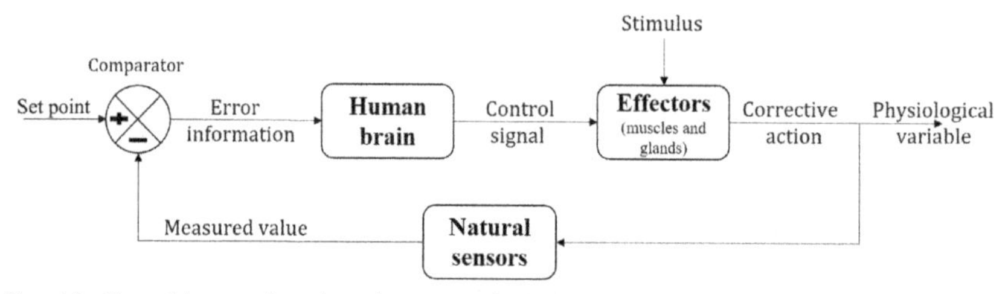

Fig. 18 Closed loop related to the control process.

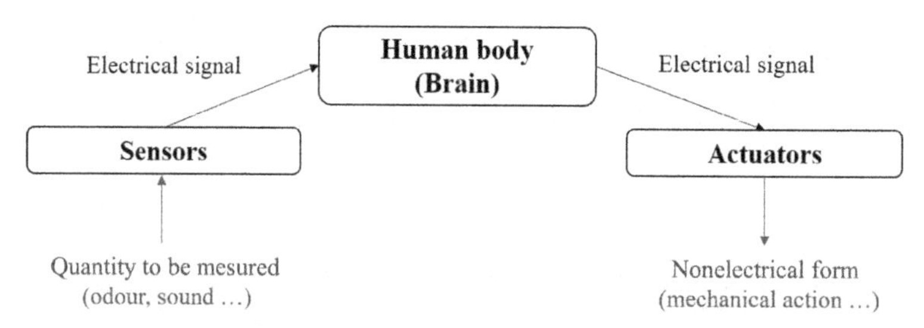

Fig. 19 The process control of human transducers.

Natural sensors and effectors are called human transducers. A transducer is a device in communication with the brain that can convert a nonelectrical action or information to an electrical signal or an electrical signal to a nonelectrical action or information. Thus, sensors convert input information to electrical signals that can be processed by the brain system. However, actuators convert electrical signals to nonelectrical information. Fig. 19 describes the process control of human transducers.

Basic sensors and effectors are summarized in Table 3. The example of postural feedback control via sensors and effectors is shown in Fig. 20 in which sensory information of body states are measured by vision, the vestibular organ and muscle spindles, and then sent to the control neural system to be processed (Kim et al., 2009). Based on an estimate of body kinematics, appropriate control plans are selected and then corresponding motor commands are produced as joint torques.

Control of physiological variables in the human body is not only achieved via negative feedback loops but can also be achieved via anticipatory controllers acting in open loops. A feed-forward controller may employ information about past and current conditions to predict the future states (Del Giudice, 2015). Contrary to negative feedback loops for which sensors and actuators are both mandatory in the control process, feed-forward control only needs

Table 3 Some basic sensors and effectors in the human body.

Human transducers	Body organs/ tissues	Senses/aptitude/action
Sensors	Eyes	Vision
	Nose	Smell
	Tongue	Taste
	Ears	Hearing/static and dynamic balance
	Skin	Touch, vibroception, thermoception, nociception, proprioception...
Effectors	Skin and muscles	Physical pressure and force
	Sweat glands	Produce sweat (increases heat loss)
	Pancreas	Produces insulin (regulates glucose in blood)
	Skeletal system	Guarantees postural balance

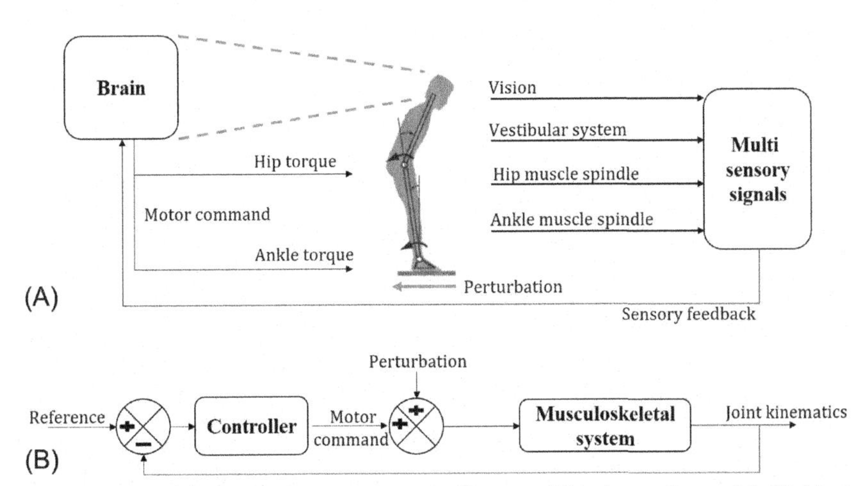

Fig. 20 Postural balance via sensors and effectors; (A) schematic model (B) block diagram. *(The figure has been adopted from Kim, S., et al., 2009. Postural feedback scaling deficits in Parkinson's disease. J. Neurophysiol. 102 (5), 2910–2920.)*

sensors to measure and anticipate external perturbations as is shown in Fig. 21. An example for voluntary movement via sensory inputs is given in Fig. 22 including feedback and feed-forward loops (Wolpert et al., 2013).

Contrary to negative loops and feed-forward controllers able to counteract changes of physiological variables from their target values, positive feedback loops amplify their initiating stimuli and move the system away from its

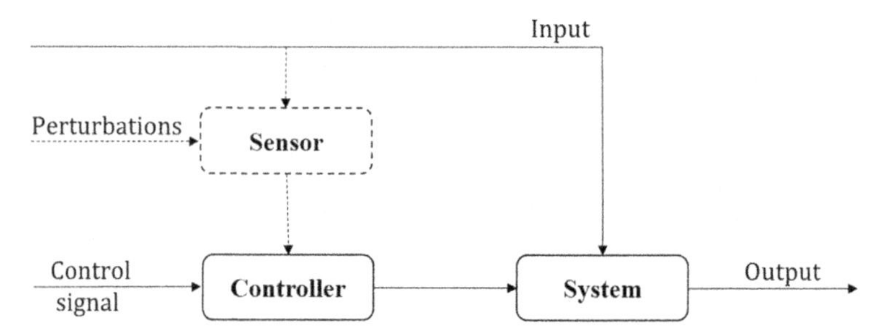

Fig. 21 Schematic representation of a feed-forward control system. *(The figure has been adopted from Del Giudice, M., 2015. Self-regulation in an evolutionary perspective. In: Handbook of Biobehavioral Approaches to Self-Regulation. Springer New York, New York, NY, pp. 25–41.)*

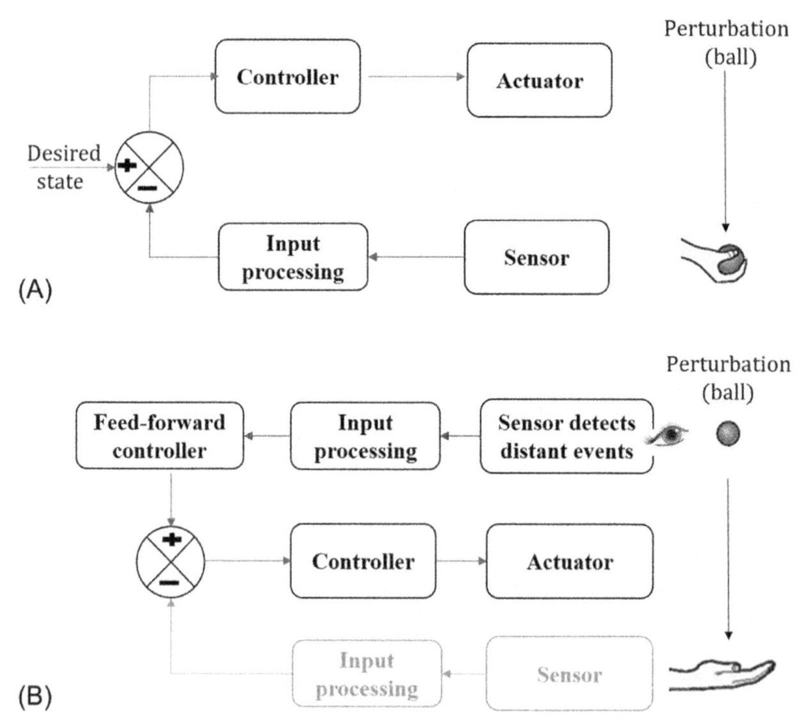

Fig. 22 Voluntary movement: (A) Feedback loop, (B) Feed-forward loop. *(The figure has been adopted from Wolpert, D.M., Pearson, K.G., Ghez, C.P.J., 2013. The organization and planning of movement. Principles of Neuroscience.)*

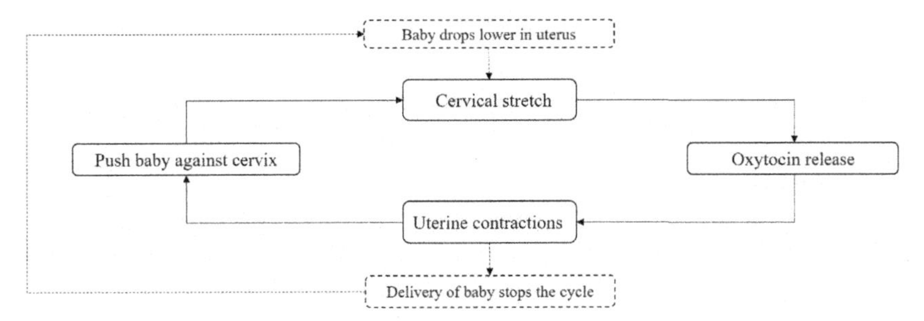

Fig. 23 Physiological positive loops: childbirth.

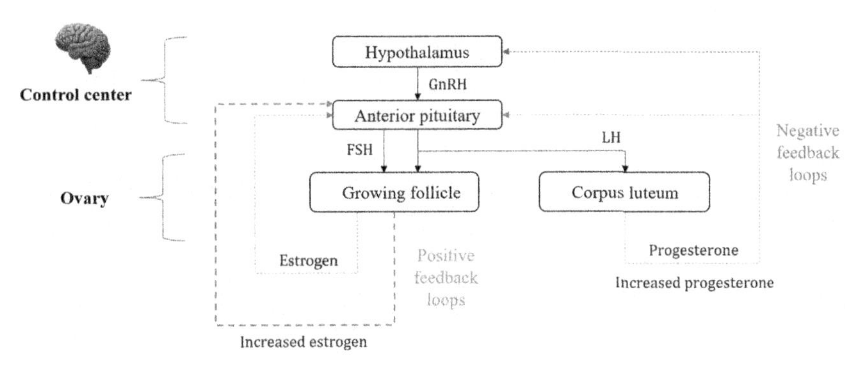

Fig. 24 Positive and negative feedback loops of the menstrual cycle. *(The figure has been adopted from Carroll, R.G., 2007. Female reproductive system. In: Elsevier's Integrated Physiology. Elsevier, pp. 177–187.)*

set point. One of the most typical examples of positive loops is childbirth, shown in Fig. 23, where contractions enhance the change until the baby comes out.

The most classic example including at the same time positive and negative loops is the menstrual cycle (Carroll, 2007). This is shown in Fig. 24. Here, note that there are three different feedback loops that are occurring. The first feedback loop, FSH and LH were secreted to start the development of the follicle. These two hormones work together to start estrogen secretion. The estrogen has a negative feedback on FSH, which causes it to stop being secreted. The second feedback loop occurs with the increased amount of estrogen that is experienced at the midpoint in the cycle. This increased amount of estrogen causes a positive feedback to occur on the LH cells in the pituitary. LH secretion rises, and ovulation occurs. The final feedback loop is a negative feedback. After ovulation, the follicle cells are transformed into the corpus luteum. The corpus luteum secretes estrogen and progesterone. This buildup of progesterone and estrogen further increases the formation of the endometrial lining. These hormones work together and send a second negative feedback to inhibit the release of FSH and LH. This causes the corpus luteum to deteriorate, slowing the production of estrogen and progesterone. The drop in these hormones signals menstruation.

As is claimed in Houk (1988), feed-forward controllers generally make persistent errors unless they are adjusted via adaptive controllers. Fig. 25 presents the model reference adaptive strategy. The reference model compares the desired and the actual outputs in order to compute the error signal. Based on the dynamic evolution of this error signal, the evaluator adjusts the characteristics of the feed-forward controller. In fact, this type of controller does not produce change in outputs in real time, but rather modifies the way the system reacts to future inputs. These properties characterize adaptive control from conventional feedback. In Houk and Rymer (2011), an adaptive strategy is explained for adaptive control of muscle length and tension. In fact, the adaptive control system contains different elements to estimate the dynamics behaviors of the controlled system. Therefore controller parameters are accordingly modified and adapted.

In conclusion, three basic strategies exist in the control process in homeostasis, namely negative feedback, feed-forward and adaptive control, as summarized in Fig. 26. This figure illustrates the structural features of these three control strategies and their combination in the same case. In fact, a feedback controller generates forcing functions by comparing desired performance with actual performance as monitored by a feedback (closed) loop. In contrast, a feed-forward controller generates commands without using continuous negative feedback (open loop). Therefore, an adaptive

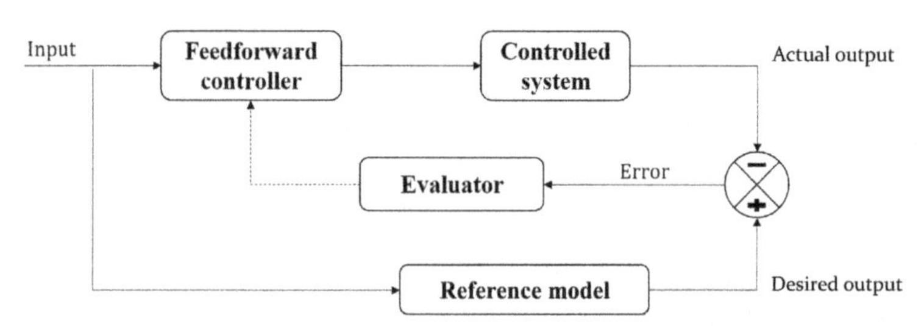

Fig. 25 Model reference of adaptive control system. *(The figure has been adopted from Houk, J.C., 1988. Control strategies in physiological systems. FASEB J. 2(2), 97–107.)*

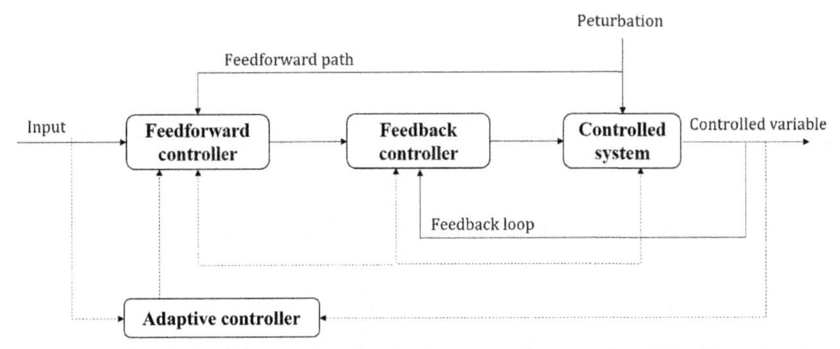

Fig. 26 Feedback, feed-forward and adaptive control strategies. *(The figure has been modified from Houk, J.C., 1988. Control strategies in physiological systems. FASEB J. 2(2), 97–107.)*

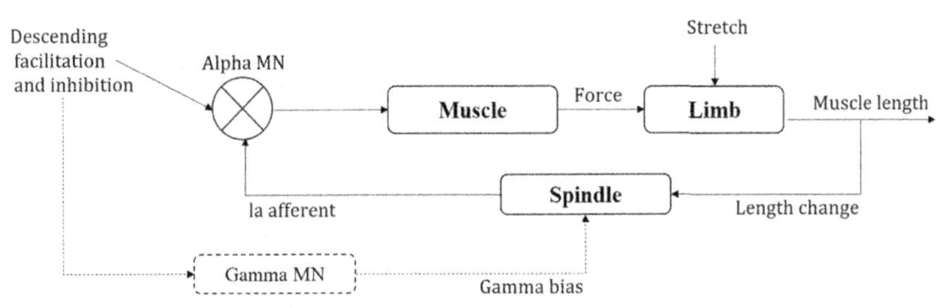

Fig. 27 Feedback vs adaptive control strategies: muscle system. *(The figure has been modified from Devasahayam, S.R., 2019. Neuromuscular control: spinal reflex and movement. In: Signals and Systems in Biomedical Engineering: Physiological Systems Modeling and Signal Processing. Springer Singapore, Singapore, pp. 387–409.)*

controller modifies the controller parameters system rather than causing immediate changes in output. Fig. 27 describes the feedback and adaptive controllers applied to the muscle system (Devasahayam, 2019).

3.4 Control therapy applications

Control therapy for physiological systems can maintain human functions and thus human life. Several control techniques have been proposed. The most familiar control approaches used for therapy strategies are optimal control, feedback control, feed-forward control and adaptive control.

3.4.1 Optimal control

A typical optimal control problem in physiological systems is to apply the adequate control laws in order to converge the controlled system on its optimal trajectory (Swan, 1981). This type of control is characterized by some degree of precision. Due to the predictive character of optimal control, these controllers are the most incorporated control algorithms in endocrine systems. Three applications are presented in this part especially for endocrine applications.

- Optimal control problem in psoriasis treatment

The dynamics behaviors of psoriasis treatment is modeled by nonlinear differential equations. This system is characterized by T-lymphocytes cells, keratinocytes cells and dendritic cells. In order to minimize the interactions between these types of cells, two optimal bounded controls are applied to the psoriasis treatment system (Grigorieva and Khailov, 2018). Finally, the applied control algorithm suppresses the weighted sum of keratinocytes concentration so that the total cost of treatment decreases.

- Optimal control problem in tumor-immune interaction system

The dynamics behaviors of tumor-immune interactions system in presence of immuno-chemotherapy is modeled by delay differential model. In order to optimize the cost associated with immuno-chemotherapy and to minimize the number of tumor cells, an optimal control is applied to the last system (Rihan et al., 2019).

- Optimal control problem in intracellular delayed HIV model

The intracellular delayed HIV model is characterized by a cytotoxic T lymphocyte (CTL) immune response. In order to reduce the number of infected cells and the viral load, two optimal controls are applied to the HIV model (Allali et al., 2018). Thus, the number of uninfected cells increases, which confirms the efficiency of drug treatment.

3.4.2 Adaptive control

A typical adaptive control problem in physiology can compensate automatically the dynamic variation in physiological systems by adjusting their features and characteristics. Thus, the controlled system has the same overall performances. In the literature, adaptive controllers are the most incorporated control algorithms in diabetes treatments.

- Adaptive control problem in Artificial Pancreas Systems

Blood glucose regulation is a complex system due to the variability of the dynamic behaviors of blood glucose concentration. In fact, meals and time-varying delays of insulin infusion can affect the control of the blood glucose concentration. Thus, the adaptive control algorithm can settle this type of challenge (Turksoy and Cinar, 2014; Nath et al., 2018).

- Adaptive control problem in convergence eye movements

The oculomotor human system contains two adaptive controllers in order to have clear and comfortable binocular vision. Thus, the oculomotor human system has a natural and autonomous adaptive capacity regulated by neural mechanisms (Erkelens et al., 2020).

3.4.3 Fuzzy logic control

In nature, most systems and concepts are naturally unpredictable and fuzzy, hence the importance of fuzzy set theory in real problems and especially in human physiology. In fact, fuzzy logic control is adequate for medicine because it is tolerant of peculiar imprecision. The fuzzy concept is based on fuzzy rules of the form IF … THEN. A fuzzy logic controller is equivalent linguistically to a PI controller (Mahfouf et al., 2001).

In the literature, fuzzy logic controllers are the most incorporated control algorithms in medicine applications. Some of them are described in the paragraphs that follow.

- Fuzzy logic control problem in anesthesia

Fuzzy logic controllers are widely adapted and used in anesthesia (Derrick et al., 1998). Generally, anesthetists fix the rules of this type of controller since they have extensive experience with patients. One example of these rules is IF "blood pressure is decreased" THEN "reduce drug infusion." More sophisticated rules are also composed.

- Fuzzy logic control problem in blood glucose regulation

Fuzzy logic controller is also used to regulate blood glucose in type 1 diabetic patients. This type of controller is based on zero order Takagi-Sugeno fuzzy logic architecture (Nath et al., 2018). This controller is characterized by two inputs and one output. The inputs are the error and the derivative in subcutaneous glucose concentration. The output of the fuzzy controller is the exogenous insulin infusion rate.

4 Future trends and challenges

One of the more complex problems is biofeedback therapy. It is considered as a contemporary challenging in preventive healthcare researches. In fact, biofeedback is a mind–body training technique that involves using visual or auditory feedback to gain control over involuntary bodily functions (Patcharatrakul et al., 2020). This may include gaining voluntary control

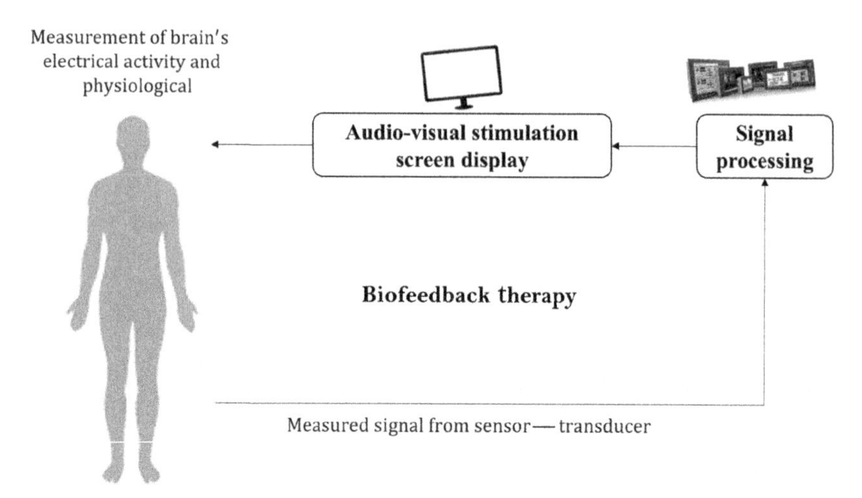

Fig. 28 Biofeedback therapy.

over such things as heart rate (Wheat and Larkin, 2010), muscle tension, blood flow, pain perception and blood pressure. This process involves being connected to a device with electrical sensors that provide feedback about specific aspects of your body, as described in Fig. 28. This technique is adopted in many clinics and hospitals (Badawi and El Saddik, 2020). Novel research aims to utilize this technique outside clinics and hospitals to benefit patients and healthy individuals.

Alternatively, in the physiological regulation, the concept of homeostasis has undergone considerable revision. This new revision is characterized by more contemporary definitions for this concept. An example of a revised concept is allostasis, which is considered as an alternative view offering novel insights relevant to efficiently regulate physiology and human behavior (Ramsay and Woods, 2014). In fact, allostasis reflects the process whereby the human physiological systems must be able to be adaptive to internal or external stimulation by adjusting one or more regulated parameters (Berntson et al., 2016). Indeed, the brain first detects stimulation and consults its database, then it computes the best response (Schulkin and Sterling, 2019). Thus, the control center (brain) adjusts in real time to the dynamic flows of energy and nutrients that reduce prediction errors for better adaptive control. For that, allostasis is now considered as a core principal in physiological regulation. However, there are still many challenges to using allostasis due to its significant computation costs.

As described previously, chaotic systems can describe physiological behaviors. Moreover, detecting chaos in the human body is a difficult challenge for biologists. In fact, basic chaos detection tools are highly sensitive to measurement noise and this problem can distort detection results. Many attempts to detect chaos in physiological systems have fallen short due to this sensitivity to measurement noise. For this reason, recent research has attempted to construct noise-robust tools for detecting chaotic behaviors in biology (Jiao et al., 2020; Toker et al., 2020).

Finally, much progress has been made on the study of artificial organs and devices. Artificial organs, such as the artificial pancreas, artificial and wearable prosthetic devices and pacemakers, improve the daily life of patients. For example, the artificial pancreas improves the regulation of blood glucose concentration, especially for type 1 diabetes patients (Boughton and Hovorka, 2019; Ginsberg and Mauseth, 2019). The artificial organ contains sensors (to measure blood glucose concentration), an insulin pump (to inject insulin in the body) and a controller (to regulate the blood glucose concentration), as described in Fig. 29.

Many patients are waiting for a heart transplantation, but unfortunately there are not enough hearts available. One of the proposed solutions for this problem is to develop a total artificial heart. The first researches were developed starting in 1960 and since then many prototypes have emerged (Cohn et al., 2015). However, the total artificial heart is not an alternative solution before transplantation due to the limited durability of the pumps. In fact, the longest recorded survival time after artificial heart transplantation was 120 days. Another solution for heart problems is the cardiac pacemaker.

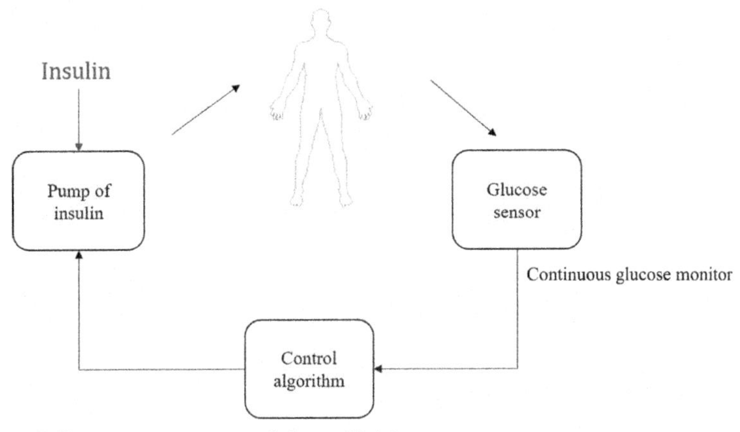

Fig. 29 Different components of the artificial pancreas.

The pacemaker is placed in the human body in order to control heart rhythms (Miller et al., 2015; Cox et al., 2018).

Further, a wearable artificial prosthetic device should be perceived by the patient as a natural body part. The main goal of artificial prosthetics is to replicate sensory-motor capabilities of the natural body parts (Novak and Riener, 2015). The loss of a body affects an individual's ability to interact with their environment. Thus, several researchers have attempted to construct artificial prosthetics in order to duplicate the function of human limbs. The main application of artificial prosthetic devices is the wearable hand prosthetic (Arjun et al., 2016; Zhang et al., 2016; Nemoto et al., 2018).

5 Conclusion

Mathematical modeling and the theory of control applied to physiological systems contribute to understanding and maintaining control of the processes of the human body. In this chapter, we reviewed the most used mathematical approaches and models in physiology. These approaches and models are supported by structural and practical identifiability as necessary tools for a comprehensive modeling. We also reviewed the dynamics of the chaos that arises in several physiological processes and presented many examples. As the natural and autonomous control process for the human body is at the heart of future medicine, we provided comprehensive info on this concept as well as different examples of homeostasis via fundamental principles of feedback control and different control approaches. Finally, we outlined a number of challenges and future trends, including biofeedback control and design and implementation of artificial organs.

References

Abdolrazaghi, M., Navidbakhsh, M., Hassani, K., 2010. Mathematical modelling and electrical analog equivalent of the human cardiovascular system. Cardiovasc. Eng. 10 (2), 45–51.

Adam, J., Bellomo, N., 2012. A Survey of Models for Tumor-Immune System Dynamics. Springer Science & Business Media.

Aidley, D.J., 1998. The Physiology of Excitable Cells. Cambridge University Press, Cambridge, UK.

Ajmera, I., et al., 2013. The impact of mathematical modeling on the understanding of diabetes and related complications. CPT Pharmacometrics Syst. Pharmacol. 2 (7), e54.

Allali, K., Harroudi, S., Torres, D.F.M., 2018. Analysis and optimal control of an intracellular delayed HIV model with CTL immune response. Math. Comput. Sci. 12 (2), 111–127.

Allman, E., Rhodes, J., 2004. Mathematical Models in Biology. An Introduction. Cambridge University Press, Cambridge, UK.

Alvarez-Arenas, A., Starkov, K., Calvo, G., 2019. Ultimate dynamics and optimal control of a multi-compartment model of tumor resistance to chemotherapy. Discrete Continuous Dyn. Syst. Ser. B 24 (5), 2017–2038.

Ambrosi, D., Quarteroni, A., Rozza, G., 2012. Modeling of Physiological Flows, Modeling, Simulation and Applications. Springer Science & Business Media.

Anderson, D., 2013. Compartmental Modeling and Tracer Kinetics. Springer Science & Business Media.

Anderson, A., Quaranta, V., 2008. Integrative mathematical oncology. Nat. Rev. Cancer 8 (3), 277.

Aon, M.A., Cortassa, S., Lloyd, D., 2011. Chaos in biochemistry and physiology. In: Encyclopedia of Molecular Cell Biology and Molecular Medicine. Wiley-VCH Verlag GmbH & Co. KGaA, Weinheim, Germany.

Aram, Z., et al., 2017. Using chaotic artificial neural networks to model memory in the brain. Commun. Nonlinear Sci. Numer. Simul. 44, 449–459.

Arjun, A., Saharan, L., Tadesse, Y., 2016. Design of a 3D printed hand prosthesis actuated by nylon 6-6 polymer based artificial muscles. In: *2016 IEEE International Conference on Automation Science and Engineering (CASE)*. IEEE, pp. 910–915.

Badawi, H.F., El Saddik, A., 2020. Biofeedback in healthcare: State of the art and meta review. In: Connected Health in Smart Cities. Springer International Publishing, Cham, pp. 113–142.

Baghdadi, G., et al., 2015. A chaotic model of sustaining attention problem in attention deficit disorder. Commun. Nonlinear Sci. Numer. Simul. 20 (1), 174–185.

Balakrishnan, N.P., Rangaiah, G.P., Samavedham, L., 2011. Review and analysis of blood glucose (BG) models for type 1 diabetic patients. Ind. Eng. Chem. Res. 50 (21), 12041–12066.

Banks, H., 2013. Modeling and Control in the Biomedical Sciences. Springer Science & Business Media.

Baxt, W.G., 1994. Complexity, chaos and human physiology: the justification for non-linear neural computational analysis. Cancer Lett. 77 (2–3), 85–93.

Bayani, A., et al., 2018. A chaotic model of migraine headache considering the dynamical transitions of this cyclic disease. EPL 123 (1), 10006.

Bekey, G., Beneken, J., 1978. Identification of biological systems: a survey. Automatica 14 (1), 41–47.

Bell, G.I., 1973. Predator-prey equations simulating an immune response. Math. Biosci. 16 (3–4), 291–314.

Bellman, R., Åström, K.J., 1970. On structural identifiability. Math. Biosci. 7 (3–4), 329–339.

Bellu, G., et al., 2007. DAISY: A new software tool to test global identifiability of biological and physiological systems. Comput. Methods Prog. Biomed. 88 (1), 52–61.

Ben Saad, A., Boubaker, O., Elhadj, Z., 2019. PD bifurcation and chaos behavior in a predator-prey model with Allee effect and seasonal perturbation. In: Recent Advances in Chaotic Systems and Synchronization. Academic Press, pp. 211–232. https://doi.org/10.1016/B978-0-12-815838-8.00011-X.

Benzinger, T.H., 1969. Heat regulation: homeostasis of central temperature in man. Physiol. Rev. 49 (4), 671–759.

Bergman, R.N., et al., 1979. Quantitative estimation of insulin sensitivity. Am. J. Physiol. Endocrinol. Metab. 236 (6), E667.

Berntson, G.G., Cacioppo, J.T., Bosch, J.A., 2016. From homeostasis to allodynamic regulation. In: Handbook of Psychophysiology. Cambridge University Press, pp. 401–426.

Bertram, R., 2011. Mathematical modeling in neuroendocrinology. Compr. Physiol. 5 (2), 911–927.

Bessonov, N., et al., 2016. Methods of blood flow modelling. Math. Model. Nat. Phenom. Edited by V. Volpert. vol. 11(1), 1–25.

Beuter, A., et al., 2003. Nonlinear Dynamics in Physiology and Medicine.
Bezzo, F., Galvanin, F., 2018. On the identifiability of physiological models: optimal design of clinical tests. In: Computer Aided Chemical Engineering. Elsevier, pp. 85–110.
Bickel, P., Doksum, K., 2001. Mathematical Statistics: Basic and Selected Topics. vol. 1. Prentice-Hall.
Bora, Ş., et al., 2019. Agent-based modeling and simulation of blood vessels in the cardiovascular system. Simulation 95 (4), 297–312.
Boubaker, O., Fourati, A., 2004. Structural identifiability of non linear systems: an overview. In: 2004 IEEE International Conference on Industrial Technology, 2004. IEEE ICIT '04. IEEE, pp. 1244–1248.
Boubaker, O., Jafari, S., 2018. Recent Advances in Chaotic Systems and Synchronization: From Theory to Real World Applications. Academic Press.
Boughton, C.K., Hovorka, R., 2019. Advances in artificial pancreas systems. Sci. Transl. Med. 11 (484), eaaw4949.
Boutayeb, A., Chetouani, A., 2006. A critical review of mathematical models and data used in diabetology. BioMed. Eng. Online 5 (1), 43.
Bronzino, J., 1999. Biomedical Engineering Handbook. CRC Press.
Brown, R.F., 1980. Compartmental system analysis: state of the art. IEEE Trans. Biomed. Eng. 27 (9), 1–11.
Brown, D., Rothery, P., 1993. Models in Biology: Mathematics, Statistics and Computing. John Wiley & Sons Ltd., Chichester, UK.
Cannon, W.B., 1929. Organization for physiological homeostasis. Physiol. Rev. 9 (3), 399–431.
Carroll, R.G., 2007. Female reproductive system. In: Elsevier's Integrated Physiology. Elsevier, pp. 177–187.
Carson, E., Cobelli, C., 2014. Modelling Methodology for Physiology and Medicine, Cap. 22. Elsevier.
Chapelot, D., Charlot, K., 2019. Physiology of energy homeostasis: models, actors, challenges and the glucoadipostatic loop. Metabolism 92, 11–25.
Cherruault, Y., 2012. Mathematical Modelling in Biomedicine: Optimal Control of Biomedical Systems. Springer Science & Business Media.
Chiarella, C., Shannon, A., 1986. An example of diabetes compartment modelling. Math. Model. 7 (9), 1239–1244.
Chis, O.-T., Banga, J.R., Balsa-Canto, E., 2011. Structural identifiability of systems biology models: a critical comparison of methods. PLoS ONE. Edited by J. Jaeger. 6 (11), e27755.
Chowell, G., et al., 2016. Mathematical models to characterize early epidemic growth: a review. Phys. Life Rev. 18, 66–97.
Clément, F., 2016. Multiscale mathematical modeling of the hypothalamo-pituitary-gonadal axis. Theriogenology 86 (1), 11–21.
Cobelli, C., Carson, E., 2008. Introduction to Modeling in Physiology and Medicine. Academic Press.
Cobelli, C., Foster, D.M., 1998. Compartmental models: theory and practice using the saam II software system. In: Advances in Experimental Medicine and Biology, pp. 79–101.
Cobelli, C., et al., 2009. Diabetes: models, signals, and control. Rev. Biomed. Eng. 2, 54–96.
Cohn, W.E., Timms, D.L., Frazier, O.H., 2015. Total artificial hearts: past, present, and future. Nat. Rev. Cardiol. 12 (10), 609–617.
Cox, J.L., et al., 2018. The maze procedure and postoperative pacemakers. Ann. Thorac. Surg. 106 (5), 1561–1569.
Davies, K.J.A., 2016. Adaptive homeostasis. Mol. Asp. Med. 49, 1–7.
De Nicolao, G., Sparacino, G., Cobelli, C., 1997. Nonparametric input estimation in physiological systems: problems, methods, and case studies. Automatica 33 (5), 851–870.

Del Giudice, M., 2015. Self-regulation in an evolutionary perspective. In: Handbook of Biobehavioral Approaches to Self-Regulation. New York, NY, Springer New York, pp. 25–41.
Derrick, J.L., Thompson, C.L., Short, T.G., 1998. The application of a modified proportional-derivative control algorithm to arterial pressure alarms in anesthesiology. J. Clin. Monit. Comput. 14 (1), 41–47.
Devasahayam, R., 2012. Signals and Systems in Biomedical Engineering: Signal Processing and Physiological Systems Modeling. Springer Science & Business Media.
Devasahayam, S.R., 2019. Neuromuscular control: spinal reflex and movement. In: Signals and Systems in Biomedical Engineering: Physiological Systems Modeling and Signal Processing. Springer Singapore, Singapore, pp. 387–409.
Dutta, S., Kushner, T., Sankaranarayanan, S., 2018. Robust data-driven control of artificial pancreas systems using neural networks. In: International Conference on Computational Methods in Systems Biology. Springer International Publishing, Cham, pp. 183–202.
Eberle, C., Ament, C., 2012. Identifiability and online estimation of diagnostic parameters with in the glucose insulin homeostasis. Biosystems 107 (3), 135–141.
Edelstein-Keshet, L., 2005. Mathematical Models in Biology. Society for Industrial and Applied Mathematics.
Elbert, T., et al., 1994. Chaos and physiology: deterministic chaos in excitable cell assemblies. Physiol. Rev. 74 (1), 1–47.
Enderle, J., Bronzino, J., 2012. Introduction to Biomedical Engineering. Academic Press, Oxford, USA.
Erkelens, I.M., et al., 2020. A differential role for the posterior cerebellum in the adaptive control of convergence eye movements. Brain Stimulation 13 (1), 215–228.
Fong, L.E., Muñoz-Rojas, A.R., Miller-Jensen, K., 2018. Advancing systems immunology through data-driven statistical analysis. Curr. Opin. Biotechnol. 52, 109–115.
Freeman, W.J., 1992. Tutorial on neurobiology: from single neurons to brain chaos. Int. J. Bifurcat. Chaos 02 (03), 451–482.
Gałach, M., 2003. Dynamics of the tumor—immune system competition—the effect of time delay. Int. J. Appl. Math. Comput. Sci. 13, 395–406.
Galvanin, F., Ballan, C.C., Barolo, M., Bezzo, F., 2013. A general model-based design of experiments approach to achieve practical identifiability of pharmacokinetic and pharmacodynamic models. J. Pharmacokinet. Pharmacodyn. 40 (4), 451–467.
Gani, A., et al., 2009. Predicting subcutaneous glucose concentration in humans: data-driven glucose modeling. IEEE Trans. Biomed. Eng. 56 (2), 246–254.
Garcia-Sevilla, F., et al., 2012a. Linear compartmental systems: II-A software to obtain the symbolic kinetic equations. J. Math. Chem. 50 (6), 1625–1648.
Garcia-Sevilla, F., et al., 2012b. Linear compartmental systems. I. kinetic analysis and derivation of their optimized symbolic equations. J. Math. Chem. 50 (6), 1598–1624.
Ghafarian, P., Jamaati, H., Hashemian, S.M., 2016. A review on human respiratory modeling. Tanaffos 15 (2), 61–69.
Giannakis, G.B., Serpedin, E., 2001. A bibliography on nonlinear system identification. Signal Process. 81 (3), 533–580.
Ginoux, J.-M., et al., 2018. Is type 1 diabetes a chaotic phenomenon? Chaos, Solitons Fractals 111, 198–205.
Ginsberg, B.H., Mauseth, R., 2019. The artificial pancreas. In: The Diabetes Textbook. Springer International Publishing, Cham, pp. 993–998.
Glass, L., 2001. Synchronization and rhythmic processes in physiology. Nature 410 (6825), 277–284.
Glass, L., Beuter, A., Larocque, D., 1988. Time delays, oscillations, and chaos in physiological control systems. Math. Biosci. 90 (1–2), 111–125.

Glynn, P., Unudurthi, S., Hund, T., 2014. Mathematical modeling of physiological systems: an essential tool for discovery. Life Sci. 111 (1), 1–5.
Goldberger, A.L., West, B.J., 1987. Chaos in physiology: health or disease? In: Chaos in Biological Systems. Springer, Boston, MA, pp. 1–4.
Goldberger, A.L., Rigney, D.R., West, B.J., 1990. Science in pictures: chaos and fractals in human physiology. Sci. Am. 262 (2), 42–49.
Grigorieva, E., Khailov, E., 2018. Optimal strategies for psoriasis treatment. Mathematical and Computational Applications 23 (3), 45–75.
Grodins, F.S., 1959. Integrative cardiovascular physiology: a mathematical synthesis of cardiac and blood vessel hemodynamics. Q. Rev. Biol. 34 (2), 93–116.
Hacısalihzade, S., 2013. Biomedical Applications of Control Engineering. Springer.
Hajizadeh, I., et al., 2018. Adaptive model predictive control for nonlinearity in biomedical applications. IFAC-PapersOnLine 51 (20), 368–373.
Heldt, T., Verghese, G.C., Mark, R.G., 2013. Mathematical Modeling of Physiological Systems. Springer, Berlin, Heidelberg, pp. 21–41.
Hester, R.L., et al., 2011. HumMod: A modeling environment for the simulation of integrative human physiology. Front. Physiol. 2, 12.
Ho, Y., 2019. Parameter Estimation for Nonlinear Mathematical Model. Springer, Singapore, pp. 69–80.
Houk, J.C., 1988. Control strategies in physiological systems. FASEB J. 2 (2), 97–107.
Houk, J.C., Rymer, W.Z., 2011. Neural control of muscle length and tension. In: Comprehensive Physiology. John Wiley & Sons, Inc., Hoboken, NJ, USA.
Huang, H., Hu, S., Sun, Y., 2019. A discrete curvature estimation based low-distortion adaptive savitzky–golay filter for ECG denoising. Sensors 19 (7), 1617.
Iii, F.J.D., et al., 2011. Control in biological systems. In: The Impact of Control Technology.
Ismail, L., et al., 2018. Circuit modeling and analysis of cardiovascular system using analog circuit analogy. In: *International Conference on Intelligent and Advanced System.*
Itik, M., Banks, S.P., 2010. Chaos in a three-dimensional cancer model. Int. J. Bifurcat. Chaos 20 (01), 71–79.
Jacquez, A., 1972. Compartmental Analysis in Biology and Medicine. Elsevier, Amsterdam.
Jiao, D., et al., 2020. The chaotic characteristics detection based on multifractal detrended fluctuation analysis of the elderly 12-lead ECG signals. Physica A: Statistical Mechanics and its Applications. North-Holland, 540, 123234.
Joe, H.M., Oh, J.H., 2019. A robust balance-control framework for the terrain-blind bipedal walking of a humanoid robot on unknown and uneven terrain. Sensors 19 (19), 4194.
Kabanikhin, S.I., et al., 2016. Identifiability of mathematical models in medical biology. Russ. J. Genet. Appl. Res. 6 (8), 838–844.
Kansal, A.R., 2004. Modeling approaches to type 2 diabetes. Diabetes Technol. Ther. 6 (1), 39–47.
Karim, A., et al., 2018. A New Automatic Epilepsy Serious Detection Method by Using Deep Learning Based on Discrete Wavelet Transform.
Khoo, M.C.K., 1999. Physiological Control Systems, Physiological Control Systems: Analysis, Simulation, and Estimation, second ed. IEEE.
Kim, S., et al., 2009. Postural feedback scaling deficits in Parkinson's disease. J. Neurophysiol. 102 (5), 2910–2920.
Kokalari, I., Karaja, T., Guerrisi, M., 2013. Review on lumped parameter method for modeling the blood flow in systemic arteries. J. Biomed. Sci. Eng. 6 (1), 16.
Kulkarni, V., 2016. Modeling and Analysis of Stochastic Systems. Chapman and Hall/CRC.
Kuttler, C., 2009. Mathematical Models in Biology.
La Perle, K.M.D., Dintzis, S.M., 2018. Endocrine system. In: Comparative Anatomy and Histology. Elsevier, pp. 251–273.

Larrabide, I., et al., 2012. HeMoLab—hemodynamics modeling laboratory: an application for modeling the human cardiovascular system. Comput. Biol. Med. 42 (10), 993–1004.
Lassoued, A., Boubaker, O., 2016. On new chaotic and hyperchaotic systems: a literature survey. Nonlinear Anal. Model. Control 21 (6), 770–789.
Leaning, M.S., et al., 1983. Modeling a complex biological system: the human cardiovascular system—1. Methodology and model description. Trans. Inst. Meas. Control. 5 (2), 71–86.
Leng, G., MacGregor, D.J., 2008. Mathematical modeling in neuroendocrinology. J. Neuroendocrinol. 20 (6), 713–718.
Li, L., 2015. Bifurcation and chaos in a discrete physiological control system. Appl. Math. Comput. 252, 397–404.
Liang, F., Liu, H., 2005. A closed-loop lumped parameter computational model for human cardiovascular system. JSME Int. J. Ser. C 48 (4), 484–493.
Liu, Y., et al., 1994. A new mathematical model of hypothalamo-pituitary-thyroid axis. Math. Comput. Model. 19 (9), 81–90.
Mackey, M., Glass, L., 1977. Oscillation and chaos in physiological control systems. Science 197 (4300), 287–289.
Mahfouf, M., Abbod, M., Linkens, D., 2001. A survey of fuzzy logic monitoring and control utilisation in medicine. Artif. Intell. Med. 21 (1–3), 27–42.
Makroglou, A., Li, J., Kuang, Y., 2006. Mathematical models and software tools for the glucose-insulin regulatory system and diabetes: an overview. Appl. Numer. Math. 56 (3), 559–573.
Mari, A., 2002. Mathematical modeling in glucose metabolism and insulin secretion. Curr. Opin. Clin. Nutr. Metab. Care 5 (5), 495–501.
Marmarelis, V.Z., 2004. Nonlinear Dynamic Modeling of Physiological Systems, John Wiley & Sons, Inc., Hoboken, NJ, USA.
Marmarelis, P.Z., Marmarelis, V.Z., 1978. Analysis of Physiological Systems: The White-Noise Approach. Springer Science & Business Media. Springer US, Boston, MA.
Mayer, H., Zaenker, K.S., an der Heiden, U., 1995. A basic mathematical model of the immune response. Chaos 5 (1), 155–161.
Miao, H., et al., 2011. On identifiability of nonlinear ODE models and applications in viral dynamics. SIAM Rev. 53 (1), 3–39.
Miller, M.A., et al., 2015. Leadless cardiac pacemakers. J. Am. Coll. Cardiol. 66 (10), 1179–1189.
Misgeld, B.J.E., et al., 2016. Estimation of insulin sensitivity in diabetic Göttingen Minipigs. Control. Eng. Pract. 55, 80–90.
Morari, M., Gentilini, A., 2001. Challenges and opportunities in process control: biomedical processes. AICHE J. 47 (10), 2140–2143.
Nath, A., et al., 2018. Blood glucose regulation in type 1 diabetic patients: an adaptive parametric compensation control-based approach. IET Syst. Biol. 12 (5), 219–225.
Nazarimehr, F., et al., 2017. Can Lyapunov exponent predict critical transitions in biological systems? Nonlinear Dyn. 88 (2), 1493–1500.
Neave, N., 2008. Hormones and Behaviour: A Psychological Approach. Cambridge University Press, Cambridge. ISBN 978-0521692014.
Nemoto, Y., Ogawa, K., Yoshikawa, M., 2018. F3Hand: a five-fingered prosthetic hand driven with curved pneumatic artificial muscles. In: *2018 40th Annual International Conference of the IEEE Engineering in Medicine and Biology Society (EMBC)*. IEEE, pp. 1668–1671.
Noble, D., 1960. Cardiac action and pacemaker potentials based on the Hodgkin-Huxley equations. Nature 188, 495–497.
Novak, D., Riener, R., 2015. A survey of sensor fusion methods in wearable robotics. Robot. Auton. Syst. 73, 155–170.
Ottesen, J., Danielsen, M., 2000. Mathematical Modeling in Medicine. ISO Press.

Ottesen, J., Olufsen, M., Larsen, J., 2004. Applied Mathematical Models in Human Physiology. Society for Industrial and Applied Mathematics.
Oxtoby, N.P., et al., 2018. Data-driven models of dominantly-inherited Alzheimer's disease progression. Brain 141 (5), 1529–1544.
Palumbo, P., et al., 2013. Mathematical modeling of the glucose–insulin system: a review. Math. Biosci. 244 (2), 69–81.
Panahi, S., et al., 2017. Modeling of epilepsy based on chaotic artificial neural network. Chaos, Solitons Fractals 105, 150–156.
Panahi, S., et al., 2019. A new chaotic network model for epilepsy. Appl. Math. Comput. 346, 395–407.
Paoletti, N., et al., 2019. Data-driven robust control for a closed-loop artificial pancreas. In: IEEE/ACM Transactions on Computational Biology and Bioinformatics. https://doi.org/10.1109/tcbb.2019.2912609.
Patcharatrakul, T., et al., 2020. Biofeedback therapy. In: Clinical and Basic Neurogastroenterology and Motility. Elsevier, pp. 517–532.
Perelson, A., 2002. Modeling viral and immune system dynamics. Nat. Rev. Immunol. 2 (1), 28.
Pia Saccomani, M., Audoly, S., D'Angiò, L., 2003. Parameter identifiability of nonlinear systems: the role of initial conditions. Automatica 39 (4), 619–632.
Pironet, A., et al., 2019. Practical identifiability analysis of a minimal cardiovascular system model. Comput. Methods Prog. Biomed. 171, 53–65.
Quarteroni, A., 2001. Modeling the cardiovascular system—a mathematical adventure: Part I. SIAM News 34 (5), 1–3.
Quarteroni, A., Formaggia, L., Veneziani, A., 2009. Cardiovascular Mathematics: Modeling and Simulation of the Circulatory System, Modeling, Simulation and Applications. Springer Science & Business Media.
Quarteroni, A., Manzoni, A., Vergara, C., 2017. The cardiovascular system: Mathematical modeling, numerical algorithms and clinical applications. Acta Numerica 26, 365–590.
Rajagopal, K., et al., 2019. Chaotic dynamics of a fractional order glucose-insulin regulatory system. Front. Inform. Technol. Electron. Eng., 1–11.
Ramsay, D.S., Woods, S.C., 2014. Clarifying the roles of homeostasis and allostasis in physiological regulation. Psychol. Rev. 121 (2), 225–247.
Raue, A., et al., 2009. Structural and practical identifiability analysis of partially observed dynamical models by exploiting the profile likelihood. Bioinformatics 25 (15), 1923–1929.
Reisman, S., et al., 2018. Biomedical Engineering Principles. CRC Press.
Ribarič, S., Kordaš, M., 2011. Teaching cardiovascular physiology with equivalent electronic circuits in a practically oriented teaching module. Adv. Physiol. Educ. 35 (2), 149–160.
Rideout, V., Beneken, J., 1975. Parameter estimation applied to physiological systems. Math. Comput. Simul. 17 (1), 23–36.
Rihan, F.A., Lakshmanan, S., Maurer, H., 2019. Optimal control of tumour-immune model with time-delay and immuno-chemotherapy. Appl. Math. Comput. 353, 147–165.
Roberts, F., 1976. Discrete Mathematical Models, With Applications to Social, Biological, and Environmental Problems. Prentice-Hall, Englewood Cliffs, NJ.
Rossler, O.E., Rossler, R., 1994. Chaos in physiology. Integr. Physiol. Behav. Sci. 29 (3), 328–333.
Rostami, Z., Mousavi, M., Rajagopal, K., Boubaker, O., Jafari, S., 2019. Chaotic solutions in a forced two-dimensional Hindmarsh-Rose neuron. In: Recent Advances in Chaotic Systems and Synchronization. Academic Press, pp. 187–209. 10.1016/B978-0-12-815838-8.00010-8.

Rupnik, M., Runovc, F., Kordaš, M., 2001. The use of equivalent electronic circuits in simulating physiological processes. IEEE Trans. Educ. 44 (4), 384–389.
Sachs, R.K., Hlatky, L.R., Hahnfeldt, P., 2001. Simple ODE models of tumor growth and anti-angiogenic or radiation treatment. Math. Comput. Model. 33 (12–13), 1297–1305.
Sanft, R., Walter, A., 2020. Exploring Mathematical Modeling in Biology Through Case Studies and Experimental Activities. Academic Press.
Sarbadhikari, S.N., Chakrabarty, K., 2001. Chaos in the brain: a short review alluding to epilepsy, depression, exercise and lateralization. Med. Eng. Phys. 23 (7), 447–457.
Sato, D., et al., 2019. A stochastic model of ion channel cluster formation in the plasma membrane. J. Gen. Physiol. 151 (9), 1116–1134.
Schiff, S.J., et al., 1994. Controlling chaos in the brain. Nature 370 (6491), 615–620.
Schulkin, J., Sterling, P., 2019. Allostasis: a brain-centered, predictive mode of physiological regulation. Trends Neurosci. Elsevier Current Trends. 42 (10), 740–752.
Sever, M., et al., 2014. The use of equivalent electronic circuits in physiology teaching. In: *Information Technology Based Proceedings of the Fifth International Conference on Higher Education and Training*, pp. 593–597.
Shabestari, P.S., et al., 2018. A new chaotic model for glucose-insulin regulatory system. Chaos, Solitons Fractals 112, 44–51.
Sharma, V., 2009. Deterministic chaos and fractal complexity in the dynamics of cardiovascular behavior: perspectives on a new frontier. Open Cardiovasc. Med. J. 3 (1), 110–123.
Shi, Y., Lawford, P., Hose, R., 2011. Review of zero-D and 1-D models of blood flow in the cardiovascular system. BioMed. Eng. OnLine 10 (1), 33.
Shi, Y., et al., 2016. Online estimation method for respiratory parameters based on a pneumatic model. In: Transactions on Computational Biology and Bioinformatics, pp. 939–946.
Shim, E.B., Sah, J.Y., Youn, C.H., 2004. Mathematical modeling of cardiovascular system dynamics using a lumped parameter method. Jpn. J. Physiol. 54 (6), 545–553.
Similowski, T., Bates, J.H.T., 1991. Two-compartment modeling of respiratory system mechanics at low frequencies: gas redistribution or tissue rheology? Eur. Respir. J. 4 (3), 353–358.
Smith, P., Cohn, W., Frazier, O., 2018. Total artificial hearts. In: Mechanical Circulatory and Respiratory Support, pp. 221–244.
Sotolongo-Costa, O., et al., 2003. Behavior of tumors under nonstationary therapy. Physica D 178 (3–4), 242–253.
Staub, J.F., et al., 2003. A nonlinear compartmental model of Sr metabolism. I. Non-steady-state kinetics and model building. Am. J. Physiol. Regul. Integr. Comp. Physiol. 284 (3), R819–R834.
Swan, G.W., 1981. Optimal control applications in biomedical engineering—a survey. Optimal Control Appl. Methods 2 (4), 311–334.
Takeuchi, Y., Iwasa, Y., Sato, K., 2007. Mathematics for Life Science and Medicine. Springer Science & Business Media.
Toker, D., et al., 2020. A simple method for detecting chaos in nature. Commun. Biol. 3 (1), 1–13.
Tong, Y.L., 1976. Parameter estimation in studying circadian rhythms. Biometrics 32 (1), 85.
Tuncer, N., et al., 2016. Structural and practical identifiability issues of immuno-epidemiological vector–host models with application to rift valley fever. Bull. Math. Biol. 78 (9), 1796–1827.
Turksoy, K., Cinar, A., 2014. Adaptive control of artificial pancreas systems—a review. J. Healthc. Eng. 5 (1), 1–22.
Valle, P.A., et al., 2018. Bounding the dynamics of a chaotic-cancer mathematical model. Math. Probl. Eng. 2018, 1–14.

Villaverde, A.F., 2019. Observability and structural identifiability of nonlinear biological systems. Complexity 2019, 1–12.
Villaverde, A.F., Barreiro, A., 2016. Identifiability of large nonlinear biochemical networks. Match 76 (2), 359–376.
Volterra, V., 1928. Variations and fluctuations of the number of individuals in animal species living together. ICES J. Mar. Sci. 3 (1), 3–51.
Wagner, C., 1996. Chaos in blood pressure control. Cardiovasc. Res. 31 (3), 380–387.
Wagner, C., 1998. Chaos in the cardiovascular system: an update. Cardiovasc. Res. 40 (2), 257–264.
Weerasinghe, H.N., et al., 2019. Mathematical models of cancer cell plasticity. J. Oncol. 2019, 1–14.
West, B.J., Zweifel, P.F., 1992. Fractal physiology and chaos in medicine. Phys. Today 45 (3), 68–70.
Westwick, D.T., Kearney, R.E., 2003. Identification of Nonlinear Physiological Systems, John Wiley & Sons, Inc., Hoboken, NJ, USA.
Wheat, A.L., Larkin, K.T., 2010. Biofeedback of heart rate variability and related physiology: a critical review. Appl. Psychophysiol. Biofeedback 35 (3), 229–242.
Wolpert, D.M., Pearson, K.G., Ghez, C.P.J., 2013. The organization and planning of movement. In: Principles of Neuroscience.
World Health Organization, 2019. World Health Statistics 2019: Monitoring Health for the SDGs, Sustainable Development goals. Geneva.
Xia, X., Moog, C.H., 2003. Identifiability of nonlinear systems with application to HIV/AIDS models. IEEE Trans. Autom. Control 48 (2), 330–336.
Yokrattanasak, J., et al., 2016. A simple, realistic stochastic model of gastric emptying. PLoS ONE, 1–8.
Zavala, E., et al., 2019. Mathematical modeling of endocrine systems. Trends Endocrinol. Metab. 30 (4), 244–257.
Zhang, T., et al., 2016. Biomechatronic design and control of an anthropomorphic artificial hand for prosthetic applications. Robotica 34 (10), 2291–2308.

CHAPTER 2

Mathematical modeling of cholesterol homeostasis

Krystian Kubica[a], **Joanna Balbus**[b]
[a]Faculty of Fundamental Problems of Technology, Department of Biomedical Engineering, Wroclaw University of Science and Technology, Wroclaw, Poland
[b]Department of Pure and Applied Mathematics, Wroclaw University of Science and Technology, Wroclaw, Poland

1 Introduction

Cholesterol is an important component of the membranes of eukaryotic cells, which determines their physical properties (Reinitzer, 1989; Alberts et al., 2010; Maxfield and van Meer, 2010; Mukherjee et al., 1998). Among the other functions, it stabilizes the so-called lipid rafts and membrane proteins (Simons and Ikonen, 1997; Edidin, 2003), while its derivatives participate in hormonal signaling (Steck and Lange, 2010; Burger et al., 2000). Cholesterol also acts as a precursor in the synthesis of adrenal steroid hormones, which include aldosterone and cortisol and about 50 steroids that are intermediates (Berg et al., 2006) as well as sex metabolites (estrogen, progesterone, testosterone) (Falkenstein et al., 2000). In addition, cholesterol acts as a substrate in the synthesis pathways of hepatic acid (Berg et al., 2006) and vitamin D3 (Zhu and Okamura, 1995), which plays an essential role in controlling the metabolism of calcium and phosphorus (Berg et al., 2006; Papadakis and McPhee, 2014). The total amount of cholesterol in the human body is approximately 0.2% of body weight (Sabine, 1977).

High blood cholesterol has been recognized as a risk factor for ischemic diseases. It is believed that an abnormal level of cholesterol is the cause of approximately 30% of ischemic heart diseases (according to the World Health Organization, https://www.who.int/gho/ncd/risk_factors/cholesterol_text/en/). For this reason, preventive measures must be followed to maintain cholesterol level within the normal physiological range, that is, $\leq$180 mg/dL of blood plasma (total cholesterol). These measures include maintaining a low-cholesterol diet in combination with an active and healthy lifestyle and/or administration of cholesterol-lowering drugs. Despite extensive research over the last 100 years, the links between

Control Theory in Biomedical Engineering
https://doi.org/10.1016/B978-0-12-821350-6.00002-0

cardiovascular diseases (CVDs) and cholesterol level in blood plasma, cholesterol content in food, and physical activity are not clearly understood. Moreover, the physiological and molecular processes underlying cholesterol homeostasis are still not definitively identified (Gold et al., 1992; Steinberg, 2004). However, it seems that there is a consensus on the relationship between plasma cholesterol level and the risk of development and/or mortality due to CVDs (Daniels et al., 2009). For ethical reasons, clinical trials and trials conducted on animals are limited, whereas epidemiological studies do not lead to unambiguous answers (McNamara, 2000). The results of studies on animals do not always correspond to processes occurring in the human body. An example of this is the research conducted on rabbits at the beginning of the 20th century in which the relationship between the formation of atherosclerotic lesions and a cholesterol-enriched diet was demonstrated. It is worth noting here that rabbits are herbivorous animals (a plant diet does not contain cholesterol) and hence their liver does not show the dependence of the rate of cholesterol synthesis on the level of cholesterol in the blood—but this relationship does exist in humans. In the case of studies on other animals, only the differences in lipoprotein profiles were observed in comparison to humans, and therefore the results of these studies make it difficult to interpret the changes occurring in the human body (McNamara, 2000).

A safe and relatively cheap way to expand the knowledge base on cholesterol homeostasis is mathematical modeling. A well-built model is a measure of the understanding of the studied process, and should not only reproduce the known reality but also give the possibility of prediction. With regard to cholesterol homeostasis, it can help in gaining a more complete understanding of the complex process and choosing optimal preventive measures and treatments for a related disorder (Goodman et al., 1973; Cobbold et al., 2001; Kervizic and Corcos, 2008; Mc Auley et al., 2012; Mc Auley and Mooney, 2015; Mishra et al., 2014; Hrydziuszko et al., 2014, 2015; Wrona et al., 2015; Toroghi et al., 2019).

Mathematical models of cholesterol homeostasis allowed for a better understanding of this complex physiological process and indicated new possibilities in identifying the regulatory mechanisms that can be controlled (Goodman et al., 1973; Cobbold et al., 2001; Kervizic and Corcos, 2008; Wattis et al., 2008; Tindall et al., 2009; Mc Auley et al., 2012; Mc Auley and Mooney, 2015; Hrydziuszko et al., 2014, 2015; Paalvast et al., 2015; Wrona et al., 2015; Pool et al., 2018). Pharmacological agents that are currently in use act at a specific point in their regulatory mechanisms, but none of them is a universal remedy due to various side effects. However, due to the development of various mathematical models, it is possible to predict the cause of the disorder in cholesterol metabolism and optimize the conditions to reduce its effects on the body.

In this study, we present a two-compartment ordinary differential equation (ODE) model of cholesterol homeostasis in the human body. The complex process was simplified into a two-compartment model in which the first compartment is blood flowing through the liver and the second compartment is the peripheral blood. Despite this simplification, we included the most important factors that affect the concentration of total cholesterol in the blood, namely, *de novo* synthesis, dietary intake, tissue demand, circulation through bile, and the kinetics of cholesterol exchange between the compartments.

2 Circulation of cholesterol in the human body

Fig. 1 shows a simplified scheme of the current state of knowledge on cholesterol homeostasis in the human body. In this scheme, we have ignored details of cholesterol absorption in the intestine, stages of its synthesis, as well

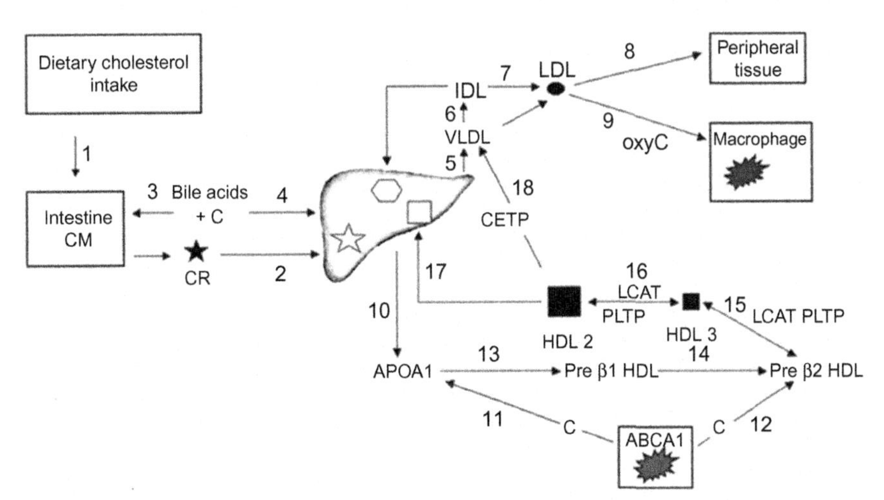

Fig. 1 A scheme of metabolic pathways involving cholesterol in the human body. *APOA1*, apolipoproteins; *C*, cholesterol; *CETP*, cholesteryl ester transfer protein; *CM*, chylomicrons; *CR*, chylomicron remnants; *HDL*, high-density lipoprotein; *IDL*, intermediate-density lipoprotein; *LCAT*, lecithin:cholesterol acyltransferase; *LDL*, low-density lipoprotein; *oxyC*, oxidized cholesterol; *PLTP*, phospholipid transferase; *VLDL*, very low density lipoprotein (Daniels et al., 2009).

as molecular aspects of binding of lipoproteins with specific receptors. However, this information can be found in a review paper (Afonso et al., 2018).

The scheme shown in Fig. 1 presents six metabolic pathways:

1. Route 1 indicates the pathway of intake of cholesterol together with food. The maximum amount of cholesterol contained in a daily dose of food is estimated to be 3 g (Sabine, 1977; McNamara, 2000). Following intake, the cholesterol with the chyme passes to the small intestine where it is absorbed into the lymphatic system in the form of chylomicrons (CM). These chylomicrons gradually release fatty acids (in the form of free fatty acids) and become aggregates called remnants (chylomicron remnants; CR). The remnants get into the bloodstream and are taken up by specific receptors in the liver (route 2). Regardless of this pathway, cholesterol is also absorbed together with the circulating bile and is transported via the portal vein to the liver.
2. Routes 3 and 4 describe the enterohepatic circulation of cholesterol along with the bile between the liver and intestines. The bile released from the liver contains up to 8% of cholesterol (Guyton and Hall, 2016), while the bile returning via the portal vein contains an additional amount of cholesterol derived from the membrane of dead intestinal epithelial cells and a part of the dietary cholesterol. Approximately 500 mg of cholesterol circulates along the selected route (Guyton and Hall, 2016). Furthermore, there is an additional factor binding cholesterol and bile, cholic acid, which is the main component of the bile. To maintain a constant pool, cholic acid is synthesized in the liver from cholesterol as needed (Guyton and Hall, 2016).
3. *De novo* synthesis of cholesterol takes place mainly in the liver (route 5) which is the main site of cholesterol synthesis. The rate of cholesterol synthesis depends on the amount of cholesterol present in the liver. The lower the concentration of cholesterol in the liver cells, the faster the rate of its synthesis (van der Wulp et al., 2013).
4. Cholesterol is released from the liver into the bloodstream in the form of lipoproteins called VLDL, which are transformed into IDL (route 6) and ultimately into LDL (route 7). Circulating with blood, LDL is captured by specific cellular receptors (controlled by the PCSK9 protein), following which it becomes one of the membrane components and participates as a substrate in the synthesis of biologically active derivatives, for example, steroid hormones (route 8). LDL is therefore a very important form of lipoproteins, although it is colloquially associated with "bad cholesterol." However, the development of atherosclerosis is closely

related to the oxidized form of LDL (oxyC). This oxidized cholesterol accumulates in macrophages (route 9), gradually transforming them into foam cells, which in turn leads to the development of atherosclerotic lesions.

5. An important part of the balanced transport of cholesterol is its reverse transport. This transport is carried out by the APOA1 protein, which is synthesized in the liver (route 10). With the help of macrophages, which have ABCA1 transport protein, APOA1 is enriched with cholesterol molecules that are mainly derived from dead cells during the process of apoptosis. Initially, the pre-β1 HDL aggregates are formed with a low-cholesterol content, which are gradually transformed into forms such as pre-β2 HDL, HDL3, and HDL2 following enrichment with cholesterol (routes 11–16). The last form is recognized by specific receptors in the liver (route 17) (Daniels et al., 2009).
6. Another important factor involved in cholesterol homeostasis is the cholesteryl ester transfer protein (CETP). This protein allows the exchange of cholesterol ester between lipoproteins (e.g., between HDL and LDL) (route 18) (Daniels et al., 2009).

3 Two-compartment model of cholesterol homeostasis

The current state of knowledge on cholesterol homeostasis, schematically shown in Fig. 1, can be summarized using a simplified mathematical model containing only two compartments: blood flowing through the liver (first compartment) and peripheral blood (second compartment) (Fig. 2).

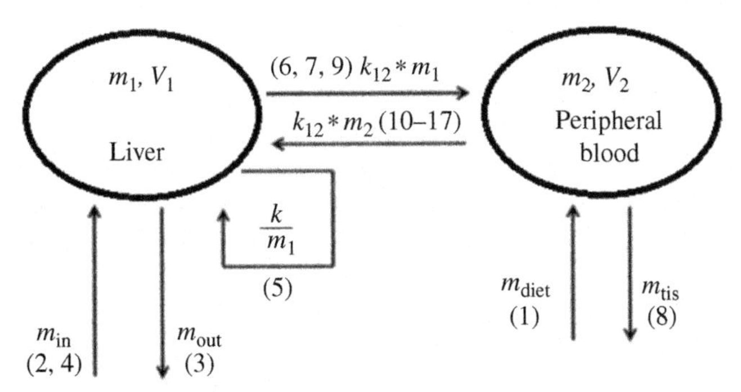

Fig. 2 Schematic representation of a two-compartment model of cholesterol homeostasis (Hrydziuszko et al., 2014). The numbers 1–17 correspond to the routes shown in Fig. 1.

The model can be expressed in mathematical terms using two differential equations: (1) and (2) (Hrydziuszko et al., 2014, 2015). The first equation expresses the rate of change in the mass of cholesterol (m_1) in the blood plasma flowing through the liver, while the second describes the rate of change in the mass of cholesterol (m_2) in the peripheral blood plasma.

$$\frac{dm_1}{dt} = \frac{k}{m_1} + k_{21}m_2 - k_{12}m_1 + m_{in} - m_{out} \tag{1}$$

$$\frac{dm_2}{dt} = -k_{21}m_2 + k_{12}m_1 - m_{tis} + m_{diet} \tag{2}$$

The amount of cholesterol in the first compartment depends on the following:

1. the rate of *de novo* synthesis of cholesterol in the liver (expressed by k/m_1, where k is the kinetic parameter), which is inversely proportional to m_1;
2. the rate of cholesterol exchange between the two compartments (expressed by $k_{12}m_1$ and $k_{21}m_2$, where k_{12} and k_{21} are the effective parameters responsible for the complex exchange of processes); and
3. the kinetics of cholesterol circulation with bile (described by parameters m_{in} and m_{out}).

In the second compartment, the amount of cholesterol depends on the following:

1. the rate of cholesterol exchange between compartments (expressed by $k_{12}m_1$ and $k_{21}m_2$);
2. the tissue demand for cholesterol (expressed by parameter m_{tis}) as a membrane component or as a substrate for the synthesis of steroid hormones and vitamin D; and
3. the rate of intake of dietary cholesterol (expressed by parameter m_{diet}).

Eqs. (1), (2) enable finding analytical stationary solutions, that is, when $dm_1/dt=0$ and $dm_2/dt=0$. These stationary solutions, marked as m_1^* and m_2^* (Eqs. 3 and 4), lead to additional restrictions on the values of the model parameters.

$$m_1^* = \frac{k}{m_{diet} + m_{in} - m_{out} - m_{tis}} \tag{3}$$

$$m_2^* = \frac{m_{diet}^2 + m_{diet}m_{in} - m_{diet}m_{out} - kk_{12} - 2m_{diet}m_{tis} - m_{in}m_{tis} + m_{out}m_{tis} + m_{tis}^2}{k_{21}\left(m_{diet} + m_{in} - m_{out} - m_{tis}\right)} \tag{4}$$

Because the mass must be positive, Eq. (3) leads to the following relationship: $m_{diet} + m_{in} > m_{out} + m_{tis}$. Similarly, Eq. (4) requires that $+m_{diet} \cdot m_{in} - m_{diet} - m_{out} k \cdot k_{12} - 2 \cdot m_{diet} \cdot m_{tis} - m_{in} \cdot m_{tis} + m_{out} \cdot m_{tis} + > 0$ because $k_{21}(m_{diet} + m_{in} - m_{out} - m_{tis})$ is greater than 0, which results from further analysis of the model solutions.

To establish the stability of the stationary point (m_1^*, m_2^*) one must compute the Jacobian matrix and study the eigenvalue of the matrix $J(m_1^*, m_2^*)$.

$$J\left(m_1^*, m_2^*\right) = \begin{bmatrix} \frac{-k}{\left(m_1^*\right)^2} - k_{12} & k_{21} \\ k_{12} & -k_{21} \end{bmatrix} \quad (5)$$

The eigenvalue of the aforementioned matrix can be obtained by the roots of the following equation:

$$\lambda^2 + \frac{k + k_{12}\left(m_1^*\right)^2 + k_{21}\left(m_1^*\right)^2}{\left(m_1^*\right)^2}\lambda + \frac{k k_{21}}{\left(m_1^*\right)^2} = 0 \quad (6)$$

The above equation has two roots that have the following form (Eq. 7):

$$\lambda_{1,2} = \frac{\mp\sqrt{\left(k_{12}\left(m_1^*\right)^2 + k_{21}\left(m_1^*\right)^2 + k\right)^2 - 4k_{21}k\left(m_1^*\right)^2} - k_{12}\left(m_1^*\right)^2 - k_{21}\left(m_1^*\right)^2 k}{2\left(m_1^*\right)^2} \quad (7)$$

The system (Eqs. 1, 2) is locally asymptotically stable if all the eigenvalues have negative real parts.

If at least one of the eigenvalues has a positive real part, then the system is unstable.

The system of Eqs. (1), (2) can be solved numerically using the Runge-Kutta method, coded in MATLAB version 2017a, solver ode45. The solutions are represented by time-dependent m_1 and m_2, that is, the mass of cholesterol contained in the blood plasma flowing through the liver and the mass of cholesterol in the peripheral blood plasma. To assess the diagnostic significance of the model, the theoretical solutions should be compared to the results of the analytical lipid profiles. Such tests are routinely carried out on the blood obtained from the peripheral vessels, and therefore their results should be compared to the results obtained for compartment II. To compare both the results (theoretical and experimental), the mass of cholesterol has to be divided by the appropriate plasma volume expressed in dL to obtain the concentration in mg/dL. For experimental verification of the solutions derived for compartment I, blood should be collected from the blood vessels of the liver, which requires significant surgical intervention.

4 Estimating the values of the model parameters

We assumed that the model refers to a 70-kg healthy subject with a total cholesterol concentration not exceeding 190 mg/dL of blood plasma. For such a person, the volume of blood plasma in the liver is estimated to be approximately 6.5 dL, while the volume of plasma in the peripheral blood is 23.5 dL. As an initial condition, we assumed that the cholesterol concentration in both the compartments is the same. According to this assumption, the initial cholesterol mass is $m_1(t_0) = 1235$ and $m_2(t_0) = 4465$ mg in compartments I and II, respectively.

In the next step, we have to determine the values of kinetic parameters. Because the rate of *de novo* cholesterol synthesis is higher when the amount of cholesterol in the liver is less, we expressed this relationship as k/m_1, where k is the kinetic constant. As the rate of cholesterol synthesis varies from 0.324 to 0.624 mg min^{-1} (Yashiro et al., 1994; Jones and Schoeller, 1990), the values of kinetic constant k vary in the range of 390–751 mg^2 min^{-1}.

Analysis of the two-compartment model in a study (Hrydziuszko et al., 2015) indicated a significant influence of the parameters describing the outflow (m_{out}) and inflow (m_{in}) of the cholesterol from and to the liver on the total cholesterol in the second compartment (peripheral blood). If parameters m_{out} and m_{in} exceeded 1.2 and 0.8 mg min^{-1}, respectively, the total cholesterol concentration was lower than 200 mg/dL. Generally, m_{out} is expected to be greater than m_{in} because about 95% of cholesterol leaving the liver with bile returns via the portal vein enriched by cholesterol derived from the membranes of dead enterocytes and partly by cholesterol in the dietary pool.

As mentioned in Section 1, almost every cell shows a cholesterol demand, which is fulfilled by the appropriate receptors that capture the cholesterol from the bloodstream. In the model, the tissue demand is described by the parameter m_{tis}, which takes a value of 0.243 mg min^{-1} (Sabine, 1977). Another source of external cholesterol is the diet based on animal products (in the diet based on plants and microorganisms, cholesterol is absent) (Sabine, 1977). The daily cholesterol content in the diet is estimated to be 3 g; however, only about up to 30% of it is absorbed in the intestine. The process of absorption of cholesterol from the diet into the plasma is delayed (because of the time required for the digestion and initial absorption by the lymphatic system) the parameter m_{diet} expressed in milligrams per minute should be equal to zero from the time the last meal was consumed (t_0) to about 400–600 min after the meal when it appears in the blood.

Complex and multistage cholesterol exchange processes occurring between compartments I and II were simplified to kinetic expressions in the form of a product of an effective kinetic constant k_{12} and k_{21} and an appropriate mass. Thus, $k_{12}m_1$ describes all the processes leading to the transport of cholesterol from compartment I to II, whereas $k_{21}m_2$ describes the reverse process. As presented in Fig. 1 LDL and HDL are the main lipoprotein fractions; LDL transports cholesterol from liver to peripheral blood, HDL transports cholesterol in opposite direction. For the stationary solutions, the expected rate of cholesterol transport in both the directions is the same; that is, $k_{12}m_1$ is equal to $k_{21}m_2$. Thus, for the normalized k_{21} value of 1 min^{-1} and according to the initial assumption that cholesterol concentrations in both the compartments are the same, the parameter $k_{12}=k_{21}V_2/V_1$, which is approximately 3.6 min^{-1}.

Now, having determined the values of parameters k, k_{21}, and k_{12}, the analysis of the stability of the model's solutions can be continued.

The system (Eqs. 1 and 2) is locally asymptotically stable if all the eigenvalues of matrix $J(\overset{*}{m}_1, \overset{*}{m}_2)$ have negative real parts.

For this condition, we can consider the following two cases:

(a) the expression under the root is negative if $(k_{12}\cdot \overset{*}{m}{}_1^2 + k_{21}\cdot \overset{*}{m}{}_1^2 + k)^2 - 4k_{21}\cdot k\cdot (\overset{*}{m}_1)^2$ is negative; and

(b) for the expression under the root to be positive, $((k_{12}\cdot(\overset{*}{m}_1)^2 + k_{21}\cdot(\overset{*}{m}_1)^2 + k)^2 - 4k_{21}\cdot(\overset{*}{m}_1)^2)^{0.5}$ has to be smaller than $k_{12}\cdot(\overset{*}{m}_1)^2 - k_{21}\cdot(\overset{*}{m}_1)^2 - k$.

For $k_{21}=1\,\text{min}^{-1}$, $k_{12}=3.6\,\text{min}^{-1}$, and k in the range of 390–751 $\text{mg}^2\,\text{min}^{-1}$, case "a" never happens, while case "b" is always satisfied for physiological masses of cholesterol in both the compartments. Thus, it can be concluded that Eqs. (3), (4) always lead to locally asymptotically stable points $(\overset{*}{m}_1, \overset{*}{m}_2)$.

5 Analysis of the solutions

Using the estimated values of the model parameters, the sensitivity of the model to changes in the values of parameters can be studied. The aim of this step is to evaluate the weight of individual parameters and to indicate the potential ways through which a disturbed system can be optimally controlled. Here, it should be noted that the results obtained are consistent with physiological knowledge. The stationary solutions (Eqs. 3, 4) should be analyzed for the condition $m_{diet} = 0$ (otherwise, $m_{diet} > 0$ corresponds to a continuous meal containing cholesterol), whereas parameters m_{in} and m_{out} describe the mean transport of cholesterol with bile, neglecting the role of the gallbladder. Analysis of the sensitivity of the model to changes in the parameters was performed based on Eq. (4). The calculated mass (m_2^*) divided by blood plasma volume (V_2) defines the concentration of total cholesterol in the peripheral blood, designated as c_2. The analysis was conducted with the following set of values: $k = 732\,\text{mg}^2\,\text{min}^{-1}$; $k_{12} = 1\,\text{min}^{-1}$; $k_{21} = 3.58\,\text{min}^{-1}$; $m_{tis} = 0.243\,\text{mg}\,\text{min}^{-1}$; $m_{in} = 0.8\,\text{mg}\,\text{min}^{-1}$; and $m_{out} = 1.2\,\text{mg}\,\text{min}^{-1}$. In each of the case shown in Fig. 3A–F, only one parameter was changed, which is shown in the description of the coordinate system.

The analysis shows that changes in the values of parameters k_{12} and k_{21} indicate qualitatively different responses: c_2 grows linearly with k_{12} exceeding 190 mg/dL for $k_{12} > 3.75\,\text{min}^{-1}$, whereas c_2 changes hyperbolically exceeding 190 mg/dL for $k_{21} < 1\,\text{min}^{-1}$. It is worth noting that the parameters k_{12} and k_{21} representing the rate of cholesterol exchange between compartments I and II appear in Eqs. (1), (2) in the same mathematical form.

The change in total cholesterol c_2 shown in Fig. 3C is in line with expectation. With age, the tissue demand for cholesterol decreases due to the decrease in the number of dividing cells as well as the decrease in the concentration of steroid hormones.

The parameter k responsible for the rate of *de novo* synthesis of cholesterol causes a linear increase in c_2 concentration along with an increase in its value. In our earlier work (Hrydziuszko et al., 2014, 2015; Wrona et al., 2015), we considered larger values of parameter k because the c_2 concentration referred to the whole blood. In this study, we reduced the range of changes in this parameter to $1000\,\text{mg}^2\,\text{min}^{-1}$ because the concentration of the cholesterol in analytical tests is referred against the volume of blood plasma.

Analysis of the model's sensitivity to the changes in the parameter (Fig. 3E and F) shows that the parameters m_{in} and m_{out} also have a significant influence on total cholesterol in the peripheral blood. Because these parameters determine the amount of cholesterol circulating along with the bile, this process of cholesterol circulation should be considered more carefully taking into account the periodic accumulation of bile in the gallbladder. If we assume that at the initial time t_0, when a meal containing cholesterol and fats is taken, the gallbladder is completely filled with bile and is expected to release its contents into the duodenum during t_0 to t_1. This period of time varies between 30 and 100 min depending on the food composition, the

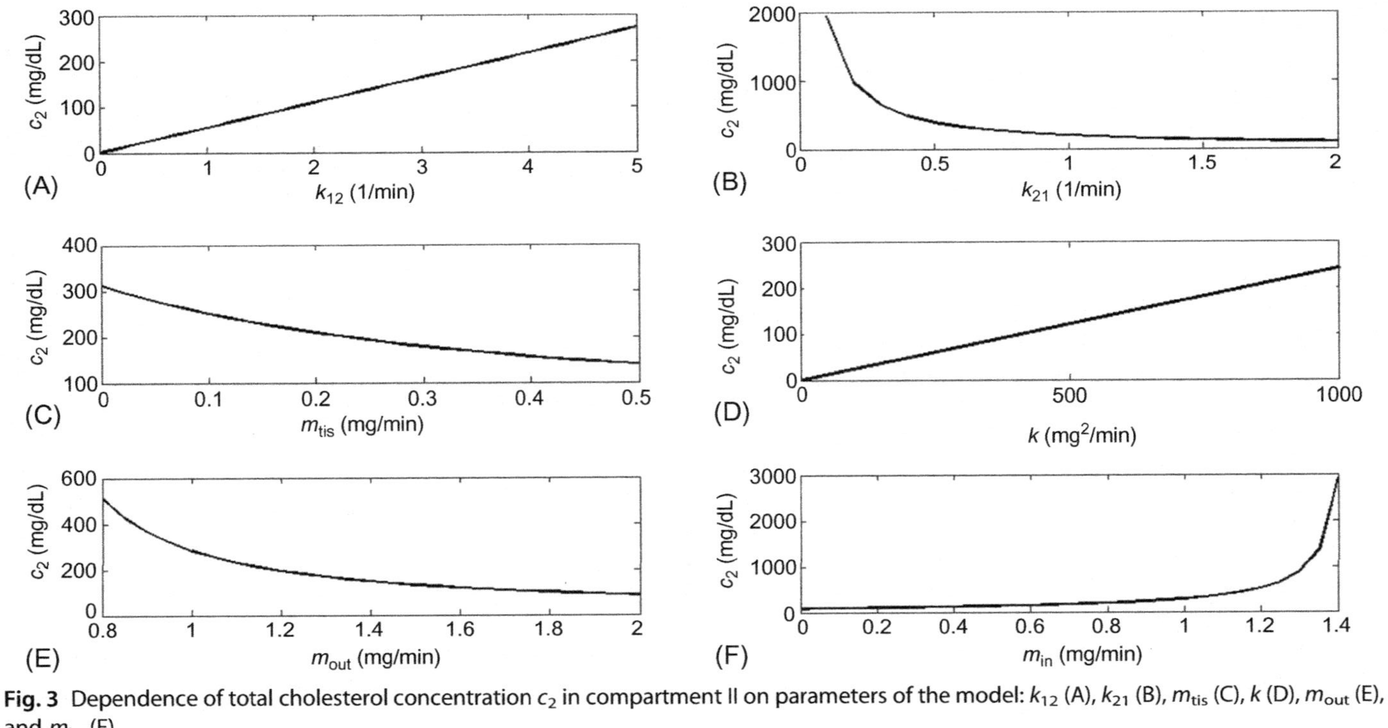

Fig. 3 Dependence of total cholesterol concentration c_2 in compartment II on parameters of the model: k_{12} (A), k_{21} (B), m_{tis} (C), k (D), m_{out} (E), and m_{in} (F).

physiological condition of the gallbladder, and the value of the body mass index (BMI) (Di Ciaula et al., 2012). After emptying, the gallbladder is refilled to its maximum volume with hepatic bile. Next, if the gallbladder is not stimulated to contract the bile accumulated within, the bile is concentrated by the active removal of water. This process is accompanied by influx of the liver bile, which can be treated as a constant component of enterohepatic circulation. Thus, the process of cholesterol transport with bile between the liver and the intestine can be described using constant parameters m_{in} and m_{out} and variable parameters M^*_{in} and M^*_{out} which change over time (associated with the gallbladder motility).

Another process that takes place over a specific time interval is the absorption of cholesterol from food, and the amount of absorption is marked as M^*_{diet}. We have already shown that such processes can be expressed by the function $A\sin^2(\omega t)$, where A is the amplitude of a given process expressed in $\mathrm{mg\,min}^{-1}$ (Żulpo et al., 2018). It is favorable to associate the angular pulsation ω with the length of the time interval in which a given process occurs by dependence (Eq. 8):

$$\omega = \frac{\pi}{t_e - t_b} \tag{8}$$

where t_b and t_e are the start and end times of a given process. Additionally, if we include a shift t_b relative to $t_0 = 0$ (the global start time of all processes), then the function $A\sin^2(\omega(t - t_b))$ takes the value 0 for $t = t_b$ and for $t = t_e$ and only once the maximum value A in the interval $(t_e - t_b)$. To determine the change in the appropriate mass described by this function in the mentioned time interval, the integral of the expression (Eq. 9) should be calculated:

$$\int_{t_b}^{t_e} A\sin^2(\omega(t - t_b))dt = A\left(\frac{t_e - t_b}{2} - \frac{1}{4\omega}(\sin(2\omega(t_e - t_b)) + \sin(2\omega(t_b - t_b)))\right) \tag{9}$$

Through this shift, the solution of Eq. (9) takes a simpler form (Eq. 10):

$$\int_{t_b}^{t_e} A\sin^2(\omega(t - t_b))dt = A\frac{t_e - t_b}{2} \tag{10}$$

To estimate the variable M^*_{out}, which is the mass of cholesterol leaving compartment I together with bile and changes over time, it is necessary to calculate the mass of bile ejected from the gallbladder. The amount of cholesterol is proportional to the mass of bile (cholesterol makes up about

8% of the dry component of bile) (Guyton and Hall, 2016). During contraction, the gallbladder releases 75% of the accumulated bile—m_{bile}. Finally, the mass of bile ejected from the gallbladder during contraction from t_0 to t_1 can be expressed as (Eq. 11):

$$\int_{t_0}^{t_1} M_b \sin^2(\omega_b(t - t_0))dt = M_b \frac{t_1 - t_0}{2} = 0.75 m_{\text{bile}} \tag{11}$$

Thus, the amount of cholesterol transported by bile into the duodenum represented by M_{out}^* is $0.08 \times 0.75 m_{\text{bile}}$.

Cholesterol carried by bile returns to the liver (first compartment) after its absorption in the small intestine, and the gallbladder is refilled with liver bile approximately three times slower compared to its emptying (Di Ciaula et al., 2012). The process of absorption of cholesterol along with bile can also be described by Eq. (8). In the case of a vegetarian diet, the amount of cholesterol returning with the bile to the liver (M_{in}^*) should be equal to the amount of cholesterol ejected along with the bile from the gallbladder reduced by losses with feces (w), that is, about 6% (Guyton and Hall, 2016) (Eq. 12):

$$M_{\text{in}}^* = M_{\text{out}}^*(1 - w) \tag{12}$$

If we define the time of absorption of cholesterol with bile as $t_3 - t_2$, then the amount of cholesterol returning to the liver (M_{in}^*) can be expressed as:

$$M_{\text{in}}^* = \int_{t_2}^{t_3} M_{\text{in}} \sin^2(\omega_{\text{in}}(t - t_2))dt = M_{\text{in}} \frac{t_3 - t_2}{2} \tag{13}$$

Based on relation (12), we can determine the amplitude (M_{in}) using Eq. (14):

$$M_{\text{in}} = M_{\text{out}} \frac{t_1 - t_0}{t_3 - t_2}(1 - w) \tag{14}$$

where

$$M_{\text{out}} = \frac{0.12}{t_1 - t_0} m_{\text{bile}} \tag{15}$$

In the case of a cholesterol-containing diet, most of the cholesterol is primarily absorbed by the lymphatic system and the expected increase in its concentration in the circulating blood is delayed compared to the time of the meal. It is worth noting here that in the intestine, the bile is also enriched by about 30% of dietary cholesterol. Thus, in this case, to estimate the amount of cholesterol returning via the portal vein to the liver, the amount ejected from the gallbladder, enrichment by 30% of the dietary amount, and the amount removed along with fecal masses should all be taken into account.

The amount of cholesterol absorbed from the diet can be calculated analogously as the other parameters M_{in}^{*} and M_{out}^{*} (Eq. 16):

$$m_{diet} = \int_{t_p}^{t_k} M_{diet} \sin^2(\omega_{diet}(t - t_{babs}))dt = M_{diet}\frac{t_{eabs} - t_{babs}}{2} \tag{16}$$

where m_{diet} is the total dose of cholesterol consumed at time t_0, and t_{babs} and t_{eabs} are the time of beginning and end of absorption, respectively.

An exemplary profile of change in cholesterol concentration in the second compartment (c_2) and the accompanying change in the rate of *de novo* cholesterol synthesis in the liver is shown in Fig. 4A and B. We calculated the concentration of cholesterol for a person weighing 70 kg with an initial mass

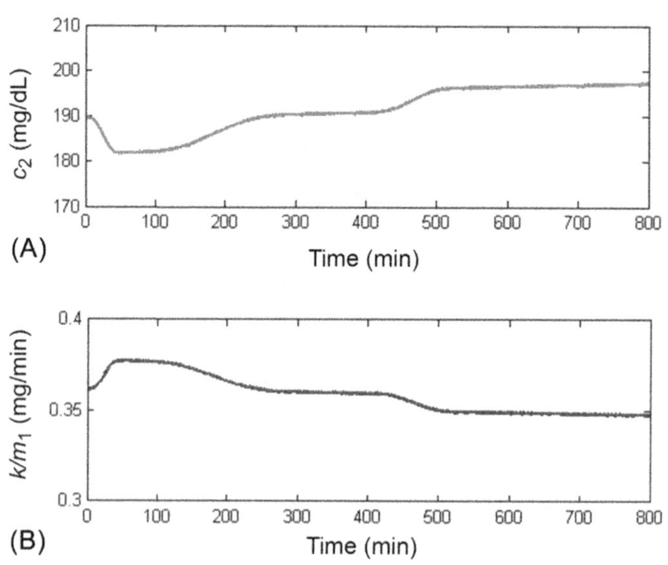

Fig. 4 Profiles of changes in total cholesterol concentration c_2 after a meal containing 1 g of cholesterol (A) and the rate of *de novo* cholesterol synthesis, k/m_1 (B). This figure presents solution for a two-compartment model with the following set of parameters: $k = 450\,\text{mg}^2\,\text{min}^{-1}$, $k_{12} = 3.58\,\text{min}^{-1}$; $k_{21} = 1\,\text{min}^{-1}$; $m_{tis} = 0.243\,\text{mg min}^{-1}$; $M_{in} = 1.96\,\text{mg min}^{-1}$; $M_{out} = 9.6\,\text{mg min}^{-1}$; $M_{diet} = 2.3\,\text{mg min}^{-1}$; $t_1 = 50\,\text{min}$; $t_2 = 70\,\text{min}$; $t_3 = 300\,\text{min}$; $t_4 = 400\,\text{min}$; and $t_5 = 530\,\text{min}$.

of cholesterol in hepatic blood plasma and in the peripheral blood plasma as equal to $m_1(t_0) = 1235$ mg and $m_2(t_0) = 4465$ mg, respectively. We assumed that at $t_0 = 0$, a meal containing 500 mg of cholesterol was consumed. Because cholesterol is insoluble in water it has to be contained in food together with fats. Therefore, such a meal results in the release of bile from the gallbladder, which carries some amount of cholesterol with an amplitude of $M_{out} = 9.6\,\text{mg}\,\text{min}^{-1}$ calculated for $m_{bile} = 4000$ mg (Mok et al., 1974) and time of release of bile $t_1 = 50$ min (Di Ciaula et al., 2012) (Eq. 15). As a result, we observed a decrease in the total concentration of cholesterol (c_2) and an appropriate increase in the kinetics of cholesterol biosynthesis. The cholesterol released into the duodenum returns with bile to the liver through the portal vein between $t_2 = 70$ min and $t_3 = 300$ min (Johnson, 2013). The amplitude of this process ($M_{in} = 1.96\,\text{mg}\,\text{min}^{-1}$) was calculated based on Eq. (12) assuming a loss of 6% (Guyton and Hall, 2016). The expected result is an increase in the level of cholesterol during this time. Finally, the cholesterol from food is absorbed between 400 and 530 min after the consumption of a meal with an amplitude of $M_{diet} = 2.3\,\text{mg}\,\text{min}^{-1}$ (Eq. 16). For this to happen, the cholesterol must first enter the small intestine, from where the majority of it is absorbed in the form of chylomicrons, which are gradually transformed into the remnants and move into the circulatory system. The expected result of dietary intake of cholesterol is an increase in its concentration (c_2) and the accompanying reduction in the rate of *de novo* cholesterol synthesis. The profile obtained for the total change in the concentration of cholesterol in the peripheral blood plasma is physiologically justified. In an experiment conducted on 26 volunteers who received a fatty meal, containing 1.0 g of fat/kg body weight and 7.0 mg/kg of cholesterol, an analogous profile was obtained (after adding up the results according to the Fridewald's rule (Friedewald et al., 1972) Tc = LDL + HDL + 0.2TG, where Tc—is total cholesterol, LDL and HDL—are the lipoprotein fractions, and TG—triglyceride) (Cohn et al., 1988).

The maintenance of normal physiological levels of blood cholesterol is recognized as an important factor in preventing the development of CVDs. Preventing adverse changes means focusing on a healthy and active lifestyle. The simple two-compartment model of cholesterol homeostasis presented in this study makes it easier to understand the mechanisms that lead to an increase in the level of blood cholesterol. This model shows that there is an increase in the concentration of cholesterol with age when the tissue demand decreases (by lowering the value of m_{tis} parameter). Disorders in any of the multistage process of cholesterol exchange between the two compartments can also lead to adverse changes in cholesterol metabolism. This effect can be observed for the increasing value of the parameter k_{12} of more than 3.6 mg min^{-1} and decreasing value of k_{21} of less than 1 mg min^{-1}.

Based on the presented model, there was a significant effect of cholesterol circulating with bile on its concentration in the blood. These relationships show the importance of diet in increasing intestinal peristalsis (e.g., diet rich in fiber). Increasing the intestinal peristalsis decreases the absorption time of cholesterol, which in the model corresponds to a decrease in the value of the parameter m_{in}. This decrease can also explain the effect of an older generation drug that binds with the components of bile and gets excreted along with the feces. When analyzing the significance of the m_{out}, the parameter describing the secretion of cholesterol in the bile, one can justify the beneficial effect of choleretic agents. At present, inhibitors of cholesterol biosynthesis (statins) are commonly used in medicine for lowering cholesterol. In terms of our model, the effect of this drug can be taken into account by reducing the value of parameter k, which is responsible for the rate of cholesterol biosynthesis.

This two-compartment model demonstrated its didactic possibilities about mechanisms of cholesterol homeostasis by using computers (Wrona et al., 2015).

We hope that in the future our simple model will be expanded to support prevention and to develop personalized therapeutic regime.

6 Summary and conclusion

The simplified two-compartment model enables studying the influence of known factors, such as the rate of *de novo* synthesis, enterohepatic circulation (Mishra et al., 2014), tissue demands, and the amount of cholesterol absorbed from food, on cholesterol homeostasis (Afonso et al., 2018). Although this model does not explicitly distinguish the lipoproteins containing cholesterol, it can be used to understand the importance of the rate of formation of the main fractions, that is, LDL and HDL. Indirectly, the role of CETP can also be examined.

In addition, the simplified two-compartment model helps in analyzing the causes of disorders of cholesterol homeostasis and the different ways to lower cholesterol concentration. In particular, it allows investigating:

- the effectiveness of therapy with drugs that inhibit the enterohepatic circulation,
- the usage of inhibitors of cholesterol synthesis (statins),
- the significance of diet components that increase the intestine peristalsis,

- the regulation of CETP activity, and
- the change in tissue requirements for cholesterol with age.

The two-compartment mathematical model developed in this study showed a significant contribution of bile in the maintenance of cholesterol homeostasis (Hrydziuszko et al., 2015). Further studies with respect to the extension of the model to include a third compartment representing the gallbladder are highly warranted. The expected response to a meal containing fat and cholesterol is contraction of the gallbladder and release of bile carrying some cholesterol into the duodenum—in terms of the model, it means that in addition to the constant component m_{out}, the variable component M^*_{out} should be considered. Thus, some cholesterol enters the intestine from the bile and the gallbladder is filled with the new hepatic bile, which again carries cholesterol from the liver. When the gallbladder is full, a constant flow (m_{out}) of the hepatic bile can be expected. Extending the model by a third compartment also creates the possibility of studying the effect of the primary component of the bile, that is, cholic acid on cholesterol homeostasis, because it is synthesized in the liver from cholesterol molecules. A well-defined three-compartment model could open the possibility of testing the susceptibility to the formation of gallstones. In addition, it will enable analyzing the effect of the dysfunction of gallbladder motility and condition of the bile ducts on cholesterol homeostasis.

References

Afonso, M.S., Machado, R.M., Lavrador, M.S., Quintao, E.C.R., Moore, K.J., Lottenberg, A.M., 2018. Molecular pathways underlying cholesterol homeostasis. Nutrients 10, 760.

Alberts, B., Bray, D., Hopkin, K., Johnson, A., Lewis, J., Raff, M., Roberts, K., Walter, P., 2010. Essential Cell Biology. Garland Science, Taylor &Francis Group, New York.

Berg, J.M., Tymoczko, J.L., Stryer, L., 2006. Biochemistry. Freeman, New York.

Burger, K., Gimpl, G., Fahrenholz, F., 2000. Regulation of receptor function by cholesterol. Cell. Mol. Life Sci. 57, 1577–1592.

Cobbold, C.A., Sherratt, J.A., Maxwell, S.R.J., 2001. Lipoprotein oxidation and its significance for atherosclerosis: a mathematical approach. Bull. Math. Biol. 64 (1), 65–95. https://doi.org/10.1006/bulm.2001.0267.

Cohn, J.S., McNamara, J.R., Cohn, S.D., Ordovas, J.M., Schaefer, E.J., 1988. Postprandial plasma lipoprotein in human subjects of different ages. J. Lipid Res. 29, 469–479.

Daniels, T.F., Killinger, K.M., Michal Jr., R.W., Jiang, Z., 2009. Lipoproteins, cholesterol homeostasis and cardiac health. Int. J. Biol. Sci. 5, 474–488.

Di Ciaula, A., Wang, D.Q., Portincasa, P., 2012. Gallbladder and gastric motility in obese newborns, preadolescents and adults. J. Gastroenterol. Hepatol. 27 (8), 1298–1305. https://doi.org/10.1111/j.1440-1746.2012.07149.x.

Edidin, M., 2003. The state of lipid rafts: from model membranes to cells. Annu. Rev. Biophys. Biomol. Struct. 32, 257–283.

Falkenstein, E., Tillmann, H., Christ, M., Feuring, M., Wehling, M., 2000. Multiple actions of steroid hormones–a focus on rapid, nongenomic effects. Pharmacol. Rev. 52, 513–555.

Friedewald, W.T., Levy, R.I., Fredrickson, D.S., 1972. Estimation of the concentration of low-density lipoprotein cholesterol in plasma, without use of the preparative ultracentrifuge. Clin. Chem. 18, 499–502.

Gold, P., Grover, S., Roncari, D.A.K., 1992. Cholesterol and Coronary Heart Disease: The Great Debate. Parthenon, Carnforth.

Goodman, D.S., Noble, R.P., Dell, R.B., 1973. Three-pool model of the long-term turnover of plasma cholesterol in man. J. Lipid Res. 14 (1973), 178–188.

Guyton, A.C., Hall, J.E., 2016. Textbook of Medicinal Physiology, 13th ed. Elsevier, Philadelphia.

Hrydziuszko, O., Wrona, A., Balbus, J., Kubica, K., 2014. Mathematical two-compartment model of human cholesterol transport in application to high blood cholesterol diagnosis and treatment. Electron. Notes Theor. Comput. Sci. 306, 19–30. https://doi.org/10.1016/j.entcs.2014.06.012.

Hrydziuszko, O., Balbus, J., Żulpo, M., Wrona, A., Kubica, K., 2015. Mathematical analyses of two-compartment model of human cholesterol circulatory transport in application to high blood cholesterol prevention, diagnosis and treatment. Chic. J. Theor. Comput. Sci. 608, 98–107. https://doi.org/10.1016/j.tcs.2015.07.057.

Johnson, R.L., 2013. Your Digestive System. Lerner Publications Company.

Jones, P.J.H., Schoeller, D.A., 1990. Evidence for diurnal periodicity in human cholesterol synthesis. J. Lipid Res. 31, 667–673.

Kervizic, G., Corcos, L., 2008. Dynamical modeling of the cholesterol regulatory pathway with Boolean networks. BMC Syst. Biol. 2, 99. https://doi.org/10.1186/1752-0509-2-99.

Maxfield, F.R., van Meer, G., 2010. Cholesterol, the central lipid of mammalian cells. Curr. Opin. Cell Biol. 22, 422–429.

Mc Auley, M.T., Mooney, K.M., 2015. Computationally modeling lipid metabolism and aging: a mini-review. Comput. Struct. Biotechnol. J. 13, 38–46. https://doi.org/10.1016/j.csbj.2014.11.006.

Mc Auley, M.T., Wilkinson, D.J., Jones, J.J.L., Kirkwood, T.B.L., 2012. A whole-body mathematical model of cholesterol metabolism and its age-associated dysregulation. BMC Syst. Biol. 6(130) https://doi.org/10.1186/1752-0509-6-130.

McNamara, D.J., 2000. Dietary cholesterol and atherosclerosis. Biochim. Biophys. Acta 1529 (1–3), 310–320.

Mishra, S., Somvanshi, P.R., Venkatesh, K.V., 2014. Control of cholesterol homeostasis by enterohepatic bile transport—the role of feedback mechanisms. RSC Adv. 4, 58964–58975.

Mok, H.Y.I., Perry, P.M., Dowling, R.H., 1974. The control of bile acid pool size. Gut 15, 247–253.

Mukherjee, S., Zha, X., Tabas, I., Maxfield, F.R., 1998. Cholesterol distribution in living cells: fluorescence imaging using dehydroergosterol as a fluorescent cholesterol analog. Biophys. J. 75, 1915–1925.

Paalvast, Y., Kuivenhoven, J., Groen, A., 2015. Evaluating computational models of cholesterol metabolism. Biochim. Biophys. Acta 1851, 1360–1376.

Papadakis, M.A., McPhee, S.J., 2014. Current Medical Diagnosis and Treatment. McGraw-Hill Education.

Pool, F., Sweby, P.K., Tindall, M.J., 2018. An integrated mathematical model of cellular cholesterol biosynthesis and lipoprotein metabolism. Processes 6, 134. https://doi.org/10.3390/pr6080134.

Reinitzer, F., 1989. Contributions to the knowledge of cholesterol. Liq. Cryst. 5, 7–18.

Sabine, J.R., 1977. Cholesterol. Mercel Dekker, New York.
Simons, K., Ikonen, E., 1997. Functional rafts in cell membranes. Nature 387, 569–572.
Steck, T.L., Lange, Y., 2010. Cell cholesterol homeostasis: mediation by active cholesterol. Trends Cell Biol. 20, 2010. https://doi.org/10.1016/j.tcb.2010.08.007.
Steinberg, D., 2004. Thematic review series: the pathogenesis of artherosclerosis. An interpretive history of the cholesterol controversy: part I. J. Lipid Res. 45, 1583–1593.
Tindall, M., Wattis, J., O'Malley, B., Pickersgill, L., Jackson, K., 2009. A continuum receptor model of hepatic lipoprotein metabolism. J. Theor. Biol. 257, 371–384.
Toroghi, M.K., Cluett, W.R., Mahadevan, R., 2019. A multi-scale model for low-density lipoprotein cholesterol (LDL-C) regulation in the human body: application to quantitative systems pharmacology. Comput. Chem. Eng. 130, 106507.
van der Wulp, M.Y.M., Verkade, H.J., Groen, A.K., 2013. Regulation of cholesterol homeostasis. Mol. Cell. Endocrinol. 368 (1–2), 1–16.
Wattis, J., O'Malley, B., Blackburn, H., Pickersgill, L., Panovska, J., Byrne, H., Jackson, K., 2008. Mathematical model for low density lipoprotein (LDL) endocytosis by hepatocytes. Bull. Math. Biol. 70, 2303–2333.
Wrona, A., Balbus, J., Hrydziuszko, O., Kubica, K., 2015. Two compartment model as a teaching tool for cholesterol homeostasis. Adv. Physiol. Educ. 39, 372–377. https://doi.org/10.1152/advan.00141.2014.
Yashiro, M., Muso, E., Matsushima, M., Nagura, R., Sawanishi, K., Sasayama, S., 1994. Two-compartment model of cholesterol kinetics for establishment of treatment strategy of LDL apheresis in nephritic hypercholesterolemia. Blood Purif. 12.
Zhu, G.D., Okamura, W.H., 1995. Synthesis of vitamin D (calciferol). Chem. Rev. 95, 1877–1952.
Żulpo, M., Balbus, J., Kuropka, P., Kubica, K., 2018. A model of gallbladder motility. Comput. Biol. Med. 93, 139–148.

CHAPTER 3

Adaptive control of artificial pancreas systems for treatment of type 1 diabetes

Iman Hajizadeh[a], Mohammad Reza Askari[a], Mert Sevil[b], Nicole Hobbs[b], Rachel Brandt[b], Mudassir Rashid[a], Ali Cinar[a,b]
[a]Department of Chemical and Biological Engineering, Illinois Institute of Technology, Chicago, IL, United States
[b]Department of Biomedical Engineering, Illinois Institute of Technology, Chicago, IL, United States

1 Introduction

The applications of systems engineering in the area of health care are numerous (Ogunnaike, 2019; Doyle et al., 2007; Goodwin et al., 2019; Parker, 2009; Parker and Doyle, 2001). They include automated insulin delivery systems for people with type 1 diabetes (T1D) (Hajizadeh et al., 2019c; Garcia-Tirado et al., 2019; Hovorka et al., 2004), modeling and control for cancer treatment (Martin, 1992; Martin and Teo, 1994), drug infusion in critical care (Yu et al., 1992; Behbehani and Cross, 1991), and developing optimal treatment for HIV infection (Hajizadeh and Shahrokhi, 2015; Pannocchia et al., 2010).

One of the most pressing medical problems in the world today is the growing epidemic of diabetes. People with T1D have β cells in the pancreas that do not produce any insulin. Consequently, people with T1D are unable to regulate their blood glucose concentration (BGC) without injecting appropriate amounts of insulin through either multiple daily injections or continuous subcutaneous insulin infusion (CSII) (Eisenbarth, 2005; Cooke and Plotnick, 2008).

Glycemic control can be enhanced in people with T1D using multivariable artificial pancreas (AP) systems that automatically compute the required amount of insulin (Turksoy et al., 2018; Hajizadeh et al., 2019c). This technology integrates a continuous glucose monitoring (CGM) sensor, a wearable device, a CSII pump, and control algorithms. Real-time measurements for feedback/feedforward control, accurate estimates of nonmeasurable physiological variables of patients, an accurate dynamic model of the human

Control Theory in Biomedical Engineering
https://doi.org/10.1016/B978-0-12-821350-6.00003-2

body, and a powerful control algorithm are required to develop a safe and reliable AP system. The AP system, tested in many simulation and clinical studies, can reduce the risks of immediate life-threatening conditions, such as severe hypoglycemia and ketoacidosis, and long-term health complications, such as cardiovascular disease, nephropathy, neuropathy, and retinopathy (Turksoy et al., 2017; Peyser et al., 2014; Esposito et al., 2018; Bekiari et al., 2018).

Complex nonlinear dynamical systems such as the metabolic processes in the human body are particularly challenging to model and control due to time-varying characteristics, nonlinear behavior, presence of stochastic and unknown disturbances, uncertain time-varying delays, and variations between and within peoples' metabolic activities. The glycemic models proposed in the literature can be categorized as physiological and data-driven models (Silvia et al., 2017). Physiological models consist of simultaneous differential equations describing the insulin and glucose interactions based on mass exchange between compartments representing various organs of depots. Data-driven models with relatively simpler structures that can characterize the relationships among the measured variables generally need less computational load. Subspace-based system identification methods can readily identify linear state-space models from multiinput, multioutput sampled data of a dynamic system. However, physiological and data-driven models with fixed parameters cannot accurately describe the dynamic behavior of BGC variations in the human body over wide ranges of real-world conditions. Therefore, the models need to be appropriately adapted online to characterize the current dynamics of the individuals and make accurate short-term predictions of glucose concentration measurements. For this purpose, adaptive system identification approaches are proposed to determine linear, time-varying models and effectively characterize the evolving glycemic dynamics, thus allowing the adaptive models to be valid over a diverse range of daily conditions. In our previous work, an adaptive-personalized modeling approach considering the effects of unannounced meals and exercise on transient glycemic dynamics was proposed and applied to 15 clinical data sets involving closed-loop experiments of the AP systems (Hajizadeh et al., 2017b, 2018a, b, d).

To minimize the risk of hypoglycemia, AP systems need safety constraints to avoid insulin overdosing. Quantifying the amount of active insulin present in the body is difficult due to lack of sensors to measure plasma insulin concentration (PIC) in the bloodstream. However, accurate PIC estimates can be obtained by using CGM measurements and infused insulin information with estimators designed based on glucose-insulin dynamic models (de Pereda et al., 2016; Eberle and Ament, 2011; Neatpisarnvanit and Boston, 2002; Hajizadeh et al., 2017a, 2018c). In our previous work, the design of adaptive-personalized PIC estimators that directly take into account the intersubject and intrasubject variabilities in glucose-insulin dynamics is investigated using different estimation techniques (Hajizadeh et al., 2017a). Clinical experimental data from subjects with T1D were used to analyze the accuracy and reliability of the PIC estimates (Hajizadeh et al., 2017a, 2018c).

The control algorithm computes the optimum amount of insulin by considering safety constraints that avoid insulin overdosing. Model predictive control (MPC) formulations are efficient techniques for AP systems as they can handle multivariable complex systems with constraints (Hajizadeh et al., 2019b, c; Hovorka et al., 2004; Clarke et al., 2009; Laguna Sanz et al., 2017; Gondhalekar et al., 2016; Chakrabarty et al., 2018; Cameron et al., 2011; Toffanin et al., 2013; Boiroux et al., 2018; El Fathi et al., 2018; Messori et al., 2018). However, MPC performance is affected by different factors such as accuracy of the model, formulation of the objective function, and system constraints. For AP systems, these factors need to be defined appropriately for effective glycemic control. In this work, accurate time-varying glycemic models are identified recursively for BGC predictions. The key controller parameters including the controller set-point, the objective function weights, and the system constraints are defined in an adaptive way to accommodate various situations faced daily by people with diabetes. Using machine-learning techniques and patients' historical data, these parameters are also modified in advance for the anticipated periods of disturbance effects such as meals and exercise (Hajizadeh et al., 2019a, d).

Motivated by the previous considerations, a personalized multivariable, multimodule artificial pancreas (PMM-AP) system is proposed to effectively control the BGC without manual user announcements for meals and exercise. The proposed PMM-AP uses physiological signals from a wearable device and estimates of unannounced meal effects (from recent glucose and insulin data) and PIC in addition to glucose measurements. A general flowchart of the proposed method is presented in Fig. 1. An adaptive-personalized PIC estimator summarized in Section 2.1 generates estimates of the PIC. To identify time-varying glycemic models, a recursive system identification technique summarized in Section 2.2 is used to characterize the time-varying glucose-insulin dynamics. Then, the identified models are employed for the design of an adaptive-learning model predictive

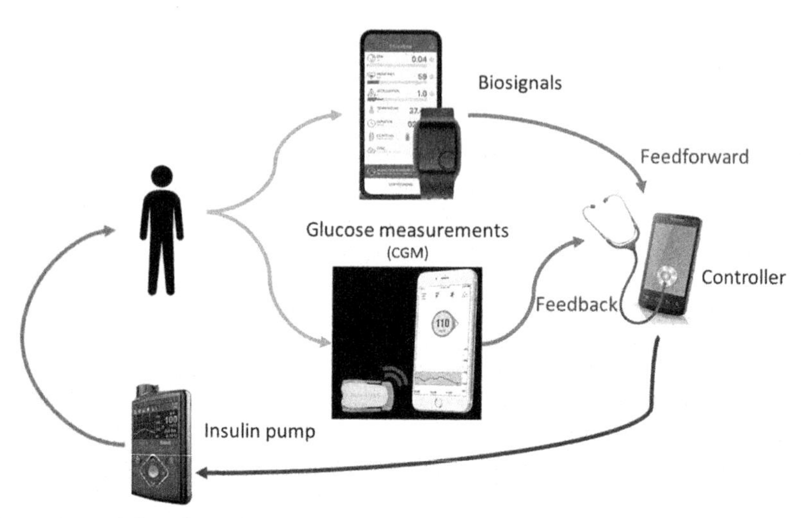

Fig. 1 General flowchart of the proposed PMM-AP system.

control (AL-MPC) formulation in Section 3. Simulation case studies using a multivariable simulator (mGIPsim) illustrate the efficacy of the proposed PMM-AP system in Section 4. Finally, concluding remarks are given in Section 5.

2 Methods

In this section, a brief overview of the adaptive-personalized PIC estimator is provided, followed by a review of the recursive system identification algorithm for the identification of linear, time-varying glycemic models. Subsequently, the AL-MPC formulation is presented. Fig. 2 illustrates the proposed PMM-AP system in which, first, an unscented Kalman filter (UKF) estimates the PIC value using the CGM data and infused insulin information. Then, the PIC and unannounced meal estimates, physiological variables and CGM data are used to identify time-varying linear state-space models. The estimated PIC and the identified state-space models are used in the AL-MPC for the insulin computation.

2.1 Adaptive-personalized PIC estimator

A glucose-insulin dynamics model, Hovorka's model, is used to design the PIC estimator (Hovorka et al., 2004; Hajizadeh et al., 2017a, 2018c). UKF algorithm (Kolås et al., 2009) is applied for the estimation of the state variables and the time-varying parameters to provide PIC estimates in real

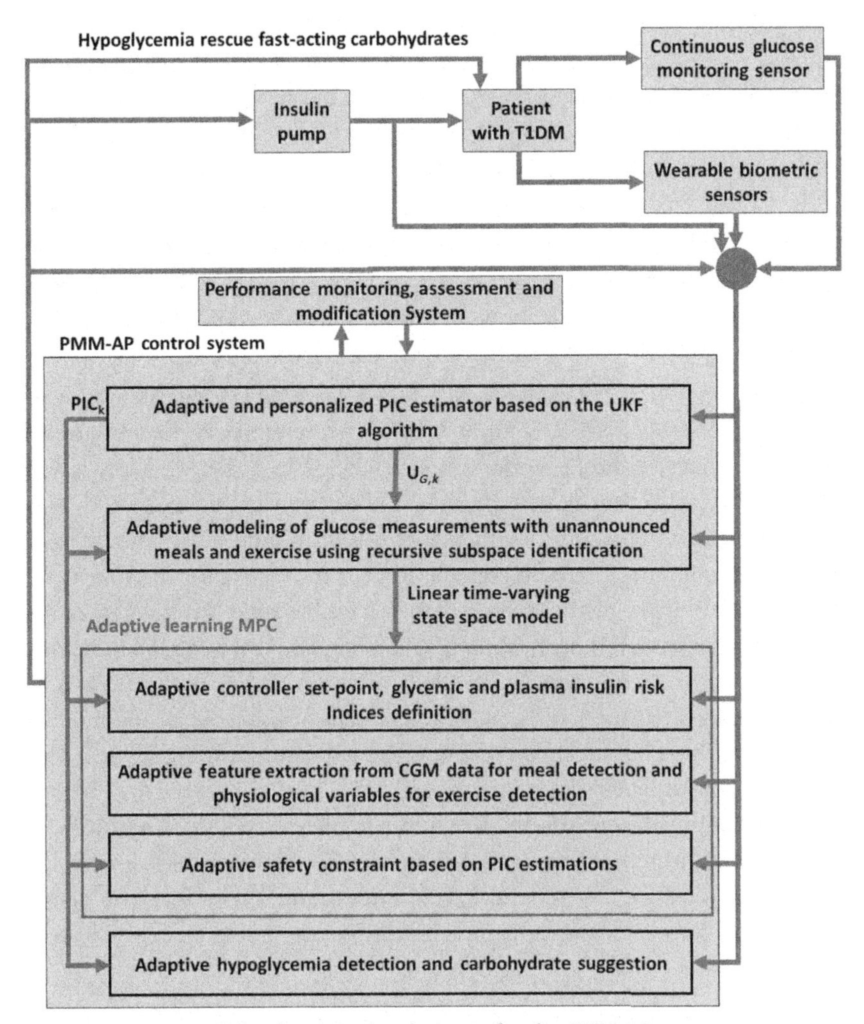

Fig. 2 A summary of the developed techniques for the PMM-AP system.

time. The designed PIC estimator is individualized by appropriately initializing some of the parameters in the insulin compartmental model using the partial least squares regression. The demographic information of individuals, such as weight, body mass index, and duration with diabetes, is used to predict the initial values for these parameters (Hajizadeh et al., 2017a, 2018c). After initialization, the time-varying parameters are estimated online. The proposed PIC estimator is able to capture the intersubject and intrasubject variabilities to provide accurate PIC estimates. Furthermore, the gut absorption rate is also included as an extended state to be estimated. The proposed nonlinear observer for the simultaneous estimation of the state variables and the time-varying parameters can be designed as

$$\hat{x}'_k = \hat{x}'^{,-}_k + K^{(f)}_k \left(y_k - \hat{y}'_k \right) \tag{1a}$$

$$\hat{y}'_k = g'\left(\hat{x}'^{,-}_k \right) \tag{1b}$$

where $\hat{x}'^{,-}_k$ is the prior estimate of the augmented state vector, $\hat{x}'_k$ denotes the estimated augmented state vector, $K^{(f)}_k$ is the observer gain obtained by the UKF algorithm, g' is the augmented output dynamic and uncertain model parameters' function, and $\hat{y}'_k$ is the estimated output (CGM) (Kolås et al., 2009; Dochain, 2003; Hajizadeh et al., 2017a, 2018c). The augmented state vector includes the states and time-varying parameters of the Hovorka's model. The states of the Hovorka's model include $[S_{1,k}\ S_{2,k}\ I_k\ x_{1,k}\ x_{2,k}\ x_{3,k}\ Q_{1,k}\ Q_{2,k}\ G_{\mathrm{sub},k}]^T$ (Hovorka et al., 2004). The state variables $S_1(t)$ and $S_2(t)$ describe the absorption rate of subcutaneously administered insulin as basal and bolus insulin and $I(t)$ represents the PIC in the bloodstream. The $Q_1(t)$ and $Q_2(t)$ describe the glucose masses in the accessible and nonaccessible compartments, respectively. The insulin action is computed by using the influence on transport and distribution ($x_1(t)$), the utilization and phosphorylation of glucose in adipose tissue ($x_2(t)$), and the endogenous glucose production in the liver ($x_3(t)$). The subcutaneous glucose concentration is $G_{\mathrm{sub}}(t)$. The time-varying parameters include $[t_{\max,\mathrm{I},\ k}\ k_{\mathrm{e},k}\ U_{\mathrm{G},k}]^T$. Considering the Hovorka's insulin compartment model, the two parameters time-to-maximum of absorption of injected insulin ($t_{\max,\mathrm{I}}(t)$) and insulin elimination from plasma ($k_{\mathrm{e}}(t)$) have a direct effect on the PIC. Furthermore, as information about meals is difficult to determine, the gut absorption rate ($U_{\mathrm{G},\ k}$) is also estimated.

2.2 Recursive subspace-based system identification

The proposed recursive system identification technique provides a time-varying stable state-space model (Hajizadeh et al., 2017b, 2018d). It updates parameters of state-space matrices online. After integrating the state-space model with a physiological insulin compartment model (Hajizadeh et al., 2018b), the final identified glycemic model for use in an AL-MPC becomes:

$$\begin{aligned} \overline{x}_{k+1} &= A_k \overline{x}_k + B_k u_k + d_k \\ \overline{y}_k &= C_k \overline{x}_k \end{aligned} \tag{2}$$

where A_k, B_k, and C_k are the system matrices, and d_k represents unmodeled/unmeasured disturbances. For the proposed model, continuous glucose measurement (CGM data) is considered as output, and infused insulin

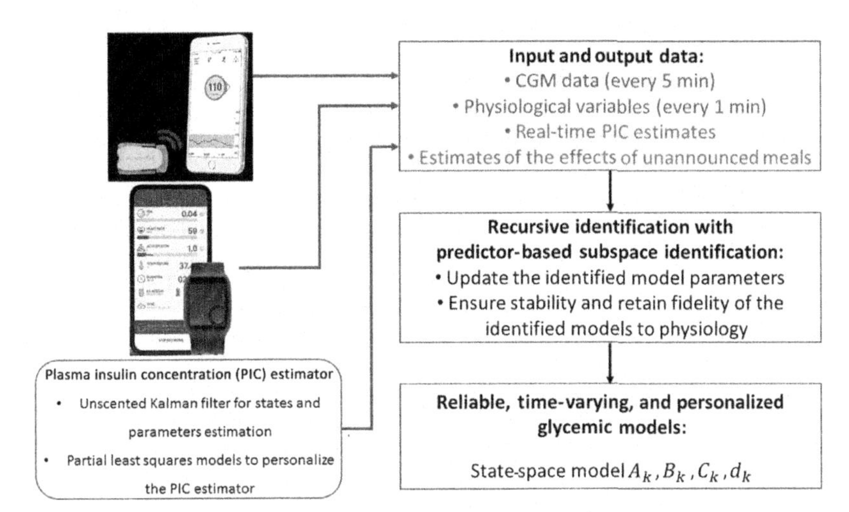

Fig. 3 A flowchart of the proposed recursive system identification technique.

information, estimates of the meal effect, and physiological variables represent inputs to indicate physical activity (PA) ($u_k = [Ins_k, Meal_k, MET_k]$). The metabolic equivalent task (MET) is the metabolic equivalent of task that represents energy expenditure, as an indicator of the intensity of PA. One of the states of the model described by Eq. (2) is the amount of insulin in the bloodstream, the PIC. The PIC safety constraints are then defined in the AL-MPC to assure that a safe amount of insulin is in the body. A general flowchart of the proposed identification technique is shown in Fig. 3. To guarantee that the underlying model can provide accurate output predictions for use in model-based predictive control algorithms, the proposed recursive subspace identification method obtains a stable time-varying state-space model of the process. This is done by incorporation of constraints on the fidelity and accuracy of the identified models, the correctness of the sign of the input-to-output gains, and the integration of heuristics to ensure the stability of the recursively identified models (Hajizadeh et al., 2018a).

3 PIC cognizant AL-MPC algorithm

In this section, we describe the glycemic and PIC risk indexes used in the AL-MPC controller objective function. The safety constraints based on the PIC and a feature extraction method for manipulating these constraints during the meal consumption periods are presented. Subsequently, the PMM-AP control formulation is presented.

3.1 Adaptive glycemic and plasma insulin risk indexes

An adaptive glycemic risk index (GRI) is used to determine the weighting matrix for penalizing the deviations of the controlled variables (outputs) from their nominal set-point values (Hajizadeh et al., 2019b, c). The GRI asymmetrically increases the set-point tracking weight when outputs diverge from the target range. Since hypoglycemic events have serious short-term implications, the set-point penalty increases rapidly in response to hypoglycemic excursions and more gradually in hyperglycemic excursions. A plasma insulin risk index (PIRI) is defined to manipulate the weighting matrix for penalizing the amount of input actuation (aggressiveness of insulin dosing) depending on the estimated PIC, thus suppressing the infusion rate if sufficient insulin is present in the bloodstream (Hajizadeh et al., 2019b, c). As it is impractical to directly consider the estimates of the PIC to define parameters of the AL-MPC due to the variability among subjects, the normalized value of the PIC is employed, which eliminates the dependency of the PIC estimates to a particular subject by standardizing with the known patient-specific basal PIC value.

3.2 Plasma insulin concentration bounds

In the proposed AL-MPC, the estimated future PIC is dynamically bounded depending on the value of the CGM measurements. For instance, if the CGM values are elevated, the bounds on the PIC are increased to ensure sufficient insulin is administered to regulate the glucose concentration. Furthermore, the PIC bounds also constrain the search space in the optimization problem, thus improving the computational tractability of the proposed AL-MPC. The PIC bounds are determined based on the CGM measurements as $\mathcal{X}_{\mathrm{PIC}} := (\mathcal{P}_{\mathrm{fasting}} + \mathcal{P}_{\mathrm{meal}})\mathcal{X}(\bar{y}_k)$, where $\mathcal{X}_{\mathrm{PIC}}$ defines the lower and upper bounds and a desired target for the normalized PIC through the predicted CGM. $\mathcal{P}_{\mathrm{meal}}$ is a parameter that modifies the PIC bounds when there is a rapid increase in CGM values. $\mathcal{P}_{\mathrm{fasting}}$ is a patient-specific parameter that defines the controller aggressiveness/conservativeness during the fasting period. These bounds and the reference target for the normalized PIC are defined as a function of the CGM value, and the AL-MPC solution should satisfy the PIC constraints while maintaining the PIC close to the desired value. The nominal PIC bounds can be determined by multiplying the normalized PIC bounds with the basal PIC value. Therefore, appropriate PIC bounds can be determined based on each subject's basal PIC value and the CGM measurement.

3.3 Feature extraction for manipulating constraints

Meal consumption can be automatically detected using qualitative descriptions of glucose time-series data, which is useful in modifying the aggressiveness of the AL-MPC (Samadi et al., 2017, 2018). In this work, features are generated from the data to describe the recent trajectory of the glycemic measurements. To this end, a p-order polynomial $y_i = f\left(t_i, \theta_k^M\right)$ with parameters θ_k^M is fitted to the most recent l glucose measurements $y_{i:i-l} := \left[y_i \; y_{i-1} \; \dots \; y_{i-l}\right]$ at each sampling time using ordinary least squares where t_i denotes the sampling index of the recent measurements. Then the derivatives of the polynomial are obtained and the first- and second-order derivatives, denoted $f^{(1)}$ and $f^{(2)}$, are analyzed to derive parameter $\mathcal{P}_{\text{meal}}$ for detecting carbohydrate consumption as

$$\mathcal{P}_{\text{meal}} = \begin{cases} \dfrac{f^{(1)}}{c_1^m} & \text{if } f^{(1)} \geq c_0^m \text{ and } f^{(2)} \geq 0 \\ \dfrac{f^{(1)}}{c_2^m} & \text{if } f^{(1)} \geq c_0^m \text{ and } f^{(2)} < 0 \\ 0 & \text{if } f^{(1)} < c_0^m \end{cases} \tag{3}$$

where c_1^m, c_2^m, and c_0^m are patient-specific threshold parameters. Detection of meals based on the $\mathcal{P}_{\text{meal}}$ parameter allows for the constraints of the AL-MPC are modified when meals are to make the controller more aggressive to suggest a sufficient insulin dose.

3.4 Adaptive-learning MPC formulation

Here, we propose a novel AL-MPC algorithm cognizant of the PIC for computing the optimal insulin infusion rate. In Fig. 2, different components of the AL-MPC are shown as the order of computations to obtain the optimum insulin doses. The proposed AL-MPC formulation employs the glycemic and PIC risk indexes that manipulate the penalty weighting matrices in the cost function. To this end, the AL-MPC computes the optimal insulin infusion over a finite horizon using the identified time-varying subspace-based models by solving at each kth sampling instance the following quadratic programming problem

$$\left\{z_i^*, v_i^*\right\}_{i=0}^{n_P} := \underset{v \in \mathcal{U},\ z \in \mathcal{Z}}{\arg\min}\ \mathcal{J}_{n_P,k}\left(\gamma_k, \{q_i\}_{i=0}^{n_P}, \{v_i\}_{i=0}^{n_P}, \{z_i\}_{i=0}^{n_P}\right)$$

$$\text{s.t.} \begin{cases} z_{i+1} = A_k z_i + B_k v_i + d_i, \ \forall i \in \mathbb{N}_0^{n_P - 1} \\ q_i = C_k z_i, \ \forall i \in \mathbb{N}_0^{n_P} \\ z_0 = \overline{x}_k \end{cases} \tag{4}$$

with the objective function

$$\begin{aligned} \mathcal{J}_{n_P,k}\left(q_i, \gamma_k, \{v_i\}_{i=0}^{n_P}, \{z_i\}_{i=0}^{n_P}\right) := & \sum_{i=0}^{n_P} (q_i - r_{k,i})^T Q_k (q_i - r_{k,i}) \\ & + \left(v_i - \mathrm{I}_{\mathrm{db,i}}\right)^T R_k \left(v_i - \mathrm{I}_{\mathrm{db,i}}\right) \\ & + e_i^{PIC} P_k e_i^{PIC} \end{aligned}$$

where $z_i \in \mathbb{R}^{\overline{n}_x}$ and $q_i \in \mathbb{R}$ denote the predicted states and outputs obtained by the model Eq. (2), respectively, $d_i \in \mathbb{R}$ is the combined effects of unmodeled disturbances, meal and exercise, the prediction/control horizon n_P, $v_i \in \mathbb{R}$ denotes the vector of constrained manipulated variable, which is the infused insulin as basal and bolus insulin, taking values in a nonempty convex set $\mathcal{U} \subseteq \mathbb{R}$ with $\mathcal{U} := \{v \in \mathbb{R} : v_{\min} \leq v \leq v_{\max}\}$, $v_{\min} \in \mathbb{R}$ and $v_{\max} \in \mathbb{R}$ denote the lower and upper bounds on the manipulated input, respectively, I_{db} is the patient-specific basal insulin rate, and $r_{k,\ i}$ is the target set-point. The index $\mathbb{N}_0^{n_P}$ represents all integers in a set as $\mathbb{N}_0^{n_P} := \{0, \ldots, n_P\}$. The nonempty convex set $\mathcal{Z} \subseteq \mathbb{R}^{\overline{n}_x}$ with $\mathcal{Z} := \left\{z \in \mathbb{R}^{\overline{n}_x} : z_{\min} \leq z \leq z_{\max}\right\}$, $z_{\min} \in \mathbb{R}^{\overline{n}_x}$ and $z_{\max} \in \mathbb{R}^{\overline{n}_x}$ denote the lower and upper bounds on the state variables, respectively, with one of the states as the estimated PIC that is constrained through the PIC bounds. The e_i^{PIC} is the deviation of the PIC from the desired PIC value. The $\overline{n}_x$ is the number of states in the model Eq. (2). Furthermore, $\overline{x}_k$ provides an initialization of the state vector, $Q_k \geq 0$, $Q_k := Q(\overline{\gamma}_k)$ is a positive semidefinite symmetric matrix used to penalize the deviations of the outputs from their nominal set-point, and $R_k > 0$, $R_k := R(\gamma_k)$ is a strictly positive definite symmetric matrix to penalize the manipulated input variables. The $r_{k,\ i}$ is the controller set-point over the prediction/control horizon, which is defined based on the current condition and historical data of the patient. For the anticipated periods of disturbances like exercise, the PIC limits are also changed to compute a safer insulin dose suitable for the exercise time. At each iteration, the quadratic programming problem in Eq. (4) is solved, and $u_k := v_0$ is the optimal solution implemented to infuse insulin over the current sampling interval with the

AL-MPC computation repeated at subsequent sampling instances using new CGM measurements, updated states, and newly calculated penalty weights of the objective function.

4 Results

The efficacy of the proposed PMM-AP is demonstrated by using a multivariable simulator (mGIPsim) developed by our research group at Illinois Institute of Technology based on a modified Hovorka's glucose-insulin dynamic model that takes into account the effects of different physical activities and meals (Rashid et al., 2019). In addition to the CGM values, the mGIPsim generates physiological variable signals reported by noninvasive wearable devices. MET is a physiological measure expressing the energy cost of PA. MET is used to express the intensity and energy expenditure of PAs. Aerobic exercises with a stationary bicycle are considered for testing the PMM-AP system. Twenty virtual subjects are simulated for 30 days with varying times and quantities of meals consumed on each day and different intensities and times of physical activities as detailed in Tables 1 and 2. The meal and PA information are not entered manually to the PMM-AP system as the PMM-AP controller is designed to regulate the BGC in the presence of significant disturbances such as unannounced meals and exercises. The energy expenditure values expressed as MET variations are computed by the simulator and used as input variables summarizing the

Table 1 Meal scenario for 30-day closed-loop experiment using mGIPsim.

Meal	Range for values	
	Time	Amount (g)
Breakfast	$[06:00, 07:00]$	$[40, 60]$
Lunch	$[12:00, 13:00]$	$[40, 60]$
Dinner	$[18:00, 19:00]$	$[40, 60]$

Table 2 Exercise scenario for 30-day closed-loop experiment using mGIPsim.

Exercise	Range for values		
	Time	Duration (min)	Power
Bicycling	$[10:00, 11:00]$	$[30, 60]$	$[50, 90]$
Bicycling	$[16:00, 17:00]$	$[30, 60]$	$[50, 90]$

physiological signals in response to physical activities in the PMM-AP system. The controller set-point is set at 110 mg/dL except during exercise when it becomes 160 mg/dL.

The quantitative evaluation of the closed-loop operation based on the proposed algorithms is presented in Table 3. The purpose of these simulations is to show that the PMM-AP is robust and reliable in handling significant disturbances to the BGC. The average percentage of time spent in the target ranges of [70, 140] mg/dL and [70, 180] mg/dL are 50.4% and 75.4% for all subjects. There is no hypoglycemic event as the BGC is never less than 70 mg/dL. The minimum and maximum observed BGC values are 73 and 279 mg/dL. The average minimum and maximum observed BGC values across all experiments during the whole simulation are 88 and 252 mg/dL, respectively. Overall, the results demonstrate that the proposed PMM-AP is able to regulate BGC effectively in presence of significant unknown disturbances caused by the diverse timing and amounts of meals and exercise specifications while mitigating severe hypoglycemic and hyperglycemic excursions. The closed-loop results for all subjects for the last day of simulations are shown in Fig. 4. The AL-mAP can proactively keep the CGM values in a safe range during exercise periods due to the learning feature of the PMM-AP controller (Hajizadeh et al., 2019a, d). This is done by using a safe (higher) controller set-point in advance, which is defined based on historical data that predicts the presence of exercise. For one select subject, the closed-loop simulation results for all 30 days are shown in Fig. 5. This figure also shows that the PMM-AP guarantees the safety and reliability of the insulin delivery system especially during exercise periods. The predicted hypoglycemic episodes warn the user to consume rescue carbohydrates about 20 min before the potential hypoglycemic episode. Overall, the PMM-AP system can also regulate the BGC with a minimum need for hypoglycemia treatments (Table 4, average 28 hypoglycemic episodes that necessitate rescue carbohydrates are predicted, about one rescue carb per day).

In this work, the insulin compartment of Hovorka's model is integrated with the recursive subspace identification technique. The adaptive and individualized PIC estimates provide accurate information on the amount of active insulin present in the body and appropriate information for the model identification. The proposed modeling technique also utilizes other additional variables such as biosignals to consider the effects of PA on BGC. Furthermore, this PIC information is used to define limits on the computed insulin doses. The PIC bounds, along with the risk indexes used

Table 3 Closed-loop simulation results for AL-MPC whole days.

Subject	Percentage of time in range					Statistics			
	<70	[70, 140]	[70, 180]	>180	>250	Mean	SD	Min	Max
S1	0.0	45.4	74.0	26.0	0.0	153.8	35.8	85	247
S2	0.0	53.9	73.8	26.2	0.7	151.6	39.2	80	261
S3	0.0	52.6	77.0	23.0	0.0	146.6	35.7	103	236
S4	0.0	57.8	81.8	18.3	0.2	143.3	36.0	80	259
S5	0.0	61.6	82.0	18.0	0.6	140.3	37.5	74	262
S6	0.0	49.1	79.8	20.2	0.0	149.3	31.5	93	232
S7	0.0	59.2	82.6	17.4	0.0	141.9	33.0	92	237
S8	0.0	58.9	80.5	19.5	0.0	141.9	34.9	91	238
S9	0.0	41.6	67.4	32.6	3.0	162.0	42.1	89	268
S10	0.0	60.0	78.9	21.1	0.1	142.8	37.2	73	254
S11	0.0	49.2	75.6	24.4	1.0	151.1	41.3	74	266
S12	0.0	54.4	76.8	23.2	0.1	148.3	37.8	83	254
S13	0.0	40.3	70.8	29.2	0.1	157.8	37.1	98	252
S14	0.0	55.2	76.9	23.1	0.3	146.9	40.0	78	260
S15	0.0	46.3	77.5	22.5	0.0	151.1	32.7	96	243
S16	0.0	57.3	83.0	17.0	0.0	143.2	30.6	88	230
S17	0.0	53.4	75.8	24.2	0.0	146.4	39.2	88	250
S18	0.0	31.4	58.8	41.2	4.4	171.0	44.1	103	275
S19	0.0	47.8	79.4	20.6	0.0	149.7	32.7	91	242
S20	0.0	33.0	58.5	41.5	3.6	170.7	43.7	105	279
Average	0.0	50.4	75.4	24.5	0.7	150.5	37.1	88	252

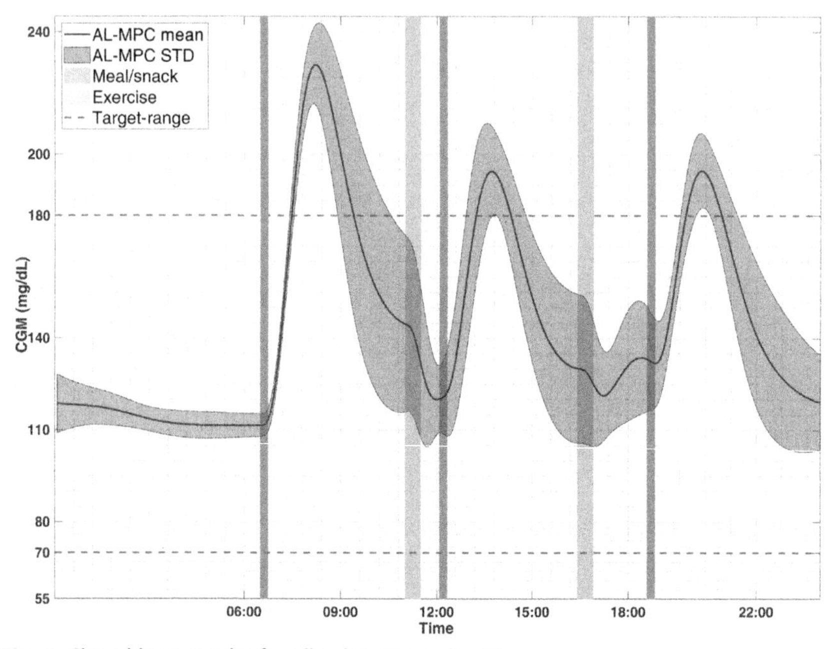

Fig. 4 Closed-loop results for all subjects on day 30.

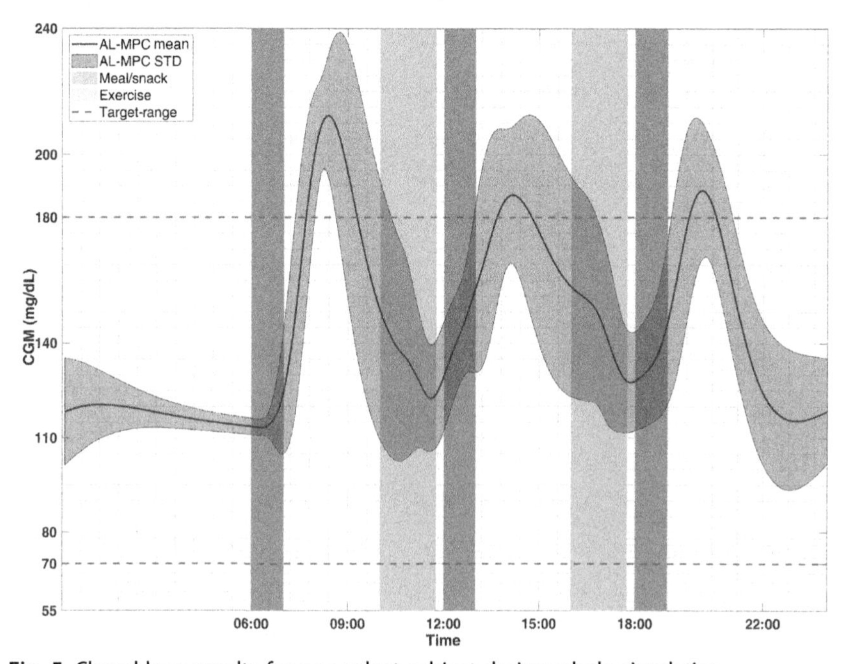

Fig. 5 Closed-loop results for one select subject during whole simulation.

Table 4 Total number of predicted hypoglycemic events and preventions by rescue carbohydrates for the whole simulation period (30 days) and the average total daily insulin (U) with AL-MPC.

Subject	Number of predicted hypo	Total daily insulin (U)
S1	11	36.9
S2	37	35.7
S3	6	30.7
S4	42	31.8
S5	80	27.9
S6	17	38.2
S7	27	40.4
S8	26	29.6
S9	10	59.7
S10	35	27.9
S11	38	30.0
S12	41	25.5
S13	4	47.5
S14	70	28.5
S15	7	43.4
S16	24	41.9
S17	61	26.1
S18	0	49.2
S19	27	43.9
S20	0	42.8
Average	28	36.9

in the AL-MPC formulation, define the aggressiveness/conservativeness of the controller. A minimum bound for the PIC is defined in the AL-MPC formulation to impose the controller to suggest a safe amount of insulin to derive the BGC toward the controller set-point value. A maximum bound is considered to avoid giving too much insulin causing hypoglycemia. A desired PIC value is also considered to reduce variability in the CGM measurements caused by variations in the PIC values.

5 Conclusions

A PMM-AP system is designed based on an AL-MPC algorithm. Accurate PIC estimates are obtained by using CGM measurements and infused insulin information with the UKF designed based on a glucose-insulin dynamic model. The proposed PIC estimator directly takes into account the intersubject and intrasubject variabilities in glucose-insulin dynamics. PIC estimates and physiological variables are combined with a recursive subspace-based system identification approach to obtain the glycemic models. The recursively identified models can describe better the dynamic behavior of BGC variations in the human body over wide ranges of real-world conditions and make accurate short-term predictions of glucose concentration measurements. The AL-MPC algorithm developed by using these adaptive models computes the optimal amount of insulin for AP systems without requiring any manual information on meal and PA specifications. Using machine-learning techniques and patients' historical data, the key control parameters are modified in advance for the anticipated periods of disturbance effects such as exercise. The proposed PMM-AP could be a reliable step toward improved glycemic control by individualizing the insulin computations and reducing the risk of postexercise hypoglycemia in the next generation of AP algorithms.

Acknowledgments

This work is supported by the National Institutes of Health under grants 1DP3DK101077-01 and 1DP3DK101075-01, and JDRF award 2-SRA-2017-506-M-B made possible by funding provided through the collaboration between JDRF and The Leona M. and Harry B. Helmsley Charitable Trust.

References

Behbehani, K., Cross, R.R., 1991. A controller for regulation of mean arterial blood pressure using optimum nitroprusside infusion rate. IEEE Trans. Biomed. Eng. 38 (6), 513–521.
Bekiari, E., Kitsios, K., Thabit, H., Tauschmann, M., Athanasiadou, E., Karagiannis, T., Haidich, A.-B., Hovorka, R., Tsapas, A., 2018. Artificial pancreas treatment for outpatients with type 1 diabetes: systematic review and meta-analysis. BMJ 361, k1310.
Boiroux, D., Duun-Henriksen, A.K., Schmidt, S., Nørgaard, K., Madsbad, S., Poulsen, N.K., Madsen, H., Jørgensen, J.B., 2018. Overnight glucose control in people with type 1 diabetes. Biomed. Signal Process. Control 39, 503–512.
Cameron, F., Bequette, B.W., Wilson, D.M., Buckingham, B.A., Lee, H., Niemeyer, G., 2011. A closed-loop artificial pancreas based on risk management. J. Diabetes Sci. Technol. 5 (2), 368–379.
Chakrabarty, A., Zavitsanou, S., Doyle, F.J., Dassau, E., 2018. Event-triggered model predictive control for embedded artificial pancreas systems. IEEE Trans. Biomed. Eng. 65 (3), 575–586.
Clarke, W.L., Anderson, S., Breton, M., Patek, S., Kashmer, L., Kovatchev, B., 2009. Closed-loop artificial pancreas using subcutaneous glucose sensing and insulin delivery and a model predictive control algorithm: the Virginia experience. J. Diabetes Sci. Technol. 3 (5), 1031–1038.
Cooke, D.W., Plotnick, L., 2008. Type 1 diabetes mellitus in pediatrics. Pediatr. Rev. 29 (11), 374–384.

de Pereda, D., Romero-Vivo, S., Ricarte, B., Rossetti, P., Ampudia-Blasco, F.J., Bondia, J., 2016. Real-time estimation of plasma insulin concentration from continuous glucose monitor measurements. Comput. Methods Biomech. Biomed. Eng. 19 (9), 934–942.
Dochain, D., 2003. State and parameter estimation in chemical and biochemical processes: a tutorial. J. Process Control 13 (8), 801–818.
Doyle, F., Jovanovic, L., Seborg, D., Parker, R.S., Bequette, B.W., Jeffrey, A.M., Xia, X., Craig, I.K., McAvoy, T., 2007. A tutorial on biomedical process control. J. Process Control 17 (7), 571–572.
Eberle, C., Ament, C., 2011. The unscented Kalman filter estimates the plasma insulin from glucose measurement. Biosystems 103 (1), 67–72.
Eisenbarth, G.S., 2005. Type 1 diabetes mellitus. Joslin's Diabetes Mellitus 14, 399–424.
El Fathi, A., Smaoui, M.R., Gingras, V., Boulet, B., Haidar, A., 2018. The artificial pancreas and meal control: an overview of postprandial glucose regulation in type 1 diabetes. IEEE Control Syst. Mag. 38 (1), 67–85.
Esposito, S., Santi, E., Mancini, G., Rogari, F., Tascini, G., Toni, G., Argentiero, A., Berioli, M.G., 2018. Efficacy and safety of the artificial pancreas in the paediatric population with type 1 diabetes. J. Transl. Med. 16 (1), 176.
Garcia-Tirado, J., Corbett, J.P., Boiroux, D., Jørgensen, J.B., Breton, M.D., 2019. Closed-loop control with unannounced exercise for adults with type 1 diabetes using the ensemble model predictive control. J. Process Control 80, 202–210.
Gondhalekar, R., Dassau, E., Doyle, F.J., 2016. Periodic zone-MPC with asymmetric costs for outpatient-ready safety of an artificial pancreas to treat type 1 diabetes. Automatica 71, 237–246.
Goodwin, G.C., Medioli, A.M., Murray, K., Sykes, R., Stephen, C., 2019. Applications of MPC in the area of health care. In: Handbook of Model Predictive Control, Springer, pp. 529–550.
Hajizadeh, I., Shahrokhi, M., 2015. Observer-based output feedback linearization control with application to HIV dynamics. Ind. Eng. Chem. Res. 54 (10), 2697–2708.
Hajizadeh, I., Rashid, M., Turksoy, K., Samadi, S., Feng, J., Frantz, N., Sevil, M., Cengiz, E., Cinar, A., 2017a. Plasma insulin estimation in people with type 1 diabetes mellitus. Ind. Eng. Chem. Res. 56 (35), 9846–9857.
Hajizadeh, I., Rashid, M., Turksoy, K., Samadi, S., Feng, J., Sevili, M., Frantz, N., Lazaro, C., Maloney, Z., Littlejohn, E., Cinar, A., 2017b. Multivariable recursive subspace identification with application to artificial pancreas systems. In: IFAC-PapersOnLine, pp. 909–914.
Hajizadeh, I., Rashid, M., Cinar, A., 2018a. Ensuring stability and fidelity of recursively identified control-relevant models. In: The 18th IFAC Symposium on System Identification (SYSID), pp. 927–932.
Hajizadeh, I., Rashid, M., Cinar, A., 2018b. Integrating compartment models with recursive system identification. In: American Control Conference (ACC), pp. 3583–3588.
Hajizadeh, I., Rashid, M., Samadi, S., Feng, J., Sevil, M., Hobbs, N., Lazaro, C., Maloney, Z., Brandt, R., Yu, X., Turksoy, K., Littlejohn, E., Cengiz, E., Cinar, A., 2018c. Adaptive and personalized plasma insulin concentration estimation for artificial pancreas systems. J. Diabetes Sci. Technol. 12 (3), 639–649.
Hajizadeh, I., Rashid, M., Turksoy, K., Samadi, S., Feng, J., Sevil, M., Hobbs, N., Lazaro, C., Maloney, Z., Littlejohn, E., Cinar, A., 2018d. Incorporating unannounced meals and exercise in adaptive learning of personalized models for multivariable artificial pancreas systems. J. Diabetes Sci. Technol. 12 (5), 953–966.
Hajizadeh, I., Hobbs, N., Samadi, S., Sevil, M., Rashid, M., Brandt, R., Askari, M.R., Maloney, Z., Cinar, A., 2019a. Controlling the AP controller: controller performance assessment and modification. J. Diabetes Sci. Technol. 13 (6), 1091–1104.
Hajizadeh, I., Rashid, M., Cinar, A., 2019b. Plasma-insulin-cognizant adaptive model predictive control for artificial pancreas systems. J. Process Control 77, 97–113.

Hajizadeh, I., Rashid, M., Samadi, S., Sevil, M., Hobbs, N., Brandt, R., Cinar, A., 2019c. Adaptive personalized multivariable artificial pancreas using plasma insulin estimates. J. Process Control 80, 26–40.
Hajizadeh, I., Samadi, S., Sevil, M., Rashid, M., Cinar, A., 2019d. Performance assessment and modification of an adaptive model predictive control for automated insulin delivery by a multivariable artificial pancreas. Ind. Eng. Chem. Res. 58 (26), 11506–11520.
Hovorka, R., Canonico, V., Chassin, L.J., Haueter, U., Massi-Benedetti, M., Federici, M.O., Pieber, T.R., Schaller, H.C., Schaupp, L., Vering, T., Wilinska, M.E., 2004. Nonlinear model predictive control of glucose concentration in subjects with type 1 diabetes. Physiol. Meas. 25 (4), 905.
Kolås, S., Foss, B.A., Schei, T.S., 2009. Constrained nonlinear state estimation based on the UKF approach. Comput. Chem. Eng. 33 (8), 1386–1401.
Laguna Sanz, A.J., Doyle III, F.J., Dassau, E., 2017. An enhanced model predictive control for the artificial pancreas using a confidence index based on residual analysis of past predictions. J. Diabetes Sci. Technol. 11 (3), 537–544.
Martin, R.B., 1992. Optimal control drug scheduling of cancer chemotherapy. Automatica 28 (6), 1113–1123.
Martin, R., Teo, K.L., 1994. Optimal Control of Drug Administration in Cancer Chemotherapy. World Scientific, Singapore.
Messori, M., Incremona, G.P., Cobelli, C., Magni, L., 2018. Individualized model predictive control for the artificial pancreas: in silico evaluation of closed-loop glucose control. IEEE Control Syst. Mag. 38 (1), 86–104.
Neatpisarnvanit, C., Boston, J.R., 2002. Estimation of plasma insulin from plasma glucose. IEEE Trans. Biomed. Eng. 49 (11), 1253–1259.
Ogunnaike, B.A., 2019. 110th anniversary: process and systems engineering perspectives on personalized medicine and the design of effective treatment of diseases. Ind. Eng. Chem. Res. 58 (44), 20357–20369.
Pannocchia, G., Laurino, M., Landi, A., 2010. A model predictive control strategy toward optimal structured treatment interruptions in anti-HIV therapy. IEEE Trans. Biomed. Eng. 57 (5), 1040–1050.
Parker, R.S., 2009. Automation and control in biomedical systems. In: Springer Handbook of Automation, Springer, pp. 1361–1378.
Parker, R.S., Doyle III, F.J., 2001. Control-relevant modeling in drug delivery. Adv. Drug Deliv. Rev. 48 (2–3), 211–228.
Peyser, T., Dassau, E., Breton, M., Skyler, J.S., 2014. The artificial pancreas: current status and future prospects in the management of diabetes. Ann. N. Y. Acad. Sci. 1311 (1), 102–123.
Rashid, M., Samadi, S., Sevil, M., Hajizadeh, I., Kolodziej, P., Hobbs, N., Maloney, Z., Brandt, R., Feng, J., Park, M., Quinn, L., Cinar, A., 2019. Simulation software for assessment of nonlinear and adaptive multivariable control algorithms: glucose-insulin dynamics in type 1 diabetes. Comput. Chem. Eng. 130, 106565.
Samadi, S., Turksoy, K., Hajizadeh, I., Feng, J., Sevil, M., Cinar, A., 2017. Meal detection and carbohydrate estimation using continuous glucose sensor data. IEEE J. Biomed. Health Inform. 21 (3), 619–627.
Samadi, S., Rashid, M., Turksoy, K., Feng, J., Hajizadeh, I., Hobbs, N., Lazaro, C., Sevil, M., Littlejohn, E., Cinar, A., 2018. Automatic detection and estimation of unannounced meals for multivariable artificial pancreas system. Diabetes Technol. Ther. 20 (3), 235–246.
Silvia, O., Josep, V., Remei, C., Joaquim, A., 2017. A review of personalized blood glucose prediction strategies for T1DM patients. Int. J. Numer. Methods Biomed. Eng. 33 (6), e2833. https://doi.org/10.1002/cnm.2833.

Toffanin, C., Messori, M., Palma, F.D., Nicolao, G.D., Cobelli, C., Magni, L., 2013. Artificial pancreas: model predictive control design from clinical experience. J. Diabetes Sci. Technol. 7 (6), 1470–1483.
Turksoy, K., Frantz, N., Quinn, L., Dumin, M., Kilkus, J., Hibner, B., Cinar, A., Littlejohn, E., 2017. Automated insulin delivery-the light at the end of the tunnel. J. Pediatr. 186, 17–28.
Turksoy, K., Littlejohn, E., Cinar, A., 2018. Multimodule, multivariable artificial pancreas for patients with type 1 diabetes: regulating glucose concentration under challenging conditions. IEEE Control Syst. Mag. 38 (1), 105–124.
Yu, C., Roy, R.J., Kaufman, H., Bequette, B.W., 1992. Multiple-model adaptive predictive control of mean arterial pressure and cardiac output. IEEE Trans. Biomed. Eng. 39 (8), 765–778.

CHAPTER 4

Modeling and optimal control of cancer-immune system

Fathalla A. Rihan[a], Nouran F. Rihan[b]
[a]Department of Mathematical Sciences, College of Science, United Arab Emirates University, Al Ain, United Arab Emirates
[b]Faculty of Pharmacy, Clinical Program, Cairo University, Cairo, Egypt

1 Introduction

Cancer is a generic term for a large group of diseases that can affect any part of the body, and is considered the second-leading cause of death worldwide. Globally, it was responsible for an estimated 9.6 million deaths in 2018, according to the International Agency for Research on Cancer (IARC) (WHO, 2018). Other terms used are malignant tumors and neoplasms. One defining feature of cancer is the rapid creation of abnormal cells that grow beyond their usual boundaries, and that can then invade adjoining parts of the body and spread to other organs. The latter process is referred to as metastasizing. Metastases are a major cause of death from cancer (Bray et al., 2018).

The interactions between cancer cells and the immune system (IS) are very complex and need sophisticated models to describe them. The IS is responsible for monitoring substances that are normally present in the body (Rihan et al., 2016). Once foreign substances exist in the body that the IS is unable to identify, it raises an alarm to initiate an attack. The substances that cause the IS response are called antigens. The IS attack includes destruction of any cells containing foreign antigens, such as pathogens or cancer cells. Pathogens are identified by certain proteins that are found on their outer surfaces that are not normally found in the human body. Immunity has two categories: *innate* (natural or nonspecific) immunity and *adaptive* (acquired or specific) immunity, which work synergically to eliminate pathogens. The *innate* IS is responsible for recognizing the molecules emitted by foreign substances. However, *adaptive* immunity is very specific; it controls responses caused by repeated exposures to the same antigens. Exposure to a particular antigen will lead to a quick and effective response to that particular antigen in the future, but not to any other antigens. Certain memory

Control Theory in Biomedical Engineering
https://doi.org/10.1016/B978-0-12-821350-6.00004-4

T lymphocytes and memory B cells will be created and be available to remember a particular species of pathogen or other antigenic substance, so that the response will be effective and rapid on subsequent exposure (Byrne et al., 2006).

Although there are great research efforts dedicated to revealing the relation between tumor cells and the IS, cancer is still considered as one of the most challenging diseases to treat. The desired outcome from cancer treatment should be destruction of all cancer cells in the body while maintaining the minimum level of healthy cells. Chemotherapy is one of the most highly adopted cancer treatment modalities; however, it was proved not to be the most convenient solution for tumor regression (de Pillis and Radunskaya, 2003; Swan, 1985). Progress is being made in attempting to eliminate tumor cells in the host by using an experimental form of immunotherapy (American Association for Cancer Research, 2016; Liu et al., 2002).

Immunotherapy (which is sometimes referred to as biological therapy) is quickly becoming one of the most important components of cancer treatments, especially in multipronged approaches (Neves and Fai Kwok, 2015). The goal of immunotherapy is to reinforce the body's own natural ability to combat cancer by enhancing the effectiveness of the IS to act against cancer cells, which involves the use of cytokines[a] with adoptive cellular immunotherapy (ACI), derived from the body or laboratory-produced versions of such substances, to improve or restore IS function (Joshi et al., 2009; Kirschner and Panetta, 1998). Although it is not entirely clear how immunotherapy treats cancer, it may work by stopping or slowing the growth of cancer cells, stopping cancer from spreading to other parts of the body, or helping the IS increase its effectiveness at eliminating cancer cells (Kuznetsov et al., 1994). There are three known categories of immunotherapy: immune response modifiers (cytokines), monoclonal antibodies[b] and vaccines.[c] The most commons cytokines are IL-2 and interferon-alpha (INF-α) (Lackie, 2010). IL-2 does not kill tumor cells directly like classical chemotherapy. Instead, IL-2 activates and stimulates the growth of immune cells,

[a] Cytokines: Protein hormones that mediate both natural and specific immunity, called interferons and interleukins. They increase the production of immune cells, can be made in the laboratory and given to patients as part of treatment for cancer.

[b] Monoclonal antibodies: a type of immunotherapy, are made in the laboratory. Monoclonal antibodies may be used alone, or they may have an attached drug or radioactive material.

[c] Cancer vaccines, another type of immunotherapy, stimulate the body's IS to destroy cancer cells.

most importantly T cells,[d] but also natural killer (NK) cells, both of which are capable of destroying cancer cells directly. The main role of ACI is that the T cells are collected from a patient and grown in the laboratory. This increases the number of T cells that are able to kill cancer cells or fight infections. These T cells are given back to the patient to help the IS to fight disease. This can be done in two ways, either by (i) a lymphokine-activated killer cell therapy (LAK therapy) or by a tumor infiltrating lymphocyte therapy (TIL) (Dunn et al., 2002). However, we should mention that the common side effects of IL-2 treatment include weight gain and low blood pressure, which can be treated with other medications (Rihan et al., 2014b; Preziosi, 2003).

Chemotherapy and radiotherapy are also common cancer therapies that have been developed to fight cancer (Lackie, 2010). The basic idea behind chemotherapy is to kill cancerous cells faster than healthy cells, while radiotherapy uses radiation to kill cancerous cells. Immunotherapy is used as a maintenance therapy following a combination of chemotherapy or radiotherapy, and in some circumstances it is used as a single agent to treat cancer (de Pillis et al., 2006, 2008). The combination is due to the fact that the chemotherapy treatment kills both cancerous and healthy cells and therefore it depletes the patient's IS, making the patient prone to dangerous infections. For this and other reasons, it is desirable to strengthen the IS after an immune-depleting course of chemotherapy. Additionally, recruiting the body's own defense to fight cancer can be a powerful treatment strategy. Therefore, maintaining a strong IS, by combining immunotherapy and chemotherapy, may be essential to successfully fighting cancer. However, the query now is how to most effectively combine cancer immunotherapy and chemotherapy. Rihan et al. (2019) study an optimal control problem (OCP) of delay differential model to describe the dynamics of tumor-immune interactions in presence of immuno-chemotherapy. The model includes constant delays in the mitotic phase to justify the time required to stimulate the effector cells and for the effector cells to develop a suitable response to the tumor cells.

Mathematical models, based on ordinary differential equations, delay differential equations (DDEs), and partial differential equations, have proven to be useful tools in analyzing and understanding IS interactions with viral,

[d] T cell: A type of white blood cell that is of key importance to the IS and is at the core of adaptive immunity, the system that tailors the body's immune response to specific pathogens. T cells are like soldiers who search out and destroy the targeted invaders.

bacterial infections and cancerous cells. Several mathematical models have been suggested to describe the interactions of tumors and the IS over time (see, e.g., the research papers; Araujo and McElwain, 2004; Bellomo et al., 2008; Chaplain, 2008; Nagy, 2005; Roose et al., 2007). Most of these papers describe the interactions between tumor cells and immune cells, tumor cells, and normal cells alone (Rihan et al., 2012), or consider the interactions of tumor-IS with chemotherapy treatment (de Pillis and Radunskaya, 2003; Swan, 1985). In this chapter, we provide a mathematical model of tumor-immune interactions in presence of chemotherapy treatment and optimal control variables. The control variables are incorporated to justify the best strategy of treatment and minimize side effects of the external treatment by reducing the production of new tumor cells while keeping the number of normal cells above the average of its carrying capacity.

This chapter is organized as follows. In Section 2, we provide a simple mathematical model with time-delay (time-lag) to represent the interaction of the IS with tumor cells. Boundedness and nonnegativity of the model solutions are also discussed. In Section 3, we extend the model to include control variables of chemotherapy treatment. Existence of optimal controls are also investigated. Section 4 presents numerical simulations and discussion to show the effectiveness of the theoretical results.

2 Mathematical models

Mathematical models provide biologists and clinicians with the tools that may guide efforts to clarify fundamental mechanisms of cancer progress and improve current strategies to stimulate the development of new ones. We first present a simple model that describes the dynamics of tumor cells, $T(t)$, and activated effector cells, $E(t)$, such as cytotoxic T cells. The model takes the form

$$\frac{dE(t)}{dt} = \sigma + \mathcal{F}(E(t), T(t)) - \mu E(t)T(t) - \delta E(t), \tag{1a}$$

$$\frac{dT(t)}{dt} = \alpha T(t)(1 - \beta T(t)) - nE(t)T(t), \tag{1b}$$

with initial conditions: $E(0) = E_0$, $T(0) = T_0$. Of course, the interaction between the effector and tumor cells leads to a reduction in the size of both populations with different rates, which are expressed by $-\mu E(t)T(t)$ and $-nE(t)T(t)$, respectively. As a result of this interaction, the immune effector cells decrease the population of tumor cells at rate n. Meanwhile, tumor cells infect some of the effector cells and therefore the population of uninfected effector cells decreases at the rate μ. $\sigma \geq 0$ is a treatment term that represents the external source of the effector cells such as ACI. Furthermore, in the absence of any tumor, the cells will die at a rate δ. The loss of tumor cells is denoted by an immune-effector cell interaction, modeled also by Michaelis-Menten kinetics to indicate the limited immune response to tumors, so that $\mathcal{F}(E, T) = \frac{\rho E(t) T(t)}{\eta + T(t)}$. In this term, ρ is the maximum immune response rate and η is the steepness of immune response. In the second equation, the rate of change of the tumor cells follows a logistic growth term $\alpha T(t)(1 - \beta T(t))$.

In model (1), one can incorporate a discrete time-lag τ to consider the time needed by the IS to develop a suitable response after recognizing the tumor cells. The new model with discrete time-lag takes the form

$$\begin{aligned}
\frac{dE(t)}{dt} &= \sigma + \frac{\rho E(t-\tau) T(t-\tau)}{\eta + T(t-\tau)} - \mu E(t-\tau) T(t-\tau) - \delta E(t), \\
\frac{dT(t)}{dt} &= r_2 T(t)(1 - \beta T(t)) - nE(t) T(t).
\end{aligned} \qquad t \geq 0 \quad (2)$$

This model is called DDEs, in which we must provide initial functions: $E(t) = \psi_1(t)$ and $E(t) = \psi_2(t)$, for all $t \in [-\tau, 0]$, instead of initial values. In model (2), the presence of tumor cells stimulates the immune response, represented by the positive nonlinear growth term for the immune cells $\rho E(t - \tau) T(t - \tau)/(\eta + T(t - \tau))$. ρ and η are positive constants, and $\tau \geq 0$ is the time-delay that presents the time needed by the IS to develop a suitable response after recognizing the tumor cells. The saturation term (Michaelis-Menten form) with the $E(t)$ compartment and logistic term with the $T(t)$ compartment are considered. The presence of the tumor cells virtually initiates the proliferation of tumor-specific effector cells to reach a saturation level parallel with the increase in the tumor populations. Hence, the recruitment function should be zero in the absence of the tumor cells, whereas it should increase monotonically toward a horizontal asymptote (Villasana and Radunskaya, 2003). Of course, the solution of DDEs model (2) should be bounded and nonnegative (Bodnar et al., 2011).

2.1 Boundedness and nonnegativity of the model solutions

To show that the solutions of model (2) are bounded and remain nonnegative in the domain of its application for sufficiently large values of time t, we recall the following lemma:

Lemma 1. *(Gronwall's lemma; Halanay, 1966, p. 9) Let* x, ψ, *and* χ *be real continuous functions defined in [*a, b*],* $\chi \geq 0$ *for* t $\in$ *[*a, b*]. We suppose that on [*a, b*] we have the inequality* $x(t) \leq \psi(t) + \int_a^t \chi(s)x(s)ds$. *Then* $x(t) \leq \psi(t) + \int_a^T \chi(s)\psi(s)e^{\left(\int_s^t \chi(\xi)d\xi\right)} ds$ *in* $[a, b]$.

Therefore, we arrive at the following proposition.

Proposition 1. *Let* $(E(t), T(t))$ *be a solution of system (2), then* $E(t) < M_1$ *and* $T(t) < M_2$ *for all sufficiently large time t, where*

$$M_1 = E(0) + \frac{\sigma}{\delta}\exp(\delta t) + \int_0^t \left[\rho e^{\delta(\tau+s)}\left(E(0) + \frac{\sigma}{\delta}e^{\delta s}\right)\exp\left(\int_s^t \rho e^{\delta(\tau+\xi)}d\xi\right)\right]ds,$$

$$M_2 = \max\left(\frac{1}{\beta}, T(0)\right).$$

Proof. Let (E, T) denotes the solution of model (2). From the second equation of system (2), we have $\frac{dT}{dt} \leq r_2 T(t)(1 - \beta T(t))$. Thus, $T(t)$ may be compared with the solution of

$$\frac{dX}{dt} = r_2 X(t)(1 - \beta X(t)), \quad \text{with } X(0) = T(0)$$

This proves that $T(t) < M_2$. From the first equation of system (2), we obtain

$$E(t) = \exp(-\delta t)\left\{E(0) + \int_0^t \left[\sigma + \frac{\rho E(s-\tau)T(s-\tau)}{\eta + T(s-\tau)} - \mu E(s-\tau)T(s-\tau)\right]\exp(\delta s)ds\right\}.$$

To show that $E(t)$ is bounded, we use the generalized Gronwall lemma. Since $\frac{T}{\eta+T} < 1$ and $\exp(-\delta t) \in (0, 1]$, we have

$$E(t) \leq E_0 + \frac{\sigma}{\delta}\exp(\delta t) + \int_0^t \rho E(s-\tau)\exp(\delta s)ds.$$

The generalized Gronwall lemma gives $E(t) < M_1$ where M_1 is uniformly bounded. It follows that if (E, T) is a solution of Eq. (2), then $(E, T) < (M_1, M_2)$ for all t. This shows that the solutions of model (2) are uniformly bounded. This completes the proof. ∎

From Eq. (1a) and the solution $T(t) = T(0)\exp\left(\int_0^t [r_2(1 - \beta T(s)) - E(s)]ds\right)$, we arrive at the following result:

Corollary 1. *If* $\frac{\rho}{\eta+T} \geq \mu$, *then the solutions (E, T) for model (2) are nonnegative for any nonnegative initial condition. However, if* $\frac{\rho}{\eta+T} < \mu$, *then there exist nonnegative initial conditions such that E(t) becomes negative in a finite time interval.*

3 Model with chemotherapy and control

We aim to design an efficient treatment protocol, where we employ the tools of optimal control theory. The formulation as an OCP allows us to:

(i) investigate the dynamical system of interacting cell populations being affected by the treatments;

(ii) optimize the application of the control such that the quantity of the treatments is optimized; and

(iii) minimize the tumor size at some of end-time.

This demonstrates how immunotherapy and chemotherapy might be combined for more effective treatment and to protect the patient from opportunistic infection, as well as fighting the cancer itself. Unlike chemotherapy, immunotherapy does not kill tumor cells directly, but it activates and stimulates the growth of immune cells, most importantly T cells, and NK cells, which are capable of destroying cancer cells directly. Therefore, the main goal of combining immuno-chemotherapy treatment is to eradicate the tumor cells, with minimum side effect, while maintaining adequate amounts of healthy tissues.

To include external chemotherapy in model (2), we should consider extra two variables namely amount of chemotherapy $u(t)$ and normal cells $N(t)$ with two control variables $v(t)$ and $w(t)$ (see Fig. 1). We also assume a homogeneity of the tumor cells and asynchronous tumor-drug interaction (Sinek et al., 2004). The modified model is

$$\frac{dE(t)}{dt} = w(t)\sigma + \frac{\rho E(t-\tau)T(t-\tau)}{\eta + T(t-\tau)} - \mu E(t-\tau)T(t-\tau) - \delta E(t) - F_1(u)E(t),$$

$$\frac{dT(t)}{dt} = r_2 T(t)(1 - \beta_1 T(t)) - nE(t)T(t) - c_1 N(t)T(t) - F_2(u)T(t),$$

$$\frac{dN(t)}{dt} = r_3 N(t)(1 - \beta_2 N(t)) - c_2 T(t)N(t) - F_3(u)N(t),$$

$$\frac{du(t)}{dt} = v(t) - d_1 u(t). \tag{3}$$

The drug kills all types of cells, with different killing rates for each type of cell: $F_i(u) = a_i(1 - e^{-u})$ is the fraction cell kill for a given amount of drug,

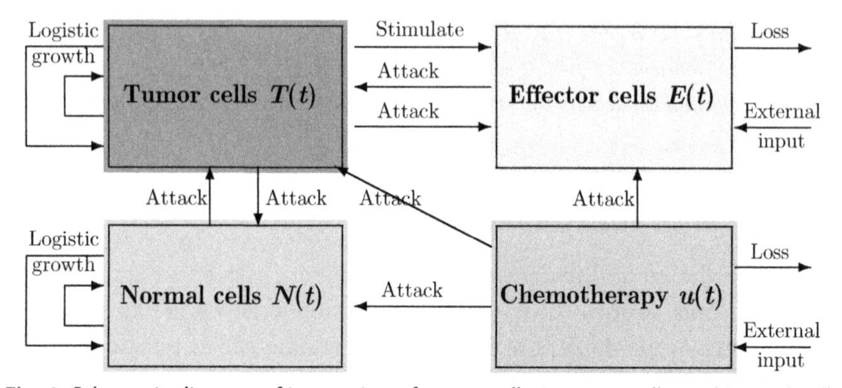

Fig. 1 Schematic diagram of interaction of tumor cells, immune cells, and normal cells in the presence of chemotherapy treatment.

$u(t)$, at the tumor site. The parameters a_1, a_2, and a_3 are the three different response coefficients. $v(t)$ represents the amount of dose that is injected into the system, while d_1 is the decay rate of the drug once it is injected. In this case, the quantity we will control directly is not $u(t)$, but $v(t)$. The tumor cells and normal cells are modeled by a logistic growth law, with parameters r_i representing the growth rate of two types of cells: $i=2$ identifies the parameter associated with the tumor, and $i=3$ identifies the one associated with the normal tissue. β_1 and β_2 are the reciprocal carrying capacities of tumor cells and host cells, respectively. The two terms $-c_1N(t)T(t)$ and $-c_2N(t)T(T)$ represent the competition between the tumor and host cells.

Let $\mathcal{C}=\mathcal{C}([-\tau,0],\mathbb{R}^4)$ be the Banach space of continuous functions mapping the interval $[-\tau, 0]$ into $\mathbb{R}^4$ with the topology of uniform convergence. It is easy to show that there exists a unique solution $(E(t), T(t), N(t), u(t))$ of system (3) with initial data $(E_0, T_0, N_0, u_0)\in\mathcal{C}$. For biological reasons, we assume that the initial data of system (3) satisfy $E_0 \geq 0$, $T_0 \geq 0$, $N_0 \geq 0$, $u_0 \geq 0$. For $\tau = 0$, the model is reduced to ODEs model developed by de Pillis and Radunskaya (2001).

The main objective in developing chemotherapy treatment, in system (3), is to reach either a tumor-free steady state or coexisting steady state in which the tumor cells' size is small, while the normal cells' size is close to its normalized carrying capacity. To keep the patient healthy while killing the tumor, our control problem consists of determining the variables $v(t)$ and $w(t)$ that will maximize the amount of effector cells and minimize the number of tumor cells. We use cost functional of the control with a constraint to keep normal cells above the average of its capacity. Therefore, our objective is to maximize the functional (see Rihan et al., 2014a)

$$\max_{v,w} J(v,w) = \int_0^{t_f} \left(E(t) - T(t) - \left[\frac{B_v}{2}[v(t)]^2 + \frac{B_w}{2}[w(t)]^2 \right] \right) dt, \quad (4)$$

where B_u, B_w are, respectively, the weight factors that describe the patient's acceptance level of chemotherapy and immunotherapy with a constraint

$$k(E, T, N, u, E_\tau, T_\tau, v) = N - 0.75 \geq 0, \quad 0 \leq t \leq t_f. \quad (5)$$

We are seeking optimal control pair (v^*, w^*) such that

$$J(v^*, w^*) = \max\{J(v,w) : (v,w) \in W\}, \quad (6)$$

where W is the control set defined by

$$\begin{aligned} W = \{(v,w) : (v,w) \text{ piecewise continuous, such that} \\ 0 \leq v(t) \leq v_{\max} < \infty, ;0 \leq w(t) \leq w_{\max} < \infty, \quad \forall t \in [0, t_f]\}. \end{aligned} \quad (7)$$

The existence of optimal controls $v^*(t)$ and $w^*(t)$ for this model is guaranteed by standard results in optimal control theory (Fleming and Rishel, 1994). Necessary conditions that the controls must satisfy are derived via Pontryagin's Maximum Principle (Pontryagin et al., 1962). The OCP given by expressions (3)–(7) is equivalent to that of minimizing the Hamiltonian $H(t)$ (see the "Appendix" section):

$$H = E - T - \frac{B_v}{2}[v(t)]^2 - \frac{B_w}{2}[w(t)]^2 + \lambda_1 \frac{dE}{dt} + \lambda_2 \frac{dT}{dt} + \lambda_3 \frac{dN}{dt} + \lambda_4 \frac{du}{dt} + \gamma k \quad (8)$$

and $\gamma \geq 0$ with $\gamma(t)k(t) = 0$, where

$$\gamma = \begin{cases} 1 & \text{if } N(t) \leq 0.75, \\ 0 & \text{otherwise.} \end{cases}$$

A standard application of Pontryagin's maximum principle leads to the following result:

Theorem 1. *Given an optimal pair $v^*(t)$ and $w^*(t)$ and corresponding solutions E^*, T^*, N^*, u^* and w^* for system (3) that minimizes $J(u(t), w(t))$ over Ω. The explicit optimal controls are connected to the existence of continuous specific functions λ_i for $i = 1, 2, 3, 4$, satisfying the adjoint system*

$$\lambda_1'(t) = -1 + \lambda_1(t)\left[\delta + a_1(1 - e^{-u^*})\right] + \lambda_2(t)nT^* + \lambda_1(t+\tau)\chi_{[0,\,t_f-\tau]}\left[\mu T^* - \frac{\rho T^*}{\eta + T^*}\right],$$

$$\lambda_2'(t) = 1 + \lambda_2\left[-r_2 + 2r_2\beta_1 T^* + nE^* + c_1 N^* + a_2(1 - e^{-u^*})\right]$$

$$+ \lambda_3 c_2 N^* + \chi_{[0,\,t_f-\tau]}\lambda_1(t+\tau)\left[\frac{\rho E^* T^*}{(\eta + T^*)^2} - \frac{\rho E^*}{\eta + T^*} + \mu E^*\right],$$

$$\lambda_3'(t) = \lambda_2 c_1 T^* - \lambda_3\left(r_3 - 2r_3\beta_2 N^* - c_2 T^* - a_3(1 - e^{-u^*})\right) - \gamma,$$

$$\lambda_4'(t) = -\lambda_1(t)a_1 e^{-u^*}E^* + \lambda_2(t)a_2 e^{-u^*}T^* + \lambda_3(t)a_3 e^{-u^*}N^* + \lambda_4(t)d_1, \tag{9}$$

with transversality conditions

$$\lambda_i(t_f) = 0,\ i = \{1,2,3,4\}\ \textit{and}\ \chi_{[0,\,t_f-\tau]} = \begin{cases} 1 & \textit{if}\ t \in [0, t_f - \tau], \\ 0 & \textit{otherwise}. \end{cases} \tag{10}$$

Furthermore, the following properties hold

$$v^* = \min\left(v_{max}, \frac{\lambda_4}{B_v}\right), \quad w^* = \min\left(w_{max}, \frac{\lambda_1 s_1}{B_w}\right). \tag{11}$$

Proof. The adjoint equations and transversality conditions can be obtained by using Pontryagin's minimum principle with delay in state (Pontryagin et al., 1962) such that

$$\begin{aligned}
\lambda_1'(t) &= -\frac{\partial H}{\partial E}(t) - \chi_{[0,\,t_f-\tau]}(t)\frac{\partial H}{\partial E_\tau}(t+\tau), \quad \lambda_1(t_f) = 0, \\
\lambda_2'(t) &= -\frac{\partial H}{\partial T}(t) - \chi_{[0,\,t_f-\tau]}(t)\frac{\partial H}{\partial T_\tau}(t+\tau), \quad \lambda_2(t_f) = 0, \\
\lambda_3'(t) &= -\frac{\partial H}{\partial N}, \quad \lambda_3(t_f) = 0, \\
\lambda_4'(t) &= -\frac{\partial H}{\partial u}, \quad \lambda_4(t_f) = 0.
\end{aligned} \tag{12}$$

The optimality system consists of the state system (3) coupled with the adjoint system (9) with initial and transversality conditions together with the characterization together with the optimal control. Therefore, the optimal system is given as follows:

$$\frac{dE^*}{dt}=w^*\sigma+\frac{\rho E^*(t-\tau)T^*(t-\tau)}{\eta+T^*(t-\tau)}-\mu E^*(t-\tau)T^*(t-\tau)-\delta E^*-a_1(1-e^{-u^*})E^*,$$

$$\frac{dT^*}{dt}=r_2T^*(1-\beta T^*)-\mu E^*(t)T^*-c_1N^*T^*-a_2(1-e^{-u^*})T^*,$$

$$\frac{dN^*}{dt}=r_3N^*(1-\beta_2N^*)-c_2T^*N^*-a_3(1-e^{-u^*})N^*,$$

$$\frac{du^*}{dt}=v^*-d_1u^*,$$

$$\lambda_1'(t)=-1+\lambda_1(t)\left[\delta+a_1(1-e^{-u^*})\right]+\lambda_2(t)nT^*+\lambda_1(t+\tau)\chi_{[0,t_f-\tau]}\left[\mu T^*-\frac{\rho T^*}{\eta+T^*}\right],$$

$$\lambda_2'(t)=1+\lambda_2\left[-r_2+2r_2\beta T^*+nE^*+c_1N^*+a_2(1-e^{-u^*})\right]+\lambda_3c_2N^*$$
$$+\chi_{[0,t_f-\tau]}\lambda_1(t+\tau)\left[\frac{\rho E^*T^*}{(\eta+T^*)^2}-\frac{\rho E^*}{\eta+T^*}+\mu E^*\right],$$

$$\lambda_3'(t)=\lambda_2c_1T^*-\lambda_3\left(r_3-2r_3\beta_2N^*-c_2T^*-a_3(1-e^{-u^*})\right)-\gamma,$$

$$\lambda_4'(t)=-\lambda_1(t)a_1e^{-u^*}E^*+\lambda_2(t)a_2e^{-u^*}T^*+\lambda_3(t)a_3e^{-u^*}N^*+\lambda_4(t)d_1,$$

$$v^*=\min\left(v_{max},\frac{\lambda_4}{B_v}\right),\quad w^*=\min\left(w_{max},\frac{\lambda_1s_1}{B_w}\right).$$

$$E^*(\theta)=\psi_1(\theta),\ T^*(\theta)=\psi_2(\theta),\ N^*(\theta)=\psi_3(\theta),\ u(\theta)=\psi_4(\theta),\quad \theta\in[-\tau,0],$$

$$\lambda_i(t_f)=0,\quad i=\{1,2,3,4\}. \tag{13}$$

□

4 Numerical simulations

The numerical approximations of the OCP are carried out using forward and backward Euler methods. Starting with an initial guess for the value of the controls on the time interval $[0,\ t_f]$, we solve the state system with control variables (3) using forward Euler' scheme. Meanwhile, the adjoint system is solved using the solutions of the state system and the transversality conditions (9) backward in time. A Pontryagin-type maximum principle is derived, for retarded OCPs with delays in the state variable when the control system is subject to a mixed control state constraint, in order to minimize the cost of treatment, reduce the tumor cell load, and keep the number of normal cells greater than 75% of its carrying capacity; see the "Appendix" section for the Matlab program.

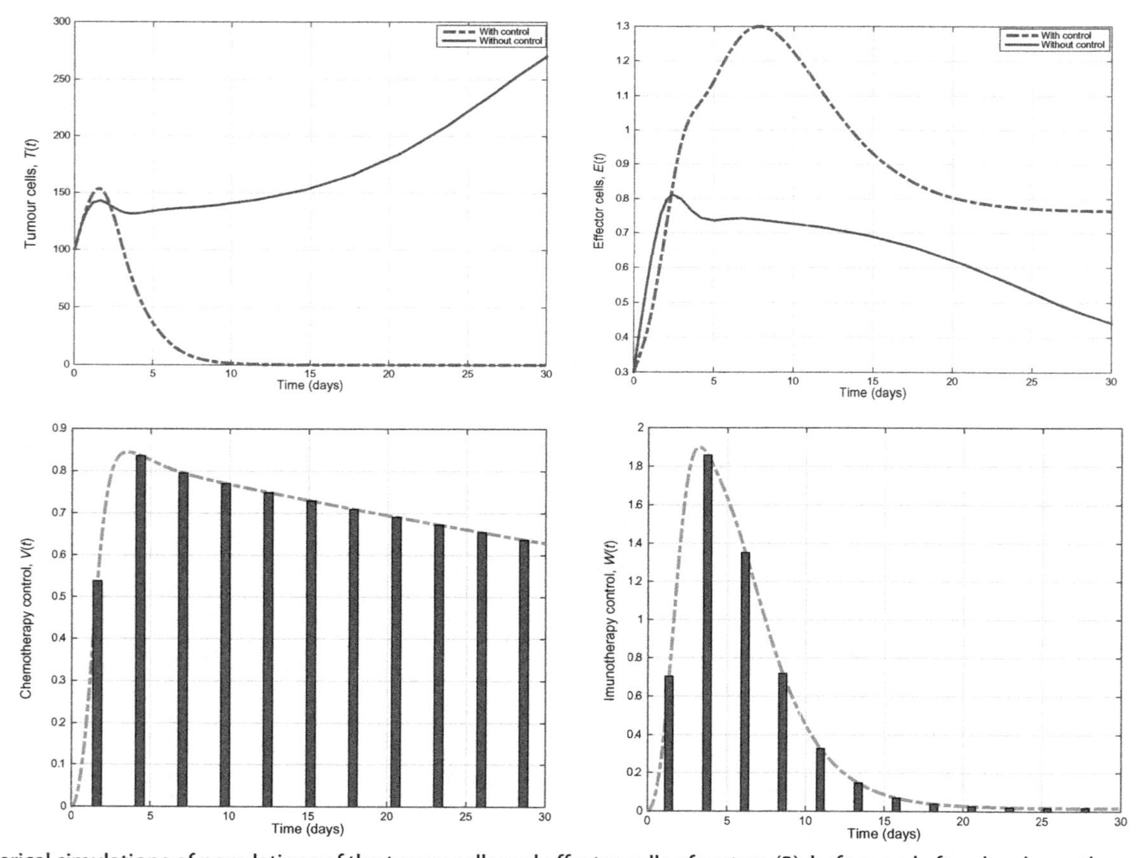

Fig. 2 Numerical simulations of populations of the tumor cells and effector cells of system (3), before and after the chemotherapy treatment with controls $v(t)$ and $w(t)$. It shows that the tumor cells' population can be eradicated by day 10.

Fig. 2 shows the impact of chemotherapy treatments (with optimal control) when we choose the parameter values in an unstable region ($\sigma = 0.2$, $\rho = 0.2$, and $\tau = 1.5$). The tumor cell population is growing over time in the absence of chemotherapy, while the presence of treatment helps the IS to keep the growth of the tumor cells under its control. The figure shows that the optimal treatment strategies reduce the tumor cell load and increase the effector cells after a few days of therapy. The numerical simulations show the rationality of the model presented, which in some degree meets the natural facts.

4.1 Numerical algorithm

Direct and indirect approaches are usually used to solve the OCPs. In the direct approach, the OCP is transformed into a nonlinear programming problem. The algorithm of solving the above OCP is roughly based on the following steps:

Step 1. Provide the initial guess for the control parameters v_0 and w_0 over the interval, and declare the parameters.

Step 2. Set the initial conditions for the state variables $y_0(t)$ with the stored values of v_0 and u_0 and solve the state system forward in time, using any DDEs solver.

Step 3. Use the transversality condition $\lambda(t_f) = 0$, the stored values v_0 and u_0 and $x(t)$ and solve the adjoint system backward in time.

Step 4. Check the control by entering the new values of the state and the adjoint state into the characterization of the optimal control.

Step 5. Verify for convergence. If values of the variables in this iteration and the latest iteration are not negligibly small, output the current values as solutions. If values are not small, return to Step 2.

5 Conclusion

In this chapter, we provided a simple mathematical model with time-lags and optimal control variables to describe the dynamics of tumor-immune interactions in presence of chemotherapy treatments. Control variables are introduced into the originally uncontrolled model and considered L^2 type objective functional to maximize the concentration of effector cells and minimize the tumor cells with minimal side effects of the chemotherapy. We showed that an optimal control exists for this problem. We derived the necessary optimality conditions as a Pontryagin-type minimum principle. We estimated the optimality system to determine the optimal control situation (i.e., the drug strategy), and predict the evolution of the tumor cells, effector cells, and normal cells of each control strategy in 30 days. The numerical simulations displayed in the figure validate the existence of optimality of the control variables and show that the immuno-chemotherapy protocol reduces the tumor load in a few months of therapy. Careful determination of the inclusion of the delay in the optimal control setting for the forward- and backward-oriented system in the numerical setting is needed.

The numerical simulations validate the existence of optimality of the control variables. The presence of chemotherapy protocol reduces the tumor load in a few days of therapy. Of course, this model can be extended to more powerful models for cancer treatment design with optimal combinations, doses, and scheduling of treatments to speed up the development of individualized therapies; see Rihan et al. (2019), Rihan and Velmurugan (2020), and Kim et al. (2018). It is also useful to investigate how a small shift (change) in the input parameters would change the stability of the tumor-free equilibrium, and detect the most significant parameter that has a major impact on the model dynamics.

Appendix

A.1 DDEs with optimal control

Mathematical modeling with DDEs is widely used for analysis and predictions in epidemiology, immunology, and physiology (Rihan, 2000; Bocharov and Rihan, 2000; Fowler and Mackey, 2002; Nelson and Perelson, 2002; Smith, 2011). Time-delays in these models take into account a dependence of the present state of the modeled system on its past history. The delay can be related to the duration of certain hidden processes like the stages of the life cycle, the time between infection of a cell and the production of new viruses, the duration of the infectious period, the immune period and so on. In real life, things are rarely so instantaneous. There is usually a propagation delay before the effects are felt. This situation can be modeled using a DDE.

$$y'(t) = f(y(t), y(t-\tau_1), y(t-\tau_2), \dots, y(t-\tau_d), t), \quad t \geq t_0, \tag{A.1}$$

where all of the time-lags, τ_i, are assumed to be none negative functions of the current time t. Because of these delay terms it is no longer sufficient to supply an initial value, at time $t = t_0$, to completely define the problem. Instead, it is necessary to define the history of the state vector, $y(t)$, sufficiently far enough back in time from t_0 to ensure that all of the delayed state terms, $y(t - \tau_i)$ are always well defined. Thus, it is necessary to supply an initial state profile of the form:

$$y(t)=\psi(t),\ t_0-\tau_{max}\leq t<t_0,\ \text{and}\ y(t_0)=y_0. \tag{A.2}$$

It should be noted that $\psi(t_{0-})$ need not be the same as y_0. This immediately introduces the possibility of a discontinuity in the state, $y(t)$.

We mention here that there are many problems in biosciences (such as epidemics, harvesting, chemostats, treatment of diseases, physiological control, vaccination) that can be addressed within an optimal control framework for systems of DDEs (Banks, 1975; Kolmanovskii and Shaikhet, 1996; Smith, 2005). However, the amount of real experience that exists with OCPs is still small. The DDE (A.1) can be converted into an OCP by adding an m-dimensional control term $u(t)$

$$y'(t)=f(y(t),y(t-\tau_1),y(t-\tau_2),\ldots,y(t-\tau_d),u(t),t) \tag{A.3}$$

and a suitable objective functional (measure): $J_0(u)$

$$\text{Minimize}\ J_0(u)=\int_0^{t_f}\mathcal{L}(y(t),y(t-\tau_1),y(t-\tau_2),\ldots,y(t-\tau_d),u(t),t)dt, \tag{A.4}$$

and subject to control constraint $a\leq u(t)\leq b$, and state constant $y(t)\leq c$, where a and b are the lower and upper bounds. The integrand, $\mathcal{L}(:)$ is called the Lagrangian of objective functional, which is continuous in $[0, t_f]$. Additional equality or inequality constraint(s) can be imposed in terms of $J_i(\mathbf{u})$.

Pontryagin's maximum principle (Pontryagin et al., 1962) gives necessary conditions that the control and the state need to satisfy, and introduces an adjoint function to affix to the differential equation to the objective functional. The necessary conditions needed to solve the OCP are derived from the so-called "Hamiltonian" H, which is given by the equation

$$H(t)=\mathcal{L}(:)+\lambda^T(t)f(:) \tag{A.5}$$

Here, $\lambda^T(t)$ is a vector of costate variables of the state variables $y(t)$, which is the solution of the equation

$$\lambda'(t)=-\frac{\partial H}{\partial y(t)}(t)-\chi_{[0,\,t_f-\tau]}\frac{\partial H}{\partial y(t-\tau)}(t+\tau)\ \ \text{(Adjoint equation)} \tag{A.6}$$

where

$$\chi_{[0,\,t_f-\tau]}=\begin{cases}1 & \text{if}\ t\in[0,t_f-\tau]\\ 0 & \text{otherwise}\end{cases}$$

Given the nonlinear Hamiltonian (A.5) in the controls v and w, the process of solving the OCP is to solve the state system (A.3) together with the adjoint Eq. (A.6) and the following conditions:

$$\begin{aligned} \frac{\partial H(t)}{\partial v} &= \frac{\partial H(t)}{\partial w} = 0 \text{ at } v^*, w^* \quad &\text{(Optimality conditions)} \\ \lambda^T(t_f) &= 0 \quad &\text{(Transversality condition)} \end{aligned} \tag{A.7}$$

OCPs are generally nonlinear and therefore generally do not have analytic solutions like the linear-quadratic OCP. As a result, it is necessary to employ numerical methods to solve the OCP (A.1)–(A.7). The numerical simulations are carried out by solving the state system (A.1) forward in time, and the adjoint system (A.6) backward in time with the given optimality and transversality conditions.

A.2 Matlab program for optimal control with DDEs

Herein, we provide the Matlab program for solving the OCP, associated with DDEs descried in this chapter.

```
program Rihan_OptimalControl
 clear all;
 clc;

 tf=30; %the final time value
 N1=10000; % the number of mish points on the whole intervel
 h=tf/N1; % the step size
 m=1.2/h; % the number of mish points in the subinyervel m=tau/h
 %the parameter values
 delta = 0.2;
 eta = 0.3;
 mu = 0.003611;
 %mu=0.00299;
 r2 = 1.03;
 r3= 1;
 b= 2*10^(-3);
 n=1.;
 c1=0.00003;
 c2=0.00000003;
 a1=0.2;
 a2=0.4;
 a3=0.1;
 d=0.01;
```

```
%v=1;
s1=0.3;
s=0.3;
roh=0.01;
B= 100;
B2=50;

%without treatment
sol = dde23('Rihan',1.2,[0.3; 100; 0.9], [0 30]);

%with treatment
% the initial values of the state variables and
%the control values at time r(i)
for i=1:m+1
r(i)=i*h-m*h;
E(i)=0.3; T(i)=100; N(i)=0.9; u(i)=0.1; v(i)=0; w(i)=0;
end
% the initail values for lambdas at the final times
for i=N1+m+1:N1+2*m+1
lambda1(i)=0;lambda2(i)=0;lambda3(i)=0;lambda4(i)=0;
end

% the forward and backword systems to be solved simultaneously
for i=m+1:N1+m
E(i+1)=E(i)+ h*(s+roh*E(i-m)*T(i-m)/(eta+T(i-m))-....
mu*E(i-m)*T(i-m)-delta*E(i)-a1*(1-exp(-u(i)))*E(i)+w(i)*s1);
T(i+1)=T(i)+h*(r2*T(i)*(1-b*T(i))-n*E(i)*T(i)-....
c1*N(i)*T(i)-a2*(1-exp(-u(i)))*T(i));
N(i+1)=N(i)+h*(r3*N(i)*(1-N(i))-c2*N(i)*T(i)-a3*(1-exp(-u(i)))
  *N(i));
u(i+1)=u(i)+h*(v(i) - d*u(i));

if ((N1-i)*h<=tf-m*h & (N1-i)*h >=0)
lambda1(N1+2*m+1-i)=lambda1(N1+2*m+2-i)-....
h*(-1+lambda1(N1+2*m+2-i)*(delta+ .....
a1*(1-exp(-u(i+1))))+lambda2(N1+2*m+2-i)*n*T(i+1)+....
lambda1(N1+2*m+2-i+m)*(mu*T(i+1)-roh*T(i+1)/(eta+T(i+1))));
lambda2(N1+2*m+1-i)=lambda2(N1+2*m+2-i)-...
h*(1+lambda2(N1+2*m+2-i)*(-r2+2*r2*...
b*T(i+1)+n*E(i+1)+c1*N(i+1)+...
a2*(1-exp(-u(i+1))))+lambda3(N1+2*m+2-i)*c2*N(i+1)+...
lambda1(N1+2*m+2-i+m)*(roh*E(i+1)*T(i+1)/(eta+T(i+1))^2 - ...
roh*E(i+1)/(eta+T(i+1))...
```

```
+mu*E(i+1)));
   else
lambda1(N1+2*m+1-i)=lambda1(N1+2*m+2-i)-...
h*(-1+lambda1(N1+2*m+2-i)*(delta+...
a1*(1-exp(-u(i+1))))+lambda2(N1+2*m+2-i)*n*T(i+1));
lambda2(N1+2*m+1-i)=lambda2(N1+2*m+2-i)-...
h*(1+lambda2(N1+2*m+2-i)*(-r2+2*r2*...
b*T(i+1)+n*E(i+1)+c1*N(i+1)+a2*(1-exp(-u(i+1))))+...
lambda3(N1+2*m+2-i)*c2*N(i+1));
    end

if(N(i+1)<=0.75)
lambda3(N1+2*m+1-i)=lambda3(N1+2*m+2-i)-...
h*(lambda2(N1+2*m+2- i )* c1 * T ( i +1)+...
lambda3 ( N1 +2* m +2- i )*(- r3 +2* r3 * N ( i +1)+...
c2 * T ( i +1)+ a3 *(1- exp (- u ( i +1))))-1);
   else

lambda3 ( N1 +2* m +1- i )= lambda3 ( N1 +2* m +2- i )-...
h *( lambda2 ( N1 +2* m +2- i )* c1 * T ( i +1)+...
lambda3 ( N1 +2* m +2- i )*(- r3 +2* r3 * N ( i +1)+...
c2 * T ( i +1)+ a3 *(1- exp (- u ( i +1)))));
 end

lambda4 ( N1 +2* m +1- i )= lambda4 ( N1 +2* m +2- i )-...
h *( lambda1 ( N1 +2* m +2- i )* a1 * E ( i +1)* exp (- u ( i +1))+...
lambda2 ( N1 +2* m +2- i )* a2 * T ( i +1)* exp (- u ( i +1))+...
lambda3 ( N1 +2* m +2- i )* a3 * N ( i +1)* exp (- u ( i +1))+ lambda4
( N1 +2* m +2- i )* d );
v ( i +1)= min (3, lambda4 ( N1 +2* m +1- i )/ B );
w ( i +1)= min (2, lambda1 ( N1 +2* m +1- i )* s1 / B2 );
r ( i +1)=( i +1)* h - m * h ;
end

figure (1)
hold off
plot ( r, E,' r -.',' LineWidth ',3)
hold on
plot ( sol . x, sol . y (1,:),' LineWidth ',2)
legend (' With Control ',' Without Control ');
xlim ([0 30])
```

```
xlabel (' Time ( days )')
ylabel (' Effector Cells, E ( t )')
grid

figure (2)
hold off
plot ( r, T,' r -.',' LineWidth ',3)
hold on
plot ( sol . x, sol . y (2,:),' LineWidth ',2)
legend (' With Control ',' Without Control ');
xlim ([0 30])
xlabel (' Time ( days )')
ylabel (' Tumour Cells, T ( t )')
grid

figure (3)
hold off
plot ( r, v,' r --',' LineWidth ',3)
hold on
bar ( r (100:900:10001), v (100:900:10001),' barwidth ',.1)
xlim ([0 30])
xlabel (' Time ( days )')
ylabel (' Chemotherapy Control, V ( t )')
grid
hold off
%
figure (4)
hold off
plot ( r, w,' r --',' LineWidth ',3)
hold on
bar ( r (100:900:10001), w (100:900:10001),' barwidth ',.1)
xlim ([0 30])
xlabel (' Time ( days )')
ylabel (' Imunotherapy Control, W ( t )')
grid
hold off
end
%

function dy = Rihan ( t, y, ylag ) % exam1f ( t, y, Z, r1, k1, r2, a, mu1,
e, k2, mu2 )
% ylag = Z (:,1);
dy = zeros (2,1);
```

```
delta =0.2;
eta = 0.3;
mu = 0.003611;
r2 = 1.03;
r3 = 1;
b = 2*10^(-3);
n =1.;
c1 =0.00003;
c2 =0.00000003;
s =0.5;
roh =0.01;

dy (1)=s+roh*ylag(1)*ylag(2)/(eta+ylag(2))-mu*ylag(1)*ylag
(2)- delta * y (1);
dy (2)= r2 * y (2)*(1 - b * y (2)) - n * y (1)* y (2)- c1 * y (3)* y (2);
dy (3)= r3 * y (3)*(1- y (3)) - c2 * y (2)* y (3);
end
```

References

American Association for Cancer Research, 2016. American Association for Cancer Research: special conference on tumor immunology and immunotherapy, Boston, MA, USA, October 20–23.

Araujo, R., McElwain, D., 2004. A history of the study of solid tumor growth: the contribution of mathematical modeling. Bull. Math. Biol. 66, 1039–1091.

Banks, H.T., 1975. Modelling and Control in Biosciences. Lecture Notes in Biomathematics, vol. 6. Springer, Berlin.

Bellomo, N., Li, N., Maini, P., 2008. On the foundations of cancer modeling: selected topics, speculations, and perspectives. Math. Mod. Methods Appl. Sci. 18, 593–646.

Bocharov, G., Rihan, F.A., 2000. Numerical modelling in biosciences using delay differential equations. J. Comput. Appl. Math. 125, 183–199.

Bodnar, M., Forys, U., Poleszczuk, J., 2011. Analysis of biochemical reactions models with delays. J. Math. Anal. Appl. 376 (1), 74–83.

Bray, F., Ferlay, J., Torre, L.A., 2018. Global Cancer Statistics 2018: Globocan estimates of incidence and mortality worldwide for 36 cancers in 185 countries. CA Cancer J. Clin. 68, 394–424.

Byrne, H., Alarcon, T., Owen, M., Webb, S., Maini, P., 2006. Modeling aspects of cancer dynamics: a review. Philos. Trans. R. Soc. A 364, 1563–1578.

Chaplain, M., 2008. Modelling aspects of cancer growth: insight from mathematical and numerical analysis and computational simulation. In: Multiscale Problems in the Life Sciences. Lecture Notes in Mathematics, vol. 1940. Springer, Berlin, pp. 147–200.

de Pillis, L.G., Radunskaya, A., 2001. A mathematical tumor model with immune resistance and drug therapy: an optimal control approach. Comput. Math. Methods Med. 3, 78–100.

de Pillis, L.G., Radunskaya, A., 2003. The dynamics of an optimally controlled tumor model: a case study. Math. Comput. Model. 37, 1221–1244.

de Pillis, L.G., Gu, W., Radunskaya, A.E., 2006. Mixed immunotherapy and chemotherapy of tumors: modeling, applications and biological interpretations. J. Theor. Biol. 238 (4), 841–862.

de Pillis, L.G., et al., 2008. Optimal control of mixed immunotherapy and chemotherapy of tumors. J. Biol. Syst. 16 (1), 51–80.

Dunn, G.P., Bruce, A., Ikeda, H., Old, L.J., Schreiber, R.D., 2002. Cancer immunoediting: from immunosurveillance to tumour scape. Nat. Immunol. 3, 991–998.

Fleming, W.H., Rishel, R.W., 1994. Deterministic and Stochasitic Optimal Control. Springer-Verlag, New York, NY.

Fowler, A.C., Mackey, M.C., 2002. Relaxation oscillations in a class of delay differential equations. SIAM J. Appl. Math. 63, 299–323.

Halanay, A., 1966. Differential Equations, Stability, Oscillations, Time Lags. Academic Press, New York, London.

Joshi, B., Wang, X., Banerjee, X., Tian, H., Matzavinos, A., Chaplain, M.A.J., 2009. On immunotherapies and cancer vaccination protocols: a mathematical modelling approach. J. Theor. Biol. 259 (4), 820–827.

Kim, R., Woods, T., Radunskaya, A., 2018. Mathematical modeling of tumor immune interactions: a closer look at the role of a PD-L1 inhibitor in cancer immunotherapy. SPORA J. Biomath. 4, 25–41.

Kirschner, D., Panetta, J.C., 1998. Modeling immunotherapy of the tumor-immune interaction. J. Math. Biol. 37, 235–252.

Kolmanovskii, V.B., Shaikhet, L.E., 1996. Control of Systems With Aftereffect. Translation of Mathematical Monographs, American Mathematical Society, USA.

Kuznetsov, V.A., Makalkin, I.A., Taylor, M.A., Perelson, A.L., 1994. Nonlinear dynamics of immunogenic tumors: parameter estimation and global bifurcation analysis. Bull. Math. Biol. 56 (2), 295–321.

Lackie, J., 2010. A Dictionary of Biomedicine. Oxford University Press, Oxford.

Liu, Y., Huang, H., Saxena, A., Xiang, J., 2002. Intratumoral coinjection of two adenoviral vectors expressing functional interleukin-18 and inducible protein-10, respectively, synergizes to facilitate regression of established tumors. Cancer Gene Ther. 9, 533–542.

Nagy, J., 2005. The ecology and evolutionary biology of cancer: a review of mathematical models of necrosis and tumor cells diversity. Math. Biosci. Eng. 2, 381–418.

Nelson, N.W., Perelson, A.S., 2002. Mathematical analysis of delay differential equation models of HIV-1 infection. Math. Biosci. 179, 73–94.

Neves, H., Fai Kwok, H., 2015. Recent advances in the field of anti-cancer immunotherapy. BBA Clin. 3, 280–288.

Pontryagin, L.S., Boltyanski, R.V., Gamkrelidge, R.V., Mischenko, E.F., 1962. The Mathematical Theory of Optimal Processes. John Wiley & Sons, New York, NY.

Preziosi, L., 2003. Cancer Modeling and Simulation. Chapman & Hall/CRC Mathematical Biology Series (Book 3).

Rihan, F.A., 2000. Numerical Treatment of Delay Differential Equation in Bioscience (Ph.D. thesis), The University of Manchester (UK).

Rihan, F., Velmurugan, G., 2020. Dynamics of delay differential models with arbitrary-derivative for tumor-immune system. Chaos Solitons Fractals 132, 109592.

Rihan, F.A., Safan, M., Abdeen, M.A., Abdel-Rahman, D.H., 2012. Mathematical modeling of tumor cell growth and immune system interactions. Int. J. Modern Phys. 95–111.

Rihan, F.A., Abdelrahman, D., Al-Maskari, F., Ibrahim, F., 2014a. A delay differential model for tumour-immune response and control with chemo-immunotherapy. Comput. Math. Methods Med. 2014, 15.

Rihan, F.A., Abdelrahman, D.H., Lakshmanan, S., Alkhajeh, A., 2014b. A time delay model of tumour-immune system interactions: global dynamics, parameter estimation, sensitivity analysis. Appl. Math. Comput. 232, 606–623.

Rihan, F.A., Hashish, A., Al-Maskari, F., Sheek-Hussein, M., Ahmed, E., Riaz, M.B., Yafia, R., 2016. Dynamics of tumor-immune system with fractional-order. J. Tumor. Res. 2 (1), 109.

Rihan, F.A., Lakshmanan, S., Maurer, H., 2019. Optimal control of tumour-immune model with time-delay and immuno-chemotherapy. Appl. Math. Comput. 353 (7), 147–165.

Roose, T., Chapman, S., Maini, P., 2007. Mathematical models of avascular tumor growth. SIAM Rev. 49, 179–208.

Sinek, J., Frieboes, H., Zheng, X., Cristini, V., 2004. Two-dimensional chemotherapy simulations demonstrate fundamental transport and tumor response limitations involving nanoparticles. Biomed. Microdevices 6 (4), 297–309.

Smith, S.E., 2005. Optimal control of delay differential equations using evolutionary algorithms. Complex. Int. 12, 1–10.

Smith, H., 2011. An Introduction to Delay Differential Equations With Applications to the Life Sciences. Springer, New York, Dordrecht, Heidelberg, London.

Swan, G., 1985. Optimal control applications in the chemotherapy of multiple myeloma. IMA J. Math. Appl. Med. Biol. 2, 139–160.

Villasana, M., Radunskaya, A., 2003. A delay differential equation model for tumour growth. J. Math. Biol. 47, 270–294.

WHO, 2018. International Agency for Research on Cancer (IARC). WHO, p. 263.

CHAPTER 5

Genetic fuzzy logic based system for arrhythmia classification

Hela Lassoued, Raouf Ketata
National Institute of Applied Sciences and Technology (INSAT), Tunis, Tunisia

1 Introduction

Recently, machine-learning approaches have had an important impact in many areas of science and technology (Angra and Sachin, 2017). Their quick evolution is providing chances to improve the quality of decision-making (Johnson et al., 2016). In fact, the world of medical research has faith in these approaches, which through experience and self-learning can resolve many issues, such as the classification and diagnosis of arrhythmias (Rajkomar et al., 2019). Accordingly, the use of a decision support system (DSS) based on machine-learning approaches becomes a necessity in medicine, especially when the diagnosis requires a lot of knowledge and experience. Moreover, a DSS is mostly used to save time and allow for rapid and efficient action. DSS systems include reasoning, evaluation, learning and many other skills of human intelligence. They guarantee not only the neutrality and objectivity of the expert but also the quality of the decision. They are also largely used by experts as a guide in order to facilitate communication between them.

Today, some cardiovascular arrhythmias such as ventricular tachycardia and ventricular fibrillation lead unexpectedly to cardiac arrest and in most cases lead to sudden death (Huynh et al., 2014). Moreover, statistics claim that cardiovascular arrhythmias are among the leading causes of death worldwide (Ettehad et al., 2016). These arrhythmias, resulting from cardiac dysfunction, are explained by the presence of certain factors, such as unhealthy eating habits, lack of physical activity, high stress, family history, age and many others (Chen et al., 2015). One way to achieve appropriate care for these arrhythmias is to rely on the electrical activity of the heart, illustrated by the electrocardiogram (ECG) signal (Chen et al., 2017). Despite the technological evolution in the field of medical instrumentation, ECG signal remains an essential examination in cardiology. However, its manual analysis requires careful inspection due to its long duration (24–48 hours). The DSS

Control Theory in Biomedical Engineering
https://doi.org/10.1016/B978-0-12-821350-6.00005-6

assistance is therefore desirable by cardiologists for classifying cardiac arrhythmias (Hijazi et al., 2016). According to the literature, the use of DSS in cardiology has demonstrated their performance compared to manual practice (Luz et al., 2016). However, developing, validating and implementing such systems for healthcare today is a challenge for current research in computer science, signal processing and medicine. Indeed, in order to ensure the accuracy and the speed of the DSS results response, optimal adjustment regarding the input hypothesis and machine-learning approaches is required.

In this context, the main purpose of this chapter is to improve the accuracy of DSS based in cardiac arrhythmias classification. Indeed, several machine-learning approaches are used for building DSS (Mincholé et al., 2019). The most common are neural networks (Vasilakos et al., 2016), fuzzy logic controllers (FLC) (Ahmadi et al., 2018), support vectors machines (Rajesh and Dhuli, 2017), recurrent neural networks (Singh et al., 2018), k-nearest neighbors (Faziludeen and Praveen, 2016), genetic algorithms (GA) (Li et al., 2017), decision trees (Kasar and Joshi, 2016), clustering algorithms (Sayilgan et al., 2017), and many others (Mondéjar-Guerra et al., 2019; Jambukia et al., 2015).

Recently, neural networks, which are inspired from the human neural system, are mostly applied for classifying certain arrhythmias (Luz et al., 2016). The most used are the multilayer perceptron (MLP) (Savalia and Vahid, 2018), the radial basis function (RBF) (Kelwade and Salankar, 2016) and the probabilistic neural network (PNN) (Gutiérrez-Gnecchi et al., 2017). The result of all the experiments done in (Lassoued and Ketata, 2018c) indicates that the RBF is the most accurate network with an accuracy of 99.9%, while the MLP has fast testing response (0.096 s) and the PNN has the speediest training response (0.070 s). However, it is concluded that there is a significant relationship not only between the neural network's performance and its configured structure but also with the input feature vector. So, several studies were conducted to evaluate neural networks with different training algorithms (Lassoued and Ketata, 2018a) or number of hidden neurons and layers (Lassoued and Ketata, 2017), or different datasets (Rajamhoana et al., 2018) and many other factors (Li et al., 2017). Nevertheless, identifying the neural network structure, by using optimization approaches, shows better performances (Lassoued and Ketata, 2018b), (Arabasadi et al., 2017). Particularly, the MLP is mostly used in deep learning, which has advanced rapidly due to its structure in which both feature extraction and classification stages are performed together (Bakator and

Radosav, 2018). The results indicate that convolutional neural networks are the most widely represented (Parvaneh et al., 2019).

In addition, the FLC that mimics experts to generate a suitable control action for a particular system has been largely integrated in DSS process. By referring to the expert knowledge, the FLC is mainly used to overcome the uncertainty and vagueness and to involve approximate reasoning for decision-making tasks. In the medical sector, and more particularly for classifying ECG signals, numerous works based on the FLC approach have been developed and evaluated (Krishnaiah et al., 2016). However, its configuration is still a challenging task. Indeed, the FLC performances depend on the adjustment of the membership functions parameters and its rule base. To overcome this problem, a number of hybrid approaches were offered in several research studies, such as fuzzy clustering techniques and fuzzy evolutionary algorithms (Rathi and Narasimhan, 2017).

Thus, in this study, a fuzzy genetic-based system for cardiac arrhythmia classification is investigated. It is applied to classify the MIT-BIH Arrhythmia Database recordings into five arrhythmia types: (1) Normal Sinus Rhythm (NSR), (2) Premature Ventricular Contraction (PVC), (3) Left Bundle Branch Block (LBBB), (4) Right Bundle Branch Block (RBBB) and (5) Paced beats (P) (Silva and Moody, 2014). Previously, we have used the neural networks for this task (Lassoued and Ketata, 2018c). However, we concluded that it is usually difficult to select the more appropriate neural network structure and to interpret its results. Accordingly, the proposed based system consists of a FLC whose membership parameters and rules number are adjusted by a GA. This system has to define the correct arrhythmia type for a previously unknown sample. Then, a comparison study between the obtained accuracies before and after the genetic optimization will be analyzed. Simultaneously, in order to access the efficiency of this study, a second comparison study between neural network approaches and the proposed FLC will be done.

This chapter is organized as follows. Section 2 describes the proposed arrhythmia classification methodology. Section 3 details the experiments and results, and the final section provides a conclusion.

2 Methodology

In this section, we describe the suggested arrhythmia classification methodology, which consists of a hybrid system between a FLC and GA. As it is described in Fig. 1, the FLC uses a pre-processing block to filter the original

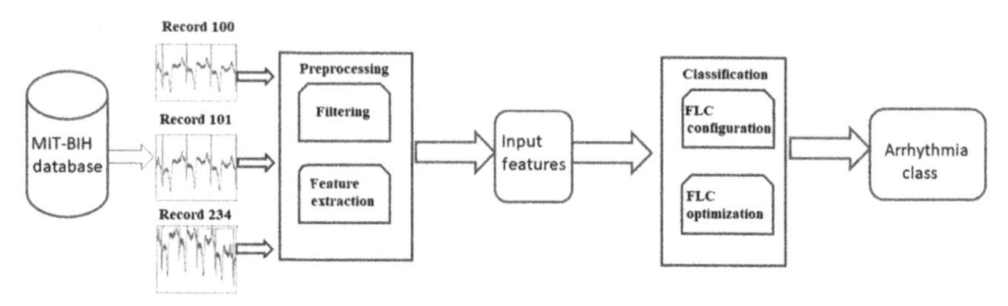

Fig. 1 Fuzzy arrhythmia classification methodology.

signal from the artifacts and to extract the features. Then, the obtained features are used as inputs for the FLC in order to classify the MIT-BIH recordings into five arrhythmias (NSR, RBBB, LBBB, PVC and P). This classifier demands two major steps including its configuration and optimization.

The proposed FLC imitates the process of a standard controller. Indeed, a standard controller, as it is described in Fig. 2A, uses the error between the output and the reference input to activate the control action. This action tries to reduce the error. If the error is reduced to zero, the output is equal to the reference input. By doing an analogy with the standard controller, the proposed FLC, as it is illustrated in Fig. 2B, uses the error between the predicted output and its corresponding target to activate the membership parameters' optimization by using a GA. Then, the FLC is updated with the optimized parameters and new outputs are generated. This process is repeated until the error reaches zero.

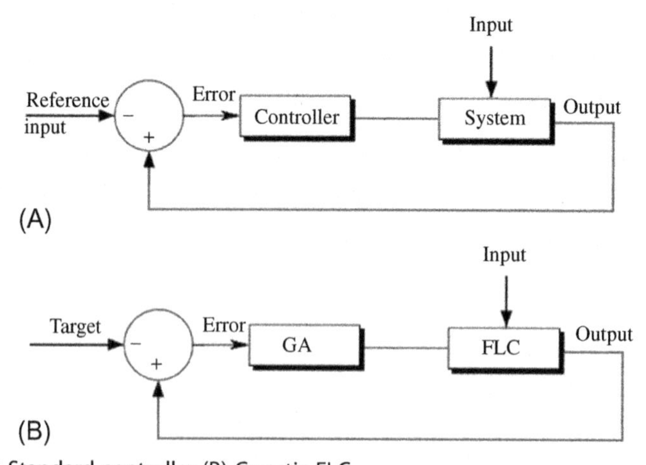

Fig. 2 (A) Standard controller (B) Genetic FLC.

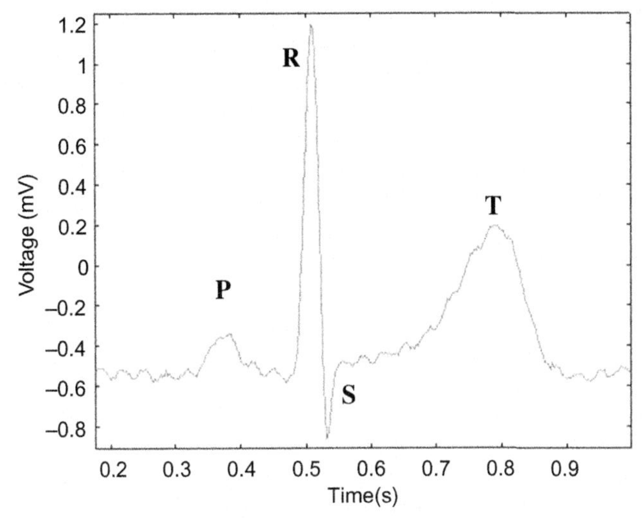

Fig. 3 Normal Sinus Rhythm heartbeat.

Fig. 3 represents a single heartbeat which consists of three electrical wave shapes. The P wave represents the contractions that pump blood in the upper chambers of the heart. It is named atrial depolarization. However, the QRS complex and the T wave represent activity in the lower chambers of the heart. They involve the ventricular depolarization and ventricular repolarization, respectively (Chen et al., 2017).

Fig. 4 shows the four arrhythmias: PVC, P, LBBB and RBBB. Fig. 4A and B describe PVC and P beats, respectively. The PVC beats correspond to premature contraction of one of the ventricles. The paced beats are generated by a pacemaker and they are detected by the enlarged RR intervals (>280 ms). However, Fig. 4C and D represent the LBBB and RBBB arrhythmias, respectively. In fact, both of them are identified by the presence of a widened QRS complex (>120 ms). Particularly, LBBB arrhythmia is detected by a wrong displacement of ST and T waves. Indeed, they are mostly opposite to the major deflection of the QRS complex. However, the RBBB arrhythmia is identified by a slurred S wave.

2.1 Preprocessing

The classification must be preceded by a pre-processing phase. It consists mainly of two main blocks: the ECG signal filtering and feature extraction.

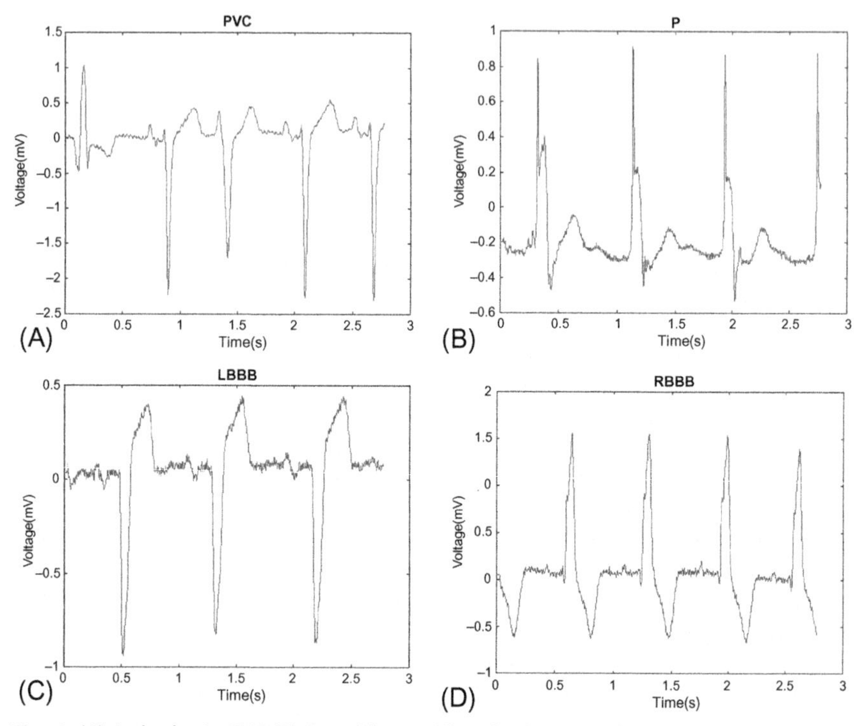

Fig. 4 (A) Arrhythmia PVC (B) Paced beats (C) Arrhythmia LBBB (D) Arrhythmia RBBB.

The extracted features are useful for cardiologists when diagnosing cardiac arrhythmias. They are mostly needed to prepare the input vector that will be used later in the classification phase (Lassoued and Ketata, 2018a).

2.1.1 ECG signal filtering

The ECG signal is measured by electrodes attached to the skin. Therefore, it is usually sensitive to different types of artifacts, such as the power line interference, the baseline wander, the electrode contact noise, the muscle contractions, the instrumentation noise and so on (Kaplan et al., 2018). In this section, we have eliminated only the baseline wander by using an appropriate low-order polynomial. Fig. 5 represents the original noisy signal. However, this signal shows a baseline shift and therefore does not represent the true amplitude. Thus, in order to remove the trend, a low-order polynomial is applied to eliminate it. Therefore, the noisy signal becomes without baseline shift, as it is shown in Fig. 6.

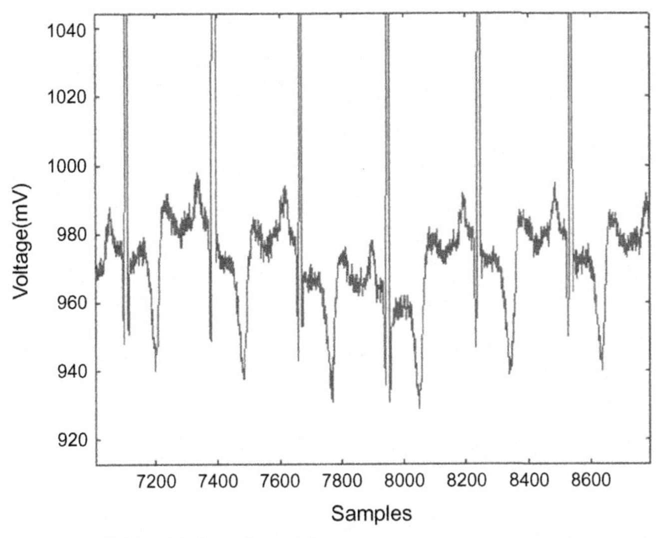

Fig. 5 Noisy signal ECG with baseline shift.

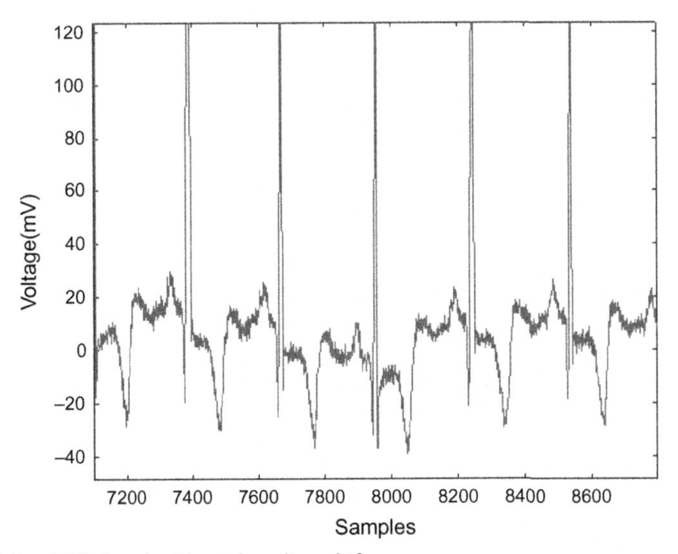

Fig. 6 Noisy ECG signal without baseline shift.

2.1.2 ECG feature extraction

After eliminating the baseline wander, the algorithm Wavedet (Martínez et al., 2004), developed in the toolbox ECG kit and carried out by Demski and Llamedo (2016), is applied to extract the morphological features. This algorithm has been applied for the detection of the ECG signal wave peaks and their durations. It uses the functions developed in the WFDB toolbox developed by Silva and Moody (2014) and it has proved its effectiveness in several researches (Soria and Martínez, 2009).

The first step in the Wavedet algorithm is to detect the QRS complexes of each heartbeat, as it is shown in Fig. 7. In fact, the QRS complex is the most repeating peak in the ECG signal. It corresponds to the depolarization of the right and left ventricles. It is mainly used to determine a patient's arrhythmias. Moreover, the high amplitude of the QRS complex facilitates its detection relative to other waves (De Chazel et al., 2004). Once the complex QRS is identified, R, S, P and T waves are deduced, as shown in Figs. 8 and 9, respectively.

Once the ECG waves have been delineated, other measures, such as duration, can be calculated. In fact, six equations (Eqs. (1)–(6)) are used to measure the duration between waves (P, Q, R, S and T) by using each wave's beginning (on) and end (off) (De Chazel et al., 2004).

$$\text{Intervalle\_PR} = \text{QRS}_{\text{on}} - \text{P}_{\text{off}} \tag{1}$$

$$\text{Intervalle\_PT} = \text{T}_{\text{on}} - \text{P}_{\text{off}} \tag{2}$$

$$\text{Intervalle\_ST} = \text{T}_{\text{on}} - \text{S}_{\text{off}} \tag{3}$$

$$\text{Intervalle\_QT} = \text{T}_{\text{on}} - \text{Q}_{\text{off}} \tag{4}$$

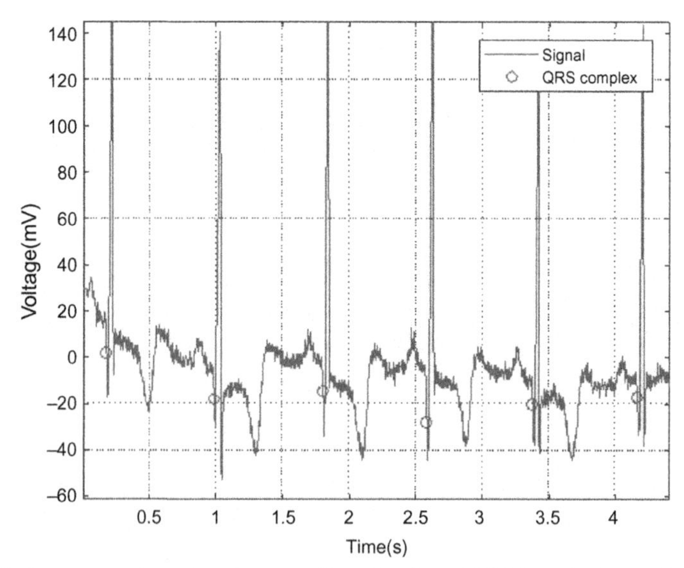

Fig. 7 QRS complex detection for recording 100 from MIT-BIH arrhythmia database.

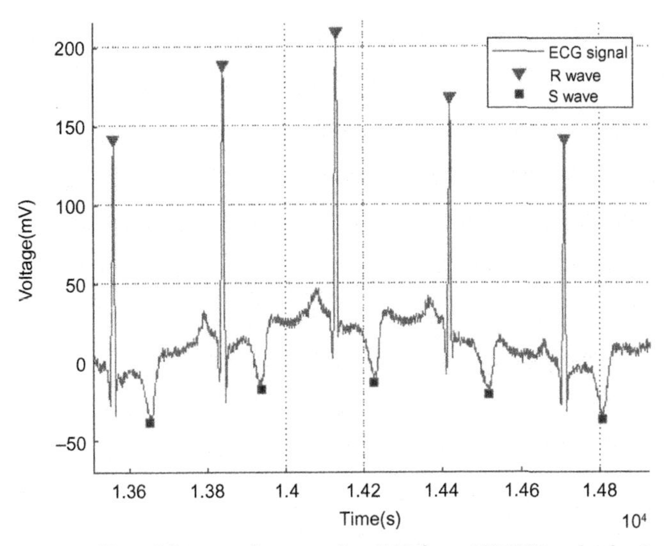

Fig. 8 Detection of R and S waves for recording 100 from MIT-BIH arrhythmia database.

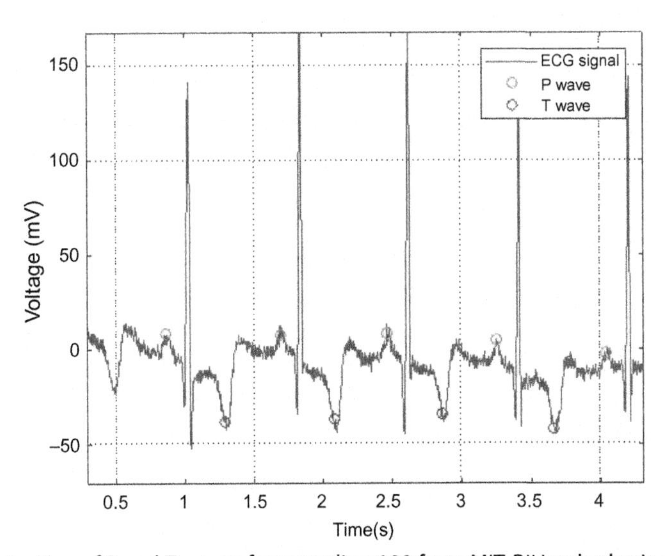

Fig. 9 Detection of P and T waves for recording 100 from MIT-BIH arrhythmia database.

$$\text{Intervalle\_TT} = \text{T}_{\text{on}} - \text{T}_{\text{off}} \tag{5}$$

$$\text{Intervalle\_QRS} = \text{QRS}_{\text{off}} - \text{QRS}_{\text{on}} \tag{6}$$

As a result, the obtained morphological features of wave duration and amplitude are used to obtain the corresponding input feature vector for each ECG recording. Then, the input feature vectors are concatenated with the target output vector in an input data matrix. This input matrix will be used to evaluate the FLC performances by comparing predicted outputs with their corresponding targets.

2.2 Fuzzy arrhythmia classification

After pre-processing, the obtained feature vectors are treated by the FLC for the patient's arrhythmia classification. However, the FLC necessitates two major steps first: configuration and optimization.

2.2.1 FLC configuration

A standard controller requires the most accurate model by using differential equations. However, a FLC does not require a mathematical model, but it uses the fuzzy sets and rules of the form (if … Then …). Fig. 10 shows the general block diagram of a FLC. It consists of four major blocks: the knowledge base, the fuzzification method, the inference mechanism and the defuzzification method.

The first block describes the knowledge base. It involves a rule base and a definition database (named Database). Accordingly, the two bases define the relationships between the premises and the corresponding consequences. The second block describes the fuzzification of the input variables, which converts the crisp inputs into fuzzy inputs by using the membership functions. Several membership functions exist. The most used are the triangular and the Gaussian functions. The third block describes the mechanism of a fuzzy inference. It can be either Mamdani or Sugeno. For the Mamdani inference mechanism, the antecedent and the consequence are both fuzzy variables. However, for the Sugeno inference, the antecedent is a fuzzy variable, but the consequence is a constant or a linear function. The last block

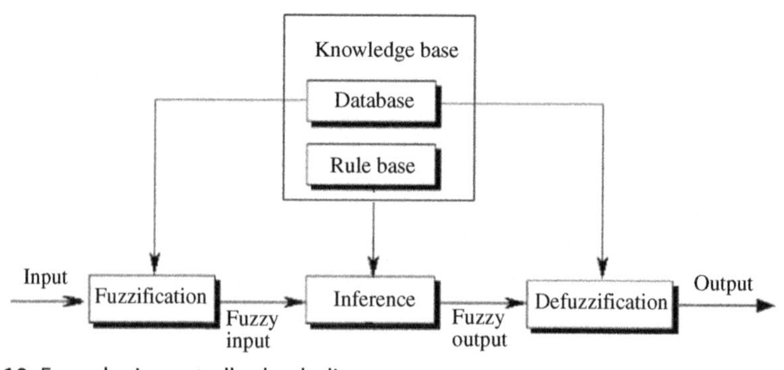

Fig. 10 Fuzzy logic controller bock diagram.

describes the defuzzification of the output variables, which converts the fuzzy outputs into crisp ones (Zadeh, 2015).

Generally, the fuzzy arrhythmia classifier has to undergo some adjustment in the knowledge base in terms of number of fuzzy variables, the membership functions types and the definition of rules. We have considered the following steps in order to configure the proposed FLC.

- Inference mechanism

We have selected the Mamdani fuzzy inference mechanism whose outputs are defined as singletons. In fact, the used defuzzification function is based on the weight of the most important fuzzy rule.

- Input/output identification

The morphological features, previously extracted in during pre-processing, are considered as the FLC inputs. Indeed, we have chosen six of them, including the mean heart rate (E1), the P wave amplitude (E2), the PR interval duration (E3), the QRS complex duration (E4), the RR interval duration (E5) and the ST interval duration (E6). In addition, for the FLC outputs, we have five arrhythmia classes (NSR, PVC, P, LBBB and RBBB).

- Discourse universe partition

We have considered for each input (E_i) three fuzzy sets as follows: the minimum (MIN), medium (MOY) and maximum (MAX). These fuzzy sets are defined on the six discourse universes $[-E_i + E_i]$. The Gaussian membership functions are used for the inputs. However, the output membership functions are singletons.

In order to define the fuzzy sets, the mean value (μ) and the standard deviation (δ) are determined for each input. They are defined in Eqs. (7) and (8), respectively. The lower limit ($\mu - 2^*\delta$), the average (μ) and the upper limit ($\mu + 2^*\delta$), are used to deduce the fuzzy sets (MIN, MOY and MAX) for the six inputs E_i (with ($i = 1 \dots 6$)). They are represented in Table 1.

Table 1 Fuzzy sets.

Input	Fuzzy sets	
E_i	MIN	$E_i < \mu - 2 * \delta$
	MOY	$\mu - 2 * \delta < E_i < \mu + 2 * \delta$
	MAX	$E_i > \mu + 2 * \delta$

$$\mu = \frac{\sum_{i=1}^{N} E_i}{N} \tag{7}$$

$$\delta = \sqrt{\frac{\sum_{i=1}^{N} (E_i - \mu)^2}{N}} \tag{8}$$

where, N refers to the number of feature vectors to be classified.

The fuzzy sets are illustrated in Fig. 11. Therefore it is clear that for each input there is a coverage rate between its fuzzy set compared to its discourse universe.

- Operator adjustment

Fuzzy logic if-then rules are formed by applying fuzzy operations to the fuzzy sets. Fuzzy operators are used to compute the rules weight. They include the intersection (AND), the union (OU), the fuzzy implication and the Cartesian product. It is recommended to adjust them carefully.

Table 2 shows the selected operators. The product and the sum of the fuzzy sets are associated with the conjunction "AND" and the disjunction "OR," respectively. The product method is used for the fuzzy implication. It is used to adjust the consequence membership function based on the antecedent values.

- Rule base construction

After achieving the input/output identification, the partition of the various discourse universes, the inference fuzzy mechanism selection and the fuzzy operators adjustment, it is possible to define the rule base. It imitates the relationship between the fuzzy sets and the five classes. Therefore, by referring to the cardiologist's knowledge, five rules haves been developed. Table 3 shows the comportment of the six features to identify the five arrhythmia classes.

Table 3 leads us to create the rule base, which is described linguistically in Table 4.

The fuzzy rules, described in Table 4, have the form of conditional expression linking the states of the input features as antecedent and the five outputs as consequences. It can be represented in another form, as it is described in Table 5.

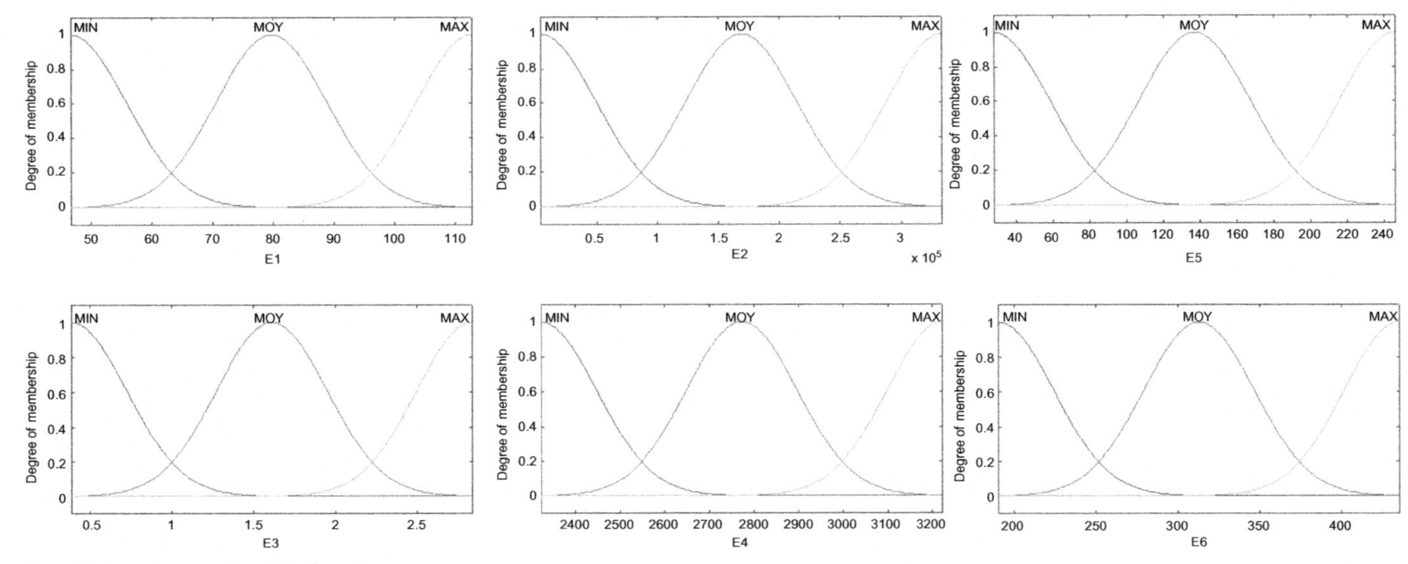

Fig. 11 Inputs membership functions.

Table 2 Adjusted operators.

Designation	Operators
Conjunction	Product
Disjunction	Sum
Implication	Product

Table 3 State of the inputs for each arrhythmia class.

Arrhythmia	E1 (bpm)	E2 (mv)	E3 (ms)	E4 (ms)	E5 (ms)	E6 (ms)
N	60–100	0.05–2.5	120–200	80–120	60–120	80–120
PVC	<60	<0.05	–	>120	–	–
P	60–100	–	–	>120	>280	–
RBBB	<60	>2.5	120–200	>120	–	>120
LBBB	<60	<0.05	120–200	>120	–	–

"–" indicates there is no feature information.

Table 4 Linguistic rule base.

Fuzzy rule	Output
If the heart rate is MOY **AND** the amplitude P is MOY **AND** The duration of the PR interval is MOY **AND** the duration of the QRS complex is MOY **AND** the duration of the RR interval is MOY **AND** the duration of the ST interval is MOY**THEN**	NSR
if the heart rate is MIN **AND** the amplitude P is MIN **AND** the duration of the QRS complex is MAX **AND** the duration of the RR interval is MAX **THEN**	PVC
If the heart rate is MOY **AND** the duration of the QRS complex is MAX **AND** the duration of the RR interval is MAX **THEN**	P
If the heart rate is MIN **AND** the amplitude P is MAX **AND** the duration of the PR interval is MOY **AND** the duration of the QRS complex is MAX **AND** the duration of the ST interval is MAX **THEN**	RBBB
If the heart rate is MIN **AND** the amplitude P is MIN **AND** the duration of the PR interval is MOY **AND** the duration of the QRS complex is MAX **THEN**	LBBB

Table 5 Rule base.

Rule no	If E1 is	AND E2 is	AND E3 is	AND E4 is	AND E5 is	AND E6 is	Then S is
R1	MOY	MOY	MOY	MOY	MOY	MOY	NSR
R2	MIN	MIN	–	MAX	–	–	PVC
R3	MOY	–	–	MAX	MAX	–	P
R4	MIN	MAX	MOY	MAX	–	MAX	RBBB
R5	MIN	MIN	MOY	MAX	–	–	LBBB

2.2.2 FLC optimization

Unlike standard controllers, the FLC configuration requires the adjustment of a greater number of parameters, as discussed in the previous section. In fact, the FLC designer has to make tuning regarding the expression of the rules, the definition of inputs and its fuzzy values, the inference mechanism, the defuzzification method and many others. Thus, configuring the FLC with the appropriate parameters is a challenging task, especially when expert knowledge is not available. Hence, optimizing the FLC parameters offers a reliable solution to this problem. We have applied the GA for the optimization of the Gaussian membership parameters and the rules number.

The GA is applied for optimization tasks and it is usually used in problems with amount parameters. By using genetic optimization, it is essential to define the chromosome representing the solution and the fitness function evaluating the produced solutions.

As it is described in Fig. 12, the GA process begins by using the initial set of solutions, named initial population, which is randomly generated and is subsequently coded into binary chromosomes. Then, the FLC is updated with each chromosome and consequently evaluated using a suitable fitness function. In this study, we have selected the Root Mean Square Error (RMSE) as a fitness function (see Eq. (9)).

$$\text{RMSE} = \sqrt{\frac{1}{N}\sum_{i=0}^{N}(t_i - o_i)^2} \tag{9}$$

Thus, the best solution is a vector that reaches the minimal value of the RMSE function. Indeed, the predicted output (o_i) is compared with its corresponding target (t_i), in order to evaluate the RMSE function. After that, in order to create the diversity of the population, selection, crossover and mutation operators are applied and a new population is evaluated (Lassoued and Ketata, 2018a). This process continues for multiple iterations

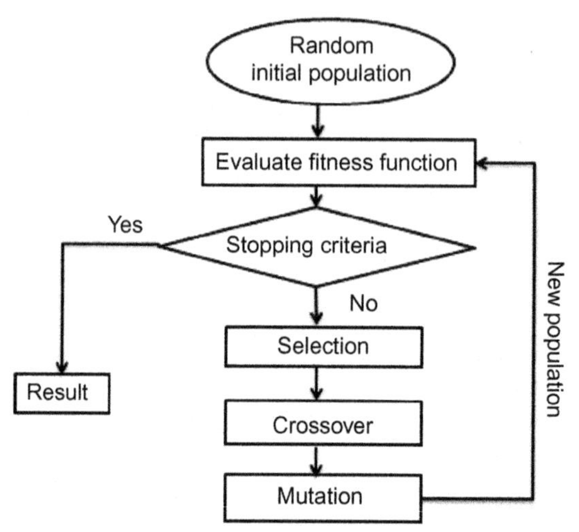

Fig. 12 Genetic algorithm process.

until the stopping criteria (the maximum generation number, the maximum number of iteration, etc.) are reached. At the end, the solution with the lowest RMSE is selected as the best one and its optimized parameters are returned.

In our case, as it is shown in Fig. 13, an example of a binary chromosome is illustrated. The chromosome represents the Gaussian parameters (μ and δ) and the number of fuzzy rules. The elitism selection is used to select which solutions are retained for further reproduction. The n-point crossover is selected to obtain new solutions from existing ones. The binary mutation is applied as introducing diversity into the solution pool by means of randomly swapping or turning off solution bits.

In our study, the GA is applied for the configured FLC with three and two Gaussian fuzzy sets. Table 6 reports the obtained performances, where MF is the membership function, Num-fuzzy_set is the number of fuzzy sets and Num_R is the number of fuzzy rules. Bold values indicate the adopted FLC performances after genetic optimization.

Table 6 shows that the minimal RMSE between the FLC outputs and their corresponding targets (RMSE = 0.619) is obtained by using three Gaussian fuzzy sets and applying 62 fuzzy rules. Accordingly, the distribution

Fig. 13 Example of a chromosome solution.

Table 6 Genetic FLC performances.

MF	Num-fuzzy-set	Num_R	RMSE
Gaussian	2	58	0.889
	3	**62**	**0.619**

of the optimized membership parameters is illustrated in Fig. 14. This figure indicates that there is a coverage rate between the input fuzzy sets compared to its discourse universe. However, for input E4, it is shown that the two fuzzy sets (MIN and MOY) are overlapped. So, for this input, only two fuzzy sets are sufficient for its fuzzification.

3 Experimental results

In this chapter, the FLC is designed to classify ECG signals into five arrhythmias. The FLC is evaluated by comparing its outputs with their corresponding targets. Accordingly, RMSE (see Eq. (9)), Accuracy (ACC), Sensitivity (Se) and Specificity (Sp) are used as performance measurements to evaluate the effectiveness of the FLC. They are expressed by Eqs. (10)–(12)

$$\mathrm{ACC} = \frac{\mathrm{NCC}}{N} \tag{10}$$

$$\mathrm{Se} = \frac{\mathrm{TP}}{\mathrm{TP} + \mathrm{FN}} \tag{11}$$

$$\mathrm{Sp} = \frac{\mathrm{TN}}{\mathrm{TN} + \mathrm{FP}} \tag{12}$$

where N is the total number of ECG recordings, NCC is the total number of correctly classified heartbeats, TP (true positive) represents the number of sick people classified as sick, FP (false positive) represents the number of nonsick people classified as sick, TN (true negative) represents the number of nonsick people classified as nonsick, and FN (false negative) represents the number of sick people classified as nonsick.

3.1 Comparison study between the performances before and after the genetic optimization

An evaluation of the FLC performances, before and after the genetic optimization, is made. Thus, by evaluating the number of correctly classified examples (TP and TN) and those badly classified (FP and FN), ACC, Se and Sp performances are deduced from the obtained confusion matrix.

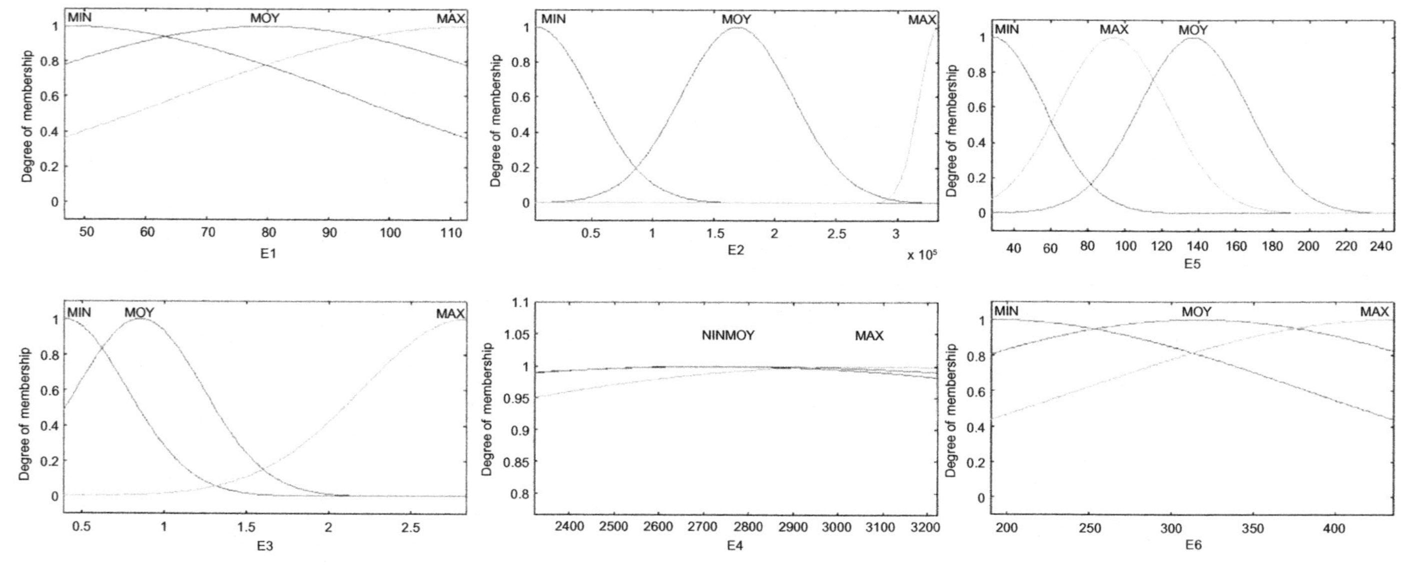

Fig. 14 Optimized membership parameters.

Table 7 Performances of the FLC before genetic optimization.

S	TP	TN	FP	FN	ACC (%)	Se (%)	Sp (%)
NSR	31	10	1	6	85.416	83.783	90.909
LBBB	3	43	1	1	**95.833**	**75.000**	**97.729**
RBBB	3	42	1	2	93.750	60.000	97.674
PVC	1	42	3	2	89.583	**33.333**	93.333
P	0	40	4	4	**83.333**	**0.000**	**90.909**
Average ACC					**89.583**		

Table 8 Performances of the FLC after the genetic optimization.

S	TP	TN	FP	FN	ACC (%)	Se (%)	Sp (%)
NSR	32	16	0	0	**99.999**	**99.999**	**99.999**
LBBB	4	31	0	1	97.058	66.666	99.999
RBBB	4	32	0	0	99.999	99.999	99.999
PVC	2	31	2	1	94.117	50.000	96.875
P	2	31	2	1	**94.117**	**50.000**	**96.875**
Average ACC					**97.054**		

Tables 7 and 8 summarize the FLC performances achieved before and after the genetic optimization, respectively. In each table, bold values indicate the highest and lowest class performances, as well as the average ACC.

On one hand, by referring to Table 7, we deduce that the configured FLC reaches an acceptable average accuracy (ACC=89.583%). Globally, it achieves good results for classifying four heartbeats types (NSR, PVC, LBBB and RBBB). In addition, we found that the maximum accuracy (ACC=95.833%) is obtained by classifying (LBBB) heartbeats with significant sensitivity (Se=75%) and specificity (Sp=97.72%). However, the performances are less efficient (ACC=83.333%, Se=0%, Sp=90.90%) by evaluating the (P) heartbeats. This indicates that the FLC has not classified any examples with arrhythmia (P). We also noticed that only 33.33% of the examples with arrhythmia (PVC) are correctly classified.

On the other hand, by referring to Table 8, we deduce that after applying the genetic optimization the FLC reaches a more accurate average accuracy (ACC=97.054%). In fact, the optimized FLC achieves good results (ACC, Se, Sp=99.999%) for classifying (NSR and RBBB) heartbeats. However, the obtained performances for the classification of the (P and PVC) arrhythmias are improved by using the optimized parameters (ACC=94.117, Se=50% and Sp=96.875). However, they are still less accurate than the other heartbeat classes (NSR, LBBB and RBBB).

Table 9 Comparison analysis with related works.

Reference	Classifier	Features	Output	ACC (%)
Lassoued and Ketata (2017)	MLP	10	2	87.000
Lassoued and Ketata (2018a)	MLP	62	5	93.800
Lassoued and Ketata (2018b)	MLP	30	5	93.700
Lassoued and Ketata (2018b)	MLP + GA	30	5	**99.999**
Lassoued and Ketata (2018c)	MLP	10	2	86.400
Lassoued and Ketata (2018c)	RBF	10	2	**99.999**
Lassoued and Ketata (2018c)	PNN	10	2	79.500
The proposed work (Genetic FLC)	FLC + GA	6	5	97.054

3.2 Comparison analysis with related works

In order to evaluate the efficacy of this work, a second comparison study with related works was conducted (see Table 9). It recapitulates the obtained accuracy by the proposed FLC and other neural network classifiers. These classifiers were previously investigated in other works using different neural network approaches (Lassoued and Ketata, 2017, 2018a, b, c). Only the proposed work deals with the fuzzy arrhythmia classification. Moreover, each classifier has used its proper methodology mainly regarding the number of features and classes (outputs). Thus, referring to Table 9, all the proposed classifiers are effective to about 99.999%–79.500% of ACC. However, we conclude that despite not achieving the most accurate results (ACC = 99.999%), the proposed FLC is considered as a precise classifier (ACC = 97.054%). It is considered more interpretable and explicitly defined than the neural network classifiers.

4 Conclusion

In this chapter, we proposed a FLC for classifying arrhythmias. We used 48 recordings from the MIT-BIH Arrhythmia Database. The proposed fuzzy based system consists of two main blocks: ECG pre-processing and fuzzy arrhythmia classification. In the first block, the elimination of the baseline shift from the ECG signals is investigated. Then, the ECG feature extraction is done. Hence, we obtained several morphological features. However, we selected only six features to evaluate the FLC. In the second block, we firstly manually configured the FLC. Then, we applied the GA for the FLC membership parameters and rules number optimization. Accordingly, a comparison study between the obtained performances before applying the genetic optimization (ACC = 89.583%) and after its application (ACC = 97.054%) is treated. This study reveals that the average accuracy (ACC=97.054%) is improved by using the optimized membership parameters and rules numbers. However, two arrhythmias (P and PVC) are not efficiently classified. Successively, in order to evaluate the efficiency of the proposed FLC, a second comparison study with related works is presented. Therefore, the reached results show the usefulness of the suggested FLC (ACC=97.054%), which may provide an effective way for earlier diagnosis of a number of arrhythmias. Moreover, the accuracy of the classifier depends not only on the FLC configuration but also on the type and number of the extracted features. Thus, we recommend using the clustering algorithms that are able to select automatically the type and number of fuzzy sets. We also recommend adding other features, such as the discrete wavelet coefficients in order to enrich the input feature vector.

References

Ahmadi, H., Gholamzadeh, M., Shahmoradi, L., Nilashi, M., Rashvand, P., 2018. Diseases diagnosis using fuzzy logic methods: a systematic and meta-analysis review. Comput. Methods Programs Biomed. 161, 145–172.

Angra, S., Sachin, A., 2017. Machine learning and its applications: a review. In: Proc of the Int. Conf. on Big Data Analytics and Computational Intelligence (ICBDAC). India, 23–25 March.

Arabasadi, Z., Alizadehsani, R., Roshanzamir, M., Moosaei, H., Yarifard, A.A., 2017. Computer aided decision making for heart disease detection using hybrid neural network-Genetic algorithm. Comput. Methods Programs Biomed. 141, 19–26.

Bakator, M., Radosav, D., 2018. Deep learning and medical diagnosis: a review of literature. Multimodal Technol. Interact. 2 (3), 47.

Chen, W., Thomas, J., Sadatsafavi, M., Fitz Gerald, J.M., 2015. Risk of cardiovascular comorbidity in patients with chronic obstructive pulmonary disease: a systematic review and meta-analysis. Lancet Respir. Med. 3 (8), 631–639.

Chen, S., Hua, W., Li, Z., Li, J., Gao, X., 2017. Heartbeat classification using projected and dynamic features of ECG signal. Biomed. Signal Process. Control 31, 165–173.

De Chazel, P., O'Dwyer, M., Reilly, R.B., 2004. Automatic classification of heartbeats using ECG morphology and heartbeat interval features. IEEE Trans. Biomed. Eng. 51 (7), 1196–1206.

Demski, A., Llamedo, S.M., 2016. ECG-kit: a Matlab toolbox for cardiovascular signal processing. J. Open Res. Softw. 4(1).

Ettehad, D., Emdin, C.A., Kiran, A., Anderson, S.G., Callender, T., Emberson, J., Rahimi, K., 2016. Blood pressure lowering for prevention of cardiovascular disease and death: a systematic review and meta-analysis. Lancet 387 (10022), 957–967.

Faziludeen, S., Praveen, S., 2016. ECG beat classification using evidential K-nearest neighbours. Procedia Comput. Sci. 89, 499–505.

Gutiérrez-Gnecchi, J.A., Morfin-Magana, R., Lorias-Espinoza, D., del Carmen Tellez-Anguiano, A., Reyes-Archundia, E., Méndez-Patiño, A., Castañeda-Miranda, R., 2017. DSP-based arrhythmia classification using wavelet transform and probabilistic neural network. Biomed. Signal Process. Control 32, 44–56.

Hijazi, S., Page, A., Kantarci, B., Soyata, T., 2016. Machine learning in cardiac health monitoring and decision support. Computer 49 (11), 38–48.

Huynh, Q.L., Reid, C.M., Chowdhury, E.K., Huq, M.M., Billah, B., Wing, L.M., Second Australian National Blood Pressure Management Committee, 2014. Prediction of cardiovascular and all-cause mortality at 10 years in the hypertensive aged population. Am. J. Hypertens. 28 (5), 649–656.

Jambukia, S.H., Vipul, K.D., Harshadkumar, B.P., 2015. Classification of ECG signals using machine learning techniques: a survey. In: Proc 2015 Int. Conf. on Adv. in Comput. Eng. and App. IEEE.

Johnson, A.E., Ghassemi, M.M., Nemati, S., Niehaus, K.E., Clifton, D.A., Clifford, G.D., 2016. Machine learning and decision support in critical care. Proc of the IEEE Institute of Electrical and Electronics Engineers, February, pp. 444–466.

Kaplan, B.S., Uysal, A.K., Sora Gunal, E., Ergin, S., Gunal, S., Gulmezoglu, M.B., 2018. A survey on ECG analysis. Biomed. Signal Process. Control 43, 216–235.

Kasar, S.L., Joshi, M.S., 2016. Analysis of multi-lead ECG signals using decision tree algorithms. Int. J. Comput. Appl. 134(16).

Kelwade, J.P., Salankar, S.S., 2016. Radial basis function neural network for prediction of cardiac arrhythmias based on heart rate time series. In: Proc of the First Int. Conf. on Control, Meas. Instrum. (CMI). India, 8-10 January.

Krishnaiah, V., Narsimha, G., Chandra, S.N., 2016. Heart disease prediction system using data mining techniques and intelligent fuzzy approach: a review. Int. J. Comput. Appl. 136 (2), 43–51.

Lassoued, H., Ketata, R., 2017. Artificial neural network classifier for heartbeat arrhythmia detection. In: *Proc. of Int. Conf. on Automatic Signal Process. (ATS), Engineering and Technology–PET*. Tunisia, 22–24 March.

Lassoued, H., Ketata, R., 2018a. ECG multi-class classification using neural network as machine learning model. In: *Proc of Int. Conf. on Adv. Syst. Electric Tech. (IC_ASET)*. Tunisia, 19–22 March.

Lassoued, H., Ketata, R., 2018b. Hybrid two stage neuro genetic system for arrhythmia diagnosis. Int. J. Comput. Sci. Netw. Secur. 18 (9), 31–42.

Lassoued, H., Ketata, R., 2018c. ECG decision support system based on feedforward neural networks. Int. J. Smart Sens. Intell. Syst. 18(11).

Li, H., Yuan, D., Ma, X., Cui, D., Cao, L., 2017. Genetic algorithm for the optimization of features and neural networks in ECG signals classification. Sci. Rep. 7, 41011.

Luz, E.J.D.S., Schwartz, W.R., Cámara-Chávez, G., Menotti, D., 2016. ECG-based heartbeat classification for arrhythmia detection: a survey. Comput. Methods Programs Biomed 127, 144–164.

Martínez, J.P., et al., 2004. A wavelet-based ECG delineator: evaluation on standard databases. IEEE Trans. Biomed. Eng. 51 (4), 570–581.

Mincholé, A., Camps, J., Lyon, A., Rodríguez, B., 2019. Machine learning in the electrocardiogram. J. Electrocardiol. 57 (Suppl.), S61–S64. https://doi. org/10.1016/j. jelectrocard.2019.08.008.

Mondéjar-Guerra, V., Novo, J., Rouco, J., Penedo, M.G., Ortega, M., 2019. Heartbeat classification fusing temporal and morphological information of ECGs via ensemble of classifiers. Biomed. Signal Process. Control 47, 41–48.

Parvaneh, S., Rubin, J., Babaeizadeh, S., Xu-Wilson, M., 2019. Cardiac arrhythmia detection using deep learning: a review. J. Electrocardiol. 57 (Suppl.), S70–S74. https://doi.org/10.1016/j.jelectrocard.2019.08.004.

Rajamhoana, S.P., Devi, C.A., Umamaheswari, K., Kiruba, R., Karunya, K., Deepika, R., 2018. Analysis of neural networks based heart disease prediction system. *Proc of the 11th Int. Conf. on Human Syst. Interact. (HSI)*. Poland, 4–6 July.

Rajesh, K.N., Dhuli, R., 2017. Classification of ECG heartbeats using nonlinear decomposition methods and support vector machine. Comput. Biol. Med. 87, 271–284.

Rajkomar, A., Jeffrey, D., Isaac, K., 2019. Machine learning in medicine. N. Engl. J. Med. 380 (14), 1347–1358.

Rathi, M., Narasimhan, B., 2017. Data mining, soft computing, machine learning and bio-inspired computing for heart disease classification/prediction—a review. Int. J. Adv. Res. Comput. Sci. Softw. Eng. 7(4).

Savalia, S., Vahid, E., 2018. Cardiac arrhythmia classification by multi-layer perceptron and convolution neural networks. Bioengineering 5 (2), 35.

Sayilgan, E., Özlem, K.C., Yalçın, İ., 2017. Use of clustering algorithms and extreme learning machine in determining arrhythmia types. In: *Proc of the 25th Conf. on Signal Process. Commun. Appl. (SIU)*. Turkey, 15-18 May.

Silva, I., Moody, G.B., 2014. An open-source toolbox for analysing and processing physionet databases in matlab and octave. J. Open Res. Softw. 2(1).

Singh, S., Pandey, S.K., Pawar, U., Janghel, R.R., 2018. Classification of ECG arrhythmia using recurrent neural networks. Procedia Comput. Sci. 132 (2018), 1290–1297.

Soria, M.L., Martínez, J.P., 2009. Analysis of multidomain features for ECG classification. In: *Proc of 36th. Ann. Comp. in Card. Conf (CinC)*. Park City.

Vasilakos, A.V., Yu, T., Yuanzhe, Y., 2016. Neural networks for computer-aided diagnosis in medicine: a review. Neurocomputing 216, 700–708.

Zadeh, L.A., 2015. Fuzzy logic: a personal perspective. Fuzzy Sets Syst. 281, 4–20.

CHAPTER 6

Modeling simple and complex handwriting based on EMG signals

Ines Chihi[a], Ernest N. Kamavuako[b], Mohamed Benrejeb[c]
[a]Laboratory of Energy Applications and Renewable Energy Efficiency (LAPER), Tunis El Manar University, Tunis, Tunisia
[b]Department of Informatics, Center for Robotics Research, King's College London, London, United Kingdom
[c]Laboratory of Research in Automation (LA.R.A), National School of Engineers of Tunis, Tunis El Manar University, Tunis, Tunisia

1 Introduction

Losing the ability to fully control an upper limb, either because of amputation or paralysis, has a profound effect on the activities of daily life. In the last decades, assistive robots and prostheses have been considered important research areas aiming to allow people with disabilities to regain functions (Mastinu et al., 2018; Adewuyi et al., 2017). Advances in mechanical technology over the past two decades have resulted in the development of advanced robotic systems. However, due to the anatomical and physiological complexity of the human hand, fully replicating all its functions, in a natural and autonomous way, has proven to be highly challenging. One of the most promising control approaches in prostheses and assistive devices is based on the electrical activities of either remnant or partially active muscles, called electromyography (EMG) signals (Kuzborskij et al., 2012; Hincapie and Kirsch, 2009). This has led to hundreds of research studies focusing on improving grasping and reaching movements (Artemiadis and Kyriakopoulos, 2010; Scheme et al., 2010; Kamavuako et al., 2013) using different approaches such as regression and pattern recognition (Chen et al., 2013; Alimi and Plamondon, 1993; Fischer and Plamondon, 2017).

The movement of the hand is produced by a complex combination of muscles from the forearm and some intrinsic muscles of the hand itself. Muscle fibers receive a nerve impulse that is converted into energy in the form of contraction by a chemical reaction at the level of myofibrils. Thus the nerve impulse results in electrical impulses whose discharge frequency depends on the required movement (Rouvière et al., 1968). EMG is the recording of muscle electrical potentials providing a window through which motor control can be investigated. EMG signals are nonstationary, stochastic in nature

Control Theory in Biomedical Engineering
https://doi.org/10.1016/B978-0-12-821350-6.00006-8

and sensitive to many disturbances. They can be affected by abrupt changing of the electrode positions, changes in electrophysiology due to sweat, changes in the impedance of the electrode, grease, muscle layers and tissue, and time (Artemiadis and Kyriakopoulos, 2008, 2011; Waris et al., 2019). All these conditions may lead to inaccurate identification of user intent, challenging the reliability of the control system (Parker et al., 2006; Hargrove et al., 2007) and necessitating some signal conditioning or modeling.

Despite the focus on EMG-based control, handwriting, which is an important means of communication, has received less attention from the assistive device point of view. Writing has inspired many researchers to propose models characterizing this biological process. Indeed, handwriting is considered a mean of communication unavoidable for academic, professional and social integration (Defazio et al., 2010). The production of graphic traces is a physical manifestation of a complex cognitive process. It contains a lot of information that can characterize a person and express the level of academic and social education and even the psychological state (angry, relaxed, etc.) of the writer and their temperamental tendencies (Viviani and Terzuolo, 1983). Writing is expressed through motions of the upper limb, and with the availability of advanced EMG recording systems EMG-driven models can allow reconstruction of individual handwriting.

Considering different kinds of handwriting, letters, geometric forms, and even numbers, this chapter reviews the most effective handwriting models using muscle activities of the upper limb. The considered structures are based on input/output relationships between pen-tip coordination moving on a 2D plane and EMG signals of the hand and/or forearm. We describe the different handwriting approaches and present the advantages and the drawbacks of each one to finally develop a summary table of the different presented black-box models. The chapter focuses on both simple movements, such as drawing numbers and geometric shapes in a single orientation, and complex movements generated in combined directions like cursive Arabic letters.

The chapter is organized as follows. After a brief history of handwriting models, we present a Kalman filter (KF)-based approach by Okorokova et al. to mimic some numbers. In the next section, we develop the approach presented by Zhang and Kamavuako. Then, in order to characterize handwriting cursive letters and complex geometric forms from two EMG forearm muscles, we present two models. The first is based on the velocity of writing developed by Murata et al. and the second is a robust interval observer model.

2 History of handwriting modeling

In 1962, Van Der Gon proposed a second-order model allowing to assimilate the hand as a mass moving on the writing surface (Van Der Gon et al., 1962). An electronic version of that model was then proposed by Mac Donald that presented the handwriting process as a dynamic mass that moves in a viscous environment (Mac Donald, 1964). The movement of this mass was described by a linear differential equation of the second order. In 1975, Yasuhara showed that the vertical movements of the pen tip are generated by the flexion movements of the forearm muscles, while horizontal displacements are generated by the abduction-adduction movements of these muscles. Then, he integrated the effect of the friction force between the tip of the pen and the writing surface, (Yasuhara, 1975) from which he identified and composed a fast writing system (Yasuhara, 1983). From this model, Iguider and Yasuhara developed, in 1995, two approaches, one for the extraction of pulsations of control (Iguider and Yasuhara, 1995) and the other, in 1996, for the recognition of cursive Arabic script (Iguider and Yasuhara, 1996). As the muscles of the forearm intervening in the handwriting act are directly located under the skin, surface electrodes were used to record the generated EMG signals. Based on this possibility, handwriting experimentation was realized at Hiroshima University to measure two forearm muscles' activities concurrently with the coordinates of some Arabic letters and geometric forms (Manabu et al., 2003). Based on this experimental approach and dataset, different mathematical approaches have been proposed to characterize the handwriting process from EMG signals.

Manabu et al. (2003) proposed a third-order model to generate writing from two EMG signals. Linderman et al. (2009) proposed a handwriting model from four EMG signals to produce handwritten numbers from zero to nine. Using the same experimental measurements, Okorokova et al. (2015) developed a hybrid model, based on the KF and the model proposed by Mac Donald (1964). Moreover, Zhang and Kamavuako (2019) developed, based on their own recording of three EMG channels, a neural network model based on an experimental approach allowing recording four geometric forms drawing on an (x, y) plane. These diverse writing characterization models have contributed to obtaining more or less satisfactory results requiring a parametric adjustment, related to each change of inputs/outputs to be modeled, limiting the efficiency of these structures, mainly due to the subjective nature of the EMG signals. In order to mitigate problems related to the complexity of this biological process, models using several kinds of graphic traces generated by one or more writers and taking into account their individual properties such as speed, fluctuation, inclination, and preferential direction have also been proposed (Chihi et al., 2015; Chihi and Benrejeb, 2018). The following sections focus on the presentation of the most important handwriting models based on different experimental data (numbers, geometric forms, Arabic letters, etc.).

3 Kalman filter-based model

In order to reconstruct handwriting traces, numbers from zero to nine, using eight channels of EMG activities of the forearm muscles, Okorokova et al. (2015) proposed an approach based on the KF, which allows the fusion of two information sources, presented by two models: dynamical and measurement models (Linderman et al., 2009; Okorokova et al., 2015). These models are defined as follows:

- Dynamical models present the physical characteristics of handwriting motion. It allows computing the dependence between the state vector s at the time t and the state vector in the past and the position of the pen moving in the plane, according to x and y directions.

$$s_t = As_{t-1} + v_t \tag{1}$$

  with:
  s_t: state vector *[6K 1]*

$$s_t = [x_t, y_t, \dot{x}_t, \dot{y}_t, \ddot{x}_t, \ddot{y}_t, \ldots, x_{t-K+1}, y_{t-K+1}, \dot{x}_{t-K+1}, \dot{y}_{t-K+1}, \ddot{x}_{t-K+1}, \ddot{y}_{t-K+1}]^T$$

  A: state transition matrix; *[6K 6K]*
  v_t: vector containing process noise *[6K 1]*
  K: samples time relative to the dynamical model

- Measurement model to compute the dependence between the state vector and eight EMGs signals inputs of this model.

$$s_t = Hz_t + w_t \tag{2}$$

  with:
  z_t: state vector *[6L 1]*
  H: measurement transformation matrix *[6K 1L]*
  w_t: vector containing measurements noise *[6L 1]*
  L: samples time relative to measurements model

We also note:

$s_t = [\boldsymbol{x}_t, \boldsymbol{y}_t, \dot{\boldsymbol{x}}_t, \dot{\boldsymbol{y}}_t, \ddot{\boldsymbol{x}}_t, \ddot{\boldsymbol{y}}_t, \ldots, \boldsymbol{x}_{t-K+1}, \boldsymbol{y}_{t-K+1}, \dot{\boldsymbol{x}}_{t-K+1}, \dot{\boldsymbol{y}}_{t-K+1}, \ddot{\boldsymbol{x}}_{t-K+1}, \ddot{\boldsymbol{y}}_{t-K+1}]^T$ is the state vector that contains the coordinates of the pen. In addition, it contains the first and second rates of change for the window of K time moments starting from t; $\boldsymbol{A}$ is the state transition matrix that describes the relationship between the state vectors at two consecutive moments; $\boldsymbol{v}_t$ contains the noise of the process.

Based on the KF, the fusion is realized by the multiplication of functions of dynamical and measurement models. The gain of the KF allows estimating the contribution of each model.

The KF is based on the fusion of two noisy information sources to estimate the dynamical system's state vector. In the Okorocova model, the first information source is a dynamical model based on the physical properties of the handwriting system. This source is presented as a multivariate autoregressive process. The parameter estimation is computed from the recorded data. The second information source is the noisy vector of the measured EMG signals. The relationship between these signals and the pen coordinate is presented using a multivariate linear regression model (Okorokova et al., 2015).

Based on eight inputs and six states, this method is suitable for real-time operations. It requires the computation and the adjustment of 48 parameters. Furthermore, the use of eight inputs increases the complexity of the model and the computation time. The KF is considered a powerful and effective method, especially for stochastic signals, but it requires some assumptions (e.g., white noise) that might be difficult to meet in practice. This may affect the response of this method in terms of accuracy, especially in real-time applications (Thomassen and Meulenbroek, 1993). Fig. 1 shows the response of the cross-validation of the KF-based model. We remark that the estimated numbers are not easily identifiable, especially for "0" and "5," which can be confused with "6" and "0," respectively. In summary, cross-validation shows the limitation of this approach despite the use of eight inputs and 48 parameters.

4 Zhang-Kamavuako model (ZK)

Zhang and Kamavuako (ZK) proposed a handwriting model to reconstruct a pen tip moving on an XP-PEN Deco 02 Digital Graphics Drawing Tablet from EMG signals recorded from three muscles (opponens pollicis, first

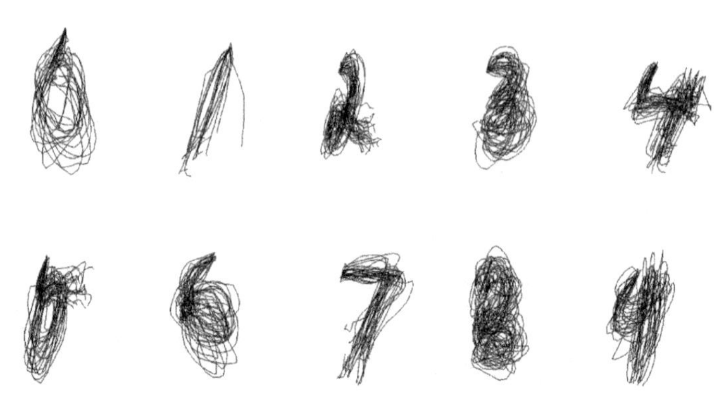

Fig. 1 Cross-validation of the handwriting model using a Kalman filter (Okorokova et al., 2015).

dorsal interosseus, and flexor carpi radialis) (Zhang and Kamavuako, 2019). The experimental procedure consisted of in-house software that collected the x and y coordinates provided by the tablet together with the three EMG channels. The considered database contains four graphic traces composed of different kinds of lines: straight, broken, and curves as shown in Fig. 2. Each trace graphic was recorded five times at a moderate speed and participants were asked to maintain the speed as constant as possible. Three features were explored, mainly the mean absolute average (MAV), root mean square (RMS), and mean power frequency (MPF), bringing the number of inputs to nine. The proposed approach is based on a back propagation (BP) neural network with nine inputs, a single hidden layer with six neurons and an output layer with two neurons. Thus the total number of weights to tune was 108, which are tuned during calibration only. This structure was chosen as a good trade-off between accuracy and speed. The four basic traces were used for training only. The ability of the model to reconstruct writing was tested by asking the subjects to perform random

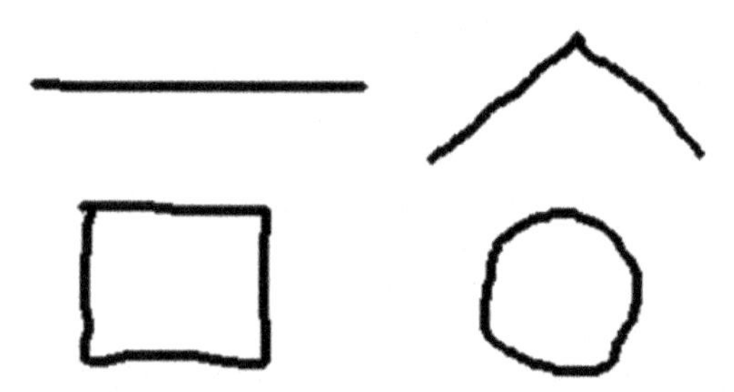

Fig. 2 The four traces used in Zhang-Kamavuako handwriting.

shapes. On average, the coefficient of determination was 0.88 ± 0.09 and 0.69 ± 0.24 for the x and y coordinates, respectively.

Results showed that the model predicted the simple curve more accurately. The prediction of complex curves such as multipeak has certain limitations and it is easy to lose part of the information, which may be related to the limited data of the training. In addition, it is almost impossible to accurately predict the coordinates of specific x and y. The prediction of the model can only reflect the shape of the trajectory. A follow-up investigation is being conducted to improve the accuracy of the model.

5 Modeling of cursive writing from two EMG signals

In this section, we present two handwriting approaches enabling the modeling of some cursive Arabic letters and different simple and complex geometric forms. The first model calculates handwriting velocities from two EMG signals. The second is considered a robust approach based on the parametric model and interval observer allowing to overcome the problem of parametric variation. For this, we start by describing the experimental data acquisition used to develop the first and the second handwriting approaches.

5.1 Experimental approach and system presentation

The generation of handwriting movement, on a 2D plane, is mainly based on two active forearm muscles activities—EMG1 and EMG2—relative to the muscles abductor pollicis longus (APL) for vertical displacement and the extensor capri ulnaris (ECU) for horizontal movements, respectively (Yasuhara, 1983; Manabu et al., 2003). In order to characterize handwriting movements, an experiment was carried out in Hiroshima City University to measure simultaneously some Arabic letters and geometric forms and two forearm EMG signals relative to APL and ECU muscles (Manabu et al., 2003). Participants wrote horizontal, vertical complexes (a combination of horizontal and vertical movements) and also rapid and slow movements (Table 1). The synchronization of the data recording was realized by sending a specific signal, a step, from the parallel interface port on the computer to the data recorder.

The following equipment was used in this experimental approach:

- Digitalized table: WACOM, KT-0405-RN
- Preamplifiers: TEAC, AR-C2EMG1
- Data recorder: TEAC, DR-C2
- Bipolar surface electrodes: MEDICOTEST, Blue Sensor N-00-S
- Computer

Fig. 3 depicts the positions of the electrodes on the forearm of a writer.

Table 1 Considered handwriting database (Chihi et al., 2018).

Description of the form	Form	Description of the form	Form
Horizontal line (1) (left/right/left)		Circle (1) (to the right)	
Horizontal line (2) (right/left/right)		Circle (2) (to the left)	
Vertical line (1) (up/down/up)		Triangle (1) (to the right)	
Vertical line (2) (down/up/down)		Triangle (2) (to the left)	

Advanced signal processing techniques were applied to extract useful information from the recorded EMG signals. These techniques filtered the fluctuation of EMG magnitudes to define new curves called integrated EMG (IEMG), represented by blue curves (gray curves in print version) in Fig. 4 (Chihi et al., 2017). We note IEMG1 and IEMG2, the integrated EMG of channel 1 and channel 2 respectively.

The type of graphic traces, considered in this experimental approach, was judiciously chosen. Indeed, based on Thomassen's analysis, which confirmed that when a person produces traits or shapes in nonpreferred directions the performance will be less precise, more unstable and is manifested in

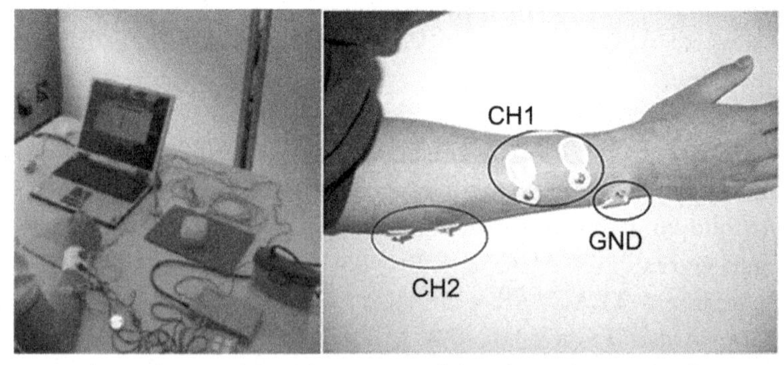

Fig. 3 Experimental approach and positions of the electrodes on the forearm (Chihi et al., 2015).

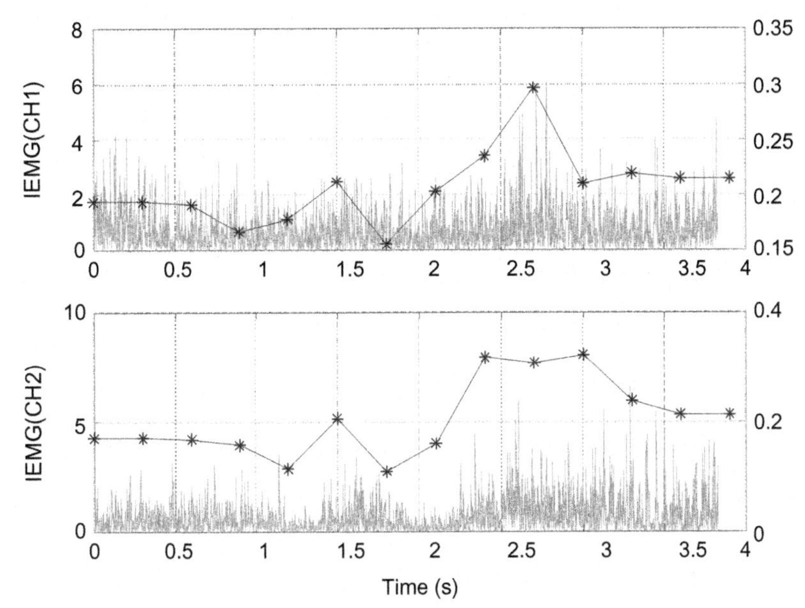

Fig. 4 Traces of data recorded during the experiment showing the integrated EMG in *blue* (*gray* in print version).

particular by the presence of tremor (Meulenbroek and Thomassen, 1991). The writer being right-handed or left-handed can cause systematic errors toward the nearest preferential orientation, which affects the production of the graphical form.

Participants in the experiment are accustomed to producing Japanese writing systems, having a direction of progression from left to right, which is different from Arab systems. In this experiment, participants wrote Arabic letters, which inevitably influenced the speed and preferential direction of the writers. The production of geometric shapes is faster.

In this context, Meunlenbroek has also shown that the generation of vertical lines is faster than horizontal lines as are the movements from top to bottom, which are also more accurate than the movements from bottom to top (Dinh et al., 2014).

These writers wrote three cursive Arabic letters (ع,س ,ح) and drew eight basic geometric shapes as shown in Table 2. Fig. 5 presents the recorded data for the Arabic letter "AIN."

Table 2 Parameters' intervals of Chihi's handwriting model (Chihi et al., 2018).

	Letter "HA"			Letter "SIN"		
	Writer 1	Writer 2	Writer 3	Writer 1	Writer 2	Writer 3
a_{ix}	[−0.2 0.4]	[−0.8 0.8]	[−1 2]	[−2.5 0.5]	[−0.3 3]	[−0.7 0.4]
b_{ix}	[−0.7 0]	[−1.2 0.8]	[−1.8 1.8]	[−1.5 0.5]	[−0.8 0.5]	[−1.5 0.5]
c_{ix}	[−30 19]	[−20 18]	[−198 300]	[−2 3]	[−18 20]	[−7 0.4]
d_{ix}	[−5 8]	[−2 3.5]	[−5 8]	[−1.2 1]	[−3 5]	[−0.4 0.4]
a_{iy}	[−0.5 0.5]	[−0.5 0.5]	[−2 1.8]	[−0.2 1]	[−1 1]	[−0.2 0.9]
b_{iy}	[−1 0.4]	[−1 0.8]	[−1.8 2.1]	[−0.8 0]	[−1 0.3]	[−1.5 0.5]
c_{iy}	[−18 20]	[−18 10]	[−198 300]	[−3 2]	[−0.8 0.8]	[−8 10]
d_{iy}	[−19 19]	[−5 5]	[−100 50]	[−3 2]	[−13 10]	[−3.5 3.5]

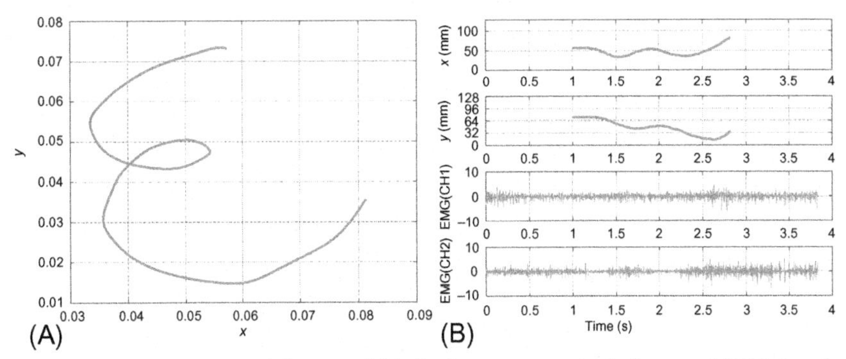

Fig. 5 Arabic letter "AIN" (A) form and (B) displacement on (*x*,*y*) plane and EMG signals.

5.2 Murata-Kosaku-Sano model (MKS)

In Manabu et al. (2003), the authors assimilate the handwriting process to a dynamic system with two EMG signal inputs and two outputs, x and y directions. After the recording of EMG signals of two forearm muscles during the writing, interpolation is used to compute two integrated EMG signals and the corresponding movement.

Indeed, the MKS model generates velocities of the pen tip moving on a digital tablet according to x and y directions. This is based on the experimental approach presented in Section 5.1.

Manabu et al. (2003) proposed a parametric model presented in Eq. (1).

$$\begin{cases} \dfrac{dx(t)}{dt} = \displaystyle\sum_{m=1}^{3} a_x(m) E_1(k-m-N_1) + \\ \qquad \displaystyle\sum_{m=1}^{3} b_x(m) E_2(k-m-N_2) \\ \dfrac{dy(t)}{dt} = \displaystyle\sum_{m=1}^{3} a_y(m) E_1(k-m-N_1) + \\ \qquad \displaystyle\sum_{m=1}^{3} b_y(m) E_2(k-m-N_2) \end{cases} \tag{3}$$

where:

t: continuous time,

a_x, a_y, b_x, b_y: parameters,

k: discrete time,

m: delay time,

N_1, N_2: dead times,

E_1, E_2: EMG signals, inputs of the model.

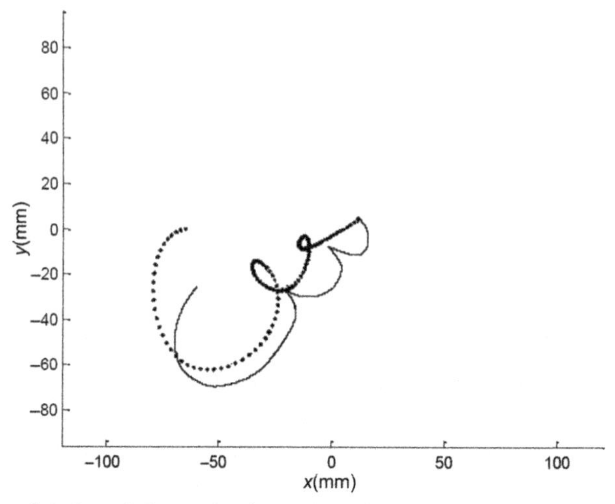

Fig. 6 Cross-validation of the Arabic letter "SIN" by MKS model (Chihi et al., 2015).

The MKS model is a parametric model that needs adjustment every time we change data, even if the data presents the same kind of graphic trace, written by the same person. Simulation results show significant error between estimated and real data especially in cross-validation (Fig. 6). We note that the solid line denotes measured data and the dotted line presents the response of the MKS model.

5.3 Interval observer for robust handwriting characterization

In the preceding section, we presented different handwriting models to characterize several kinds of graphic traces: numbers, geometric forms, letters, and so on. However, these models are limited in terms of accuracy and need parametric adjustment each time we change the data. This can be explained by the stochastic nature of the EMG signals. Based on the experimental approach previously presented in Chihi et al. (2017), a linear fourth-order model to generate handwriting traces according to x and y directions was developed to characterize the handwriting process from only two EMG signals of the forearm. This model is given by the following equations:

$$
\begin{aligned}
x_e(k) &= \sum_{i=1}^{4} \hat{a}_{ix}\, y_e(k-i) + \sum_{i=1}^{4} \hat{b}_{ix}\, x_e(k-i) + \sum_{i=1}^{5} \hat{c}_{ix}\, e_1(k-i+1) + \sum_{i=1}^{5} \hat{d}_{ix}\, e_2(k-i+1) \\
y_e(k) &= \sum_{i=1}^{4} \hat{a}_{iy}\, x_e(k-i) + \sum_{i=1}^{4} \hat{b}_{iy}\, y_e(k-i) + \sum_{i=1}^{5} \hat{c}_{iy}\, e_1(k-i+1) + \sum_{i=1}^{5} \hat{d}_{iy}\, e_2(k-i+1)
\end{aligned}
\tag{4}
$$

with:

x_e and y_e: estimated outputs according to x and y coordinates,

e_1,e_2: electromyography signals, inputs of the model,

a_{ix},b_{ix},c_{ix},d_{ix}: parameters according to x_e,

a_{iy},b_{iy},c_{iy},d_{iy}: parameters according to y_e.

The recursive least squares algorithm is used to estimate the handwriting model parameters (Eqs. 3–5). Parameter estimation is based on the minimization of the sum of squares of the difference between the observed and the computed values (Kim et al., 2015, 2016; Umberto et al., 2017).

$$\hat{\theta}(k) = \hat{\theta}(k-1) + P(k) \sum_{i=n+1}^{k} \gamma(i)\Psi(i) \tag{5}$$

$$P(k) = P(k-1) - \frac{P(k-1)\Psi(k)\Psi^T(k)P(k-1)}{1 + \Psi^T(k)P(k-1)\Psi(k)} \tag{6}$$

$$\varepsilon(k) = \gamma(k) - \hat{\theta}(k-1)\Psi(k) \tag{7}$$

with:

$\hat{\theta}(k)$: estimated parameters,

$P(k)$: adaptation matrix, also called inverse correlation matrix of the input signal,

$\gamma(k)$: outputs of the system,

$\psi(k)$: observation matrix,

$\varepsilon(k)$: error of estimation.

Fig. 7 presents the cross-validation of the handwriting model (Eq. 2). The blue line (dark gray line in print version) is used to present experimental data and the dotted purple line (dotted light gray line in print version) presents the model response. The cross-validation consists of keeping the structure and parameters of a well-defined model and applying new input-output data that have not been used for parameter estimation.

This principle was applied to validate the developed models both within and between writers using either similar or different traces. Multiwriter validation with the same kind of graphic trace provided the highest validation result (Gene et al., 1979; Prabir, 1989; Sylvain and Matthieu, 2016).

It is important to note that the model structure is fixed whatever the graphic trace, however, this approach is based on nonphysical parameters, a_{ix}, b_{ix}, c_{ix}, d_{ix}, a_{iy}, b_{iy}, c_{iy}, and d_{iy}. Therefore, parametric adjustment is required each time we change input-output data, even if it is the same person and the same type of graphic trace to be estimated. This leads to the

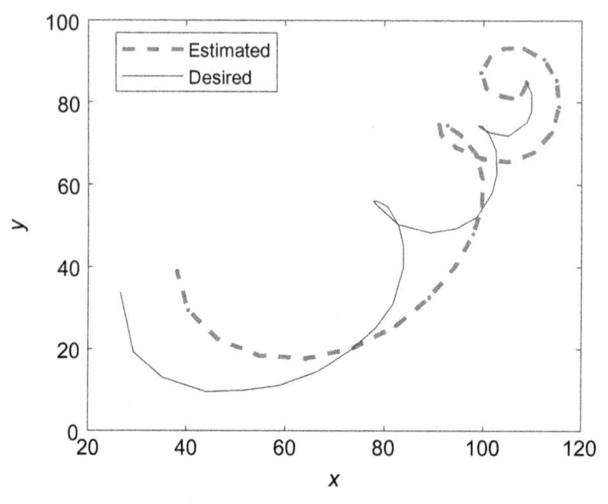

Fig. 7 The response of cross-validation of the model (4): Arabic letter SIN.

following question: *How can we model different kinds of graphic traces without adjustment of the model parameters?*

To answer this question, we propose, in this section, an interval observer to estimate manuscript coordinates in the two-dimensional plane, using the parametric model (Eq. 2). This estimator is based on two Luenberger observers, which contain two parameter intervals, upper and lower. Indeed, we studied parameter variation of different models characterizing the letters "SIN" and "HA" (Ito and Dinh, 2018; Chihi et al., 2018). This study led us to show a significant difference between parameter intervals even if it is computed to characterize the same kind of letter written by the same writer. Besides, this study allowed classifying the model parameters according to two intervals, upper and lower (see Table 2). In order to explain the proposed approach, we start by studying parametric variations of two Arabic letters "HA" and "SIN." These writings are different; the letter "SIN" is constituted by three primitives, similar to arcs of a circle, two small and one bigger. The second letter "HA" is composed of two primitives, a line quasihorizontal (left/right) and a large arc as shown in Figs. 9 and 10.

As already mentioned, each participant wrote each graphic form many times. For three writers, we study, in a first step, the parametric variations of two letters ("HA" and "SIN") estimated using model (2).

$$\begin{cases} \widetilde{X}(k+1) = A\widetilde{X}(k) + BU(k) + K\left(Y - \widetilde{Y}\right) \\ \widetilde{Y}(k) = C\widetilde{X}(k) \end{cases} \tag{8}$$

$$\begin{cases} \widetilde{X}(k+1) = A\widetilde{X}(k) + BU(k) + Ke^{norm}(k) \\ \widetilde{Y}(k) = C\widetilde{X}(k) \end{cases}$$
$$e^{norm}(k) = \frac{e(k)}{\sum e(k)}, e(k) = Y(k) - \widetilde{Y}(k) \tag{9}$$

with:

$\widetilde{X}$: state vector,
$\widetilde{Y}$: output vector,
e: error between real and estimated outputs,
e^{norm}: normalized error,
A: state matrix,
B: input matrix,
C: output matrix,
K: observer gain.

Fig. 8 presents the principle of the proposed handwriting interval observer based on the model given in Eq. (2). Using Eqs. (3)–(6), we developed upper and lower observers to estimate each shape. Finally, based on Eq. (7), we combine the two observers to develop the proposed handwriting interval

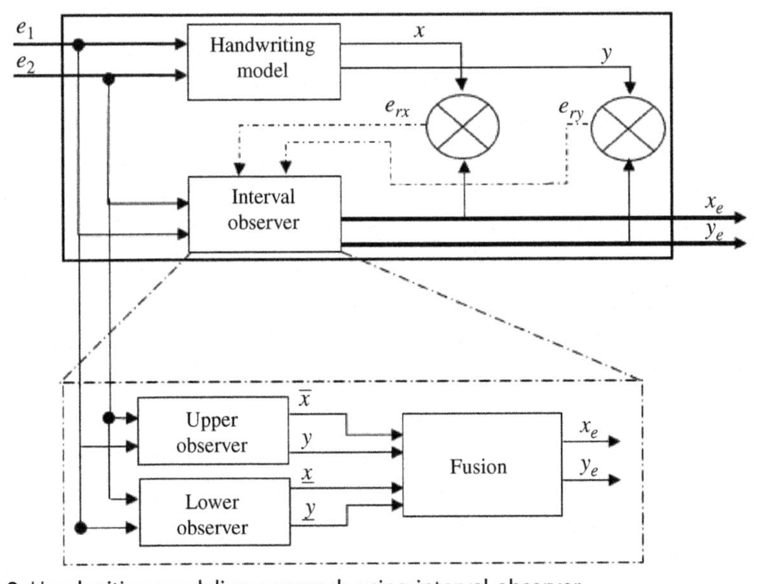

Fig. 8 Handwriting modeling approach using interval observer.

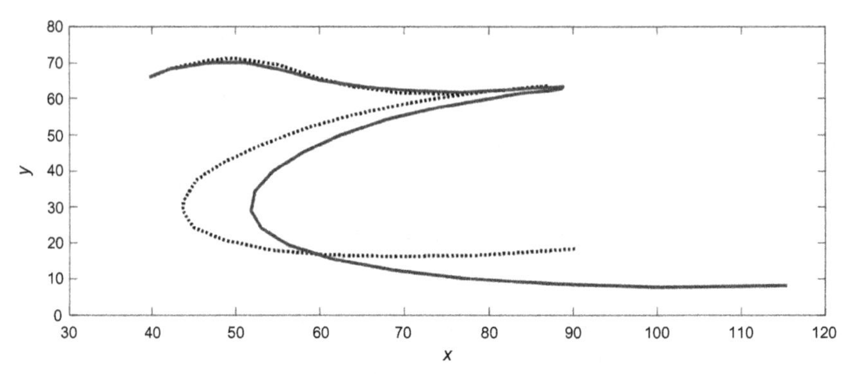

Fig. 9 Letter "HA"—one-writer validation: desired response and interval observer response.

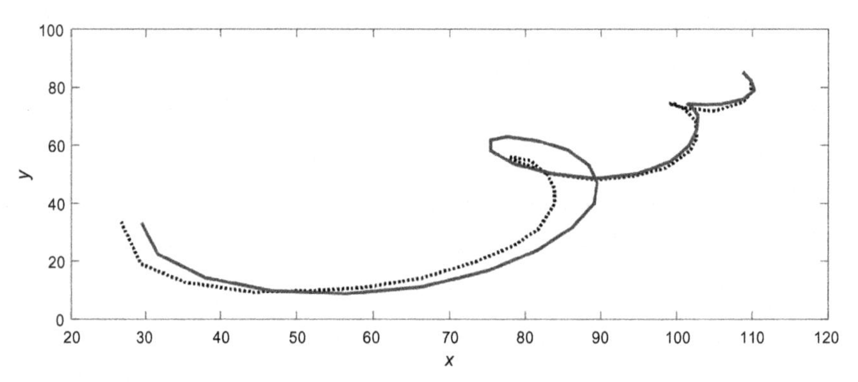

Fig. 10 Letter "SIN"—multiwriter validation: desired response and interval observer response.

estimator. Cross-validation of the proposed approach is presented in Figs. 9 and 10 with considerable accuracy of the handwriting structure using the principle of the interval observer.

In order to obtain a better estimation accuracy, we developed the interval observer, deduced from two observers—upper and lower—that contains superior and inferior variations of the parameters (Ito and Dinh, 2018, Chihi et al., 2018).

- **Upper observer $\overline{\overline{X}}$:**

$$\begin{cases} \overline{\overline{X}}(k+1) = A_s\overline{\overline{X}}(k) + B_s\overline{U}(k) + K_s\left(\overline{Y} - \overline{\overline{Y}}\right) \\ \overline{\overline{Y}}(k) = C_s\overline{\overline{X}}(k) \end{cases} \tag{10}$$

- **Lower observer $\underline{\tilde{X}}$:**

$$\begin{cases} \underline{\tilde{X}}(k+1) = A_i\underline{\tilde{X}}(k) + B_i\underline{U}(k) + K_i\left(\underline{Y} - \underline{\tilde{Y}}\right) \\ \underline{\tilde{Y}}(k) = C_i\underline{\tilde{X}}(k) \end{cases} \tag{11}$$

$$\begin{cases} X_e(\mathrm{k}) = \alpha\overline{\tilde{X}}(k) + \beta\underline{\tilde{X}}(k) \\ Y_e(k) = \alpha\overline{\tilde{Y}}(k) + \beta\underline{\tilde{Y}}(k) \end{cases} \tag{12}$$

with:

$$\alpha \in [0, 1]$$
$$\beta \in [0, 1], \alpha + \beta = 1$$

The estimated coordinates are then calculated using the relationships from Eqs. (3)–(7).

For Figs. 9 and 10, the dotted blue line (dotted gray line in print version) shows the experimental data and the solid blue line (solid gray line in print version) is relative to the outputs of the model.

6 Discussion

In this chapter, we presented different models that have been proposed to characterize handwriting from EMG signals. The first model is based on the KF to reproduce numbers zero to nine from eight EMG channels. This complex model requires the estimation of 48 parameters and fixing specific hypotheses that may not be respected in real-time development. Besides, this approach is constituted by the fusion of two models, which increases the computational time. The results of this approach show a significant error between estimated and desired data. The second model is based on the neural network approach to estimate some geometric shapes from three EMG signals. The results of this approach are acceptable, however, it shows that the model needs refinement. This model is characterized by a moderate computational time during training only because of the single hidden layer. In order to model handwriting of Arabic letters and geometric forms from only two EMG signals recorded from the forearm muscles, Manabu et al. (2003) proposed a dynamic third-order model to characterize pen-tip movements on an (x, y) plane. However, the model is limited in terms of accuracy and computational time.

Unlike these handwriting models, the advantage of the modeling approach based on interval observer is to reduce the number of inputs

Table 3 Synthesis on the properties of the studied handwriting models.

	Data description	One-writer validation	Multiwriter validation	Order	Number of parameters	Accuracy	Complexity	Computing time
KF model	Numbers	Medium	Medium	6	48	Medium	High	High
ZK model	Shapes	Medium	Low	NA	108	Medium	High	Medium
MKS model	Geometric forms and Arabic letters	Medium	Low	3	11	Medium	Medium	Medium
Model (2)		Medium	Low	4	18	Medium	Low	Low
Interval observer model		High	High	4	14	High	Medium	Medium

and to use only two EMG signals of the most active forearm muscles. Besides, this approach allows producing simple and complex writing, especially cursive Arabic letters, containing combing movements. The model based on the interval observer mitigates the needs for parameter adjustment for new data. According to Table 3, the interval observer approach allows ameliorating the estimation of handwriting from EMG signals. Indeed, validation results show good accuracy in both one-writer and multiwriter cases (Table 3).

7 Conclusion

In this chapter, we presented different approaches that allow the characterization of handwriting from muscle activities. The developed models generate different kinds of writing: numbers, geometric forms and cursive Arabic letters. These traces have different characteristics, like speed, complexity, orientation, and so on. The proposed approaches are appropriate for many fields and practical applications such as bio-engineering, biorobotics, intelligent therapy and military applications.

References

Adewuyi, A.A., Hargrove, L.J., Kuiken, T.A., 2017. Resolving the effect of wrist position on myoelectric pattern recognition control. J. Neuroeng. Rehabil. 1, 14–39.

Alimi, M.A., Plamondon, R., 1993. Parameter analysis of handwriting strokes generation models. In: Proc. International Conference on Handwriting, pp. 4–6.

Artemiadis, P.K., Kyriakopoulos, K.J., 2008. Assessment of muscle fatigue using a probabilistic framework for an EMG-based robot control scenario. In: 8th IEEE International Conference on Bioinformatics and BioEngineering, Athens, pp. 1–6.

Artemiadis, P.K., Kyriakopoulos, K.J., 2010. An EMG-based robot control scheme robust to time-varying EMG signal features. IEEE Trans. Inf. Technol. Biomed. 14 (3), 582–588.

Artemiadis, P.K., Kyriakopoulos, K.J.A., 2011. Switching regime model for the EMG-based control of a robot arm. IEEE Trans. Syst. Man Cybern. B (Cybern.) 41 (1), 53–63.

Chen, X., Zhang, D., Zhu, X., 2013. Application of a self-enhancing 22 classification method to electromyography pattern recognition for multifunctional prosthesis control. J. Neuroeng. Rehabil. 10 (44), 1–13.

Chihi, I., Benrejeb, M., 2018. Online fault detection approach of unpredictable inputs: application to handwriting system. Complexity 2018, 12.

Chihi, I., Abdelkrim, A., Benrejeb, M., 2015. Multi-model approach to characterize human handwriting motion. Biol. Cybern. 110 (1), 17–30.

Chihi, I., Abdelkrim, A., Benrejeb, M., 2017. Internal model control to characterize human handwriting motion. Arab J. Appl. Sci. 14 (6), 861–869.

Chihi, I., Sidhom, L., Maamri, O., 2018. Robust handwriting estimator from two forearm muscles activities. Int. J. Appl. Eng. Res. 13 (23), 16213–16219.

Defazio, J., Jones, J., Tennant, F., Anne Hook, S., 2010. Academic literacy: the importance and impact of writing across the curriculum – a case study. J. Scholarsh. Teach. Learn. 10 (2), 34–47.

Dinh, T.N., Mazenc, F., Niculescu, S.I., 2014. Interval observer composed of observers for nonlinear systems. In: 13th European Control Conference, pp. 660–665.

Fischer, A., Plamondon, R., 2017. Signature verification based on the kinematic theory of rapid human movements. IEEE Trans. Hum.-Mach. Syst. 47 (2), 169–180.

Gene, H.G., Michael, H., Grace, W., 1979. Generalized cross-validation as a method for choosing a good ridge parameter. Technometrics 21 (2), 215–223.

Hargrove, L.J., Englehart, K., Hudgins, B., 2007. A comparison of surface and intramuscular myoelectric signal classification. IEEE Trans. Biomed. Eng. 54 (5), 847–853.

Hincapie, J.G., Kirsch, R.F., 2009. Feasibility of EMG-based neural network controller for an upper extremity neuroprosthesis. IEEE Trans. Neural Syst. Rehabil. Eng. 17 (1), 80–90.

Iguider, Y., Yasuhara, M., 1995. Extracting control pulses of handwriting movement. Trans. Soc. Instrum. Control Eng. 31 (8), 1175–1184.

Iguider, Y., Yasuhara, M., 1996. An active recognition pulses of handwriting isolated Arabic characters. Trans. Soc. Inst. Control Eng. 32 (8), 1267–1276.

Ito, H., Dinh, T.N., 2018. Interval observers for global feedback control of nonlinear systems with robustness with respect to disturbances. Eur. J. Control 39, 68–77.

Kamavuako, E.N., Rosenvang, J.C., Bøg, M.F., Smidstrup, A., Erkocevic, E., Niemeier, M.J., Jensen, W., Farina, D., 2013. Influence of the feature space on the estimation of hand grasping force from intramuscular EMG. Biomed. Signal Process. Control 8 (1), 1–5.

Kim, Y., Kim, Y.H., Lee, S., 2015. Multivariable nonlinear identification of smart buildings. Mech. Syst. Signal Process. 62 (63), 254–271.

Kim, Y., Kim, J.M., Kim, Y.H., 2016. System identification of smart buildings under ambient excitations. J Meas 87, 294–302.

Kuzborskij, A., Gijsberts, A., Caputo, B., 2012. On the challenge of classifying 52 hand movements from surface electromyography. In: Annual International Conference of the IEEE Engineering in Medicine and Biology Society, San Diego, CA, pp. 4931–4937.

Linderman, M., Lebedev, M.A., Erlichman, J.S., 2009. Recognition of handwriting from electromyography. PLoS ONE 4 (8), e6791.

Mac Donald, J.S., 1964. Experimental Studies of Handwriting Signals. (Ph.D. dissertation) Massachusetts Institute of Technology, Cambridge.

Manabu, S., Kosaku, T., Murata, Y., 2003. Modeling of human handwriting motion by electromyographic signals on forearm muscles. In: CCCT'03.

Mastinu, E., Ahlberg, J., Lendaro, E., Hermansson, L., Håkansson, B., Ortiz-Catalan, M., 2018. An alternative myoelectric pattern recognition approach for the control of hand prostheses: a case study of use in daily life by a dysmelia subject. IEEE J. Transl. Eng. Health Med. 6, 2168–2372.

Meulenbroek, R.G.J., Thomassen, A.J.W.M., 1991. Stroke-direction preferences in drawing and handwriting. Hum. Mov. Sci. 10, 247–270.

Okorokova, E., Lebedev, M., Linderman, M., Ossadtchi, A., 2015. A dynamical model improves reconstruction of handwriting from multichannel electromyographic recordings. Front. Neurosci. 9 (517), 389–404.

Parker, P., Englehart, K., Hudgins, B., 2006. Myoelectric signal processing for control of powered limb prosthesis. J. Electromyogr. Kinesiol. 16 (6), 541–548.

Prabir, B., 1989. A comparative study of ordinary cross-validation, v-fold cross-validation and the repeated learning-testing methods. Biometrika 76 (3), 503–514.

Rouvière, H., Delmas, A., Delmas, V., 1968. Anatomie Humaine Descriptive, topographique et fonctionnelle. vol. 3. Editions Masson, p. 354.

Scheme, E., Fougner, A., Stavdahl, O., Chan, A., Englehart, K., 2010. Examining the adverse effects of limb position on pattern recognition based myoelectric control. In: Proceedings of the 32nd Annual International Conference of the IEEE Engineering in Medicine and Biology Society, Buenos Aires, Argentina, pp. 6337–6340.
Sylvain, A., Matthieu, L., 2016. Choice of V for V-fold cross-validation in least-squares density estimation. J. Mach. Learn. Res. 17 (208), 1–50.
Thomassen, A.J.W.M., Meulenbroek, R.G.J., 1993. Effects of manipulation horizontal progression in handwriting. Acta Psychol. 82, 329–352.
Umberto, B., Yeesock, K., Philip, P., Changwon, K., 2017. System identification, health monitoring, and control design of smart structures and materials. Adv. Mech. Eng. 9 (4), 1–2.
Van Der Gon, D., Thuring, J.P., Strackee, J., 1962. A handwriting simulator. Phys. Med. Biol. 407–414.
Viviani, P., Terzuolo, C.A., 1983. The organization of movement in handwriting and typing. Lang. Prod. 2, 103–146.
Waris, A., Mendez, I., Englehart, K., Jensen, W., Kamavuako, E.N., 2019. On the robustness of real-time myoelectric control investigations: a multiday Fitts' law approach. J. Neural Eng. 16 (2), 026003.
Yasuhara, M., 1975. Experimental studies of handwriting process. Rep. Univ. Electro-Commun. Japan 25 (2), 233–254.
Yasuhara, M., 1983. Identification and decomposition of fast handwriting process. IEEE Trans. Circ. Syst. 30 (11), 828–832.
Zhang, D., Kamavuako, E.N., 2019. Estimation of Handwriting Trajectory From EMG Signals. (Individual master project) Department of Engineering, King's College London.

PART II

Applications in medical robotics

CHAPTER 7

Medical robotics

Olfa Boubaker
University of Carthage, National Institute of Applied Sciences and Technology, Tunis, Tunisia

1 Introduction

Medical robots are robotic machines utilized in health sciences. They can be categorized into three main classes (Cianchetti et al., 2018): (1) medical devices including surgery robotic devices, diagnosis and drug delivery devices, (2) assistive robotics including wearable robots and rehabilitation devices, and (3) robots mimicking the human body including prostheses, artificial organs, and body-part simulators. A variety of other classification approaches for medical robotics are proposed in the vast literature and will be discussed in this work (Taylor, 1997; Cleary and Nguyen, 2001; Hockstein et al., 2007; Dogangil et al., 2010; Enayati et al., 2016; Yang et al., 2017a; George Thuruthel et al., 2018).

During the last three decades, medical robots have been increasingly used to perform a growing number of health tasks. As such, they show promising future potential for use in a wide range of health issues (Preising et al., 1991; Dario et al., 1996; Speich and Rosen, 2008; Okamura et al., 2010; Bogue, 2011; Ferrigno et al., 2011; Kramme et al., 2011; Troccaz, 2013; Schweikard and Ernst, 2015; Cianchetti et al., 2018).

In the late 1980s, the first commercial surgical robots were built (Hockstein et al., 2007) and the first commercial myoelectric prostheses were being used in rehabilitation centers around the word (Zuo and Olson, 2014). The first prototypes date back several decades. Although the history of medical robotics is short, a review of the literature shows a comprehensive bibliography revealing the wealth and maturity of the domain.

Compared to manual machines in healthcare, medical robotic systems offer a wide range of advantages. They are flexible and can be programmed to perform a number of tasks. They are more versatile and cost effective. Further, they can eliminate human fatigue as well as improve the precision and capabilities of physicians. In order to design medical robots, a fundamental knowledge of biological systems is needed because these machines should be more accurate than other robotic systems and should contain

Control Theory in Biomedical Engineering
https://doi.org/10.1016/B978-0-12-821350-6.00007-X

sensing systems that provide them with safety features (Troccaz et al., 1995; Davies, 1996; Fei et al., 2001). Although medical robotics offer many advantages, it is recognized by many authors that the acceptance of robots in healthcare has been slowed by safety concerns (Preising et al., 1991). To be accepted and widely deployed, medical robots have to provide real advantages, including reduction of access trauma, faster recovery, scar limitation, cost reduction, eases of use, human-machine communication capabilities, and so on.

This chapter gives a comprehensive review of the current literature on medical robotics and highlights the different classification approaches related to medical robots. It also stresses fundamental requirements like safety. Finally, the chapter presents the main applications of medical robots such as in surgery, rehabilitation, and training.

2 Literature review

There are almost 4721 publications related to medical robotics; 513 papers were published in 2019 alone (see Fig. 1).

As shown in Fig. 2, the leading countries with the most publications on medical robotics are the United States with 1498 publications (31.73%), China with 534 (11.31%), and Japan with 452 (9.57%).

The field of medical robotics is distinguished by a rich bibliography, including at least two handbooks (Kramme et al., 2011; Segil, 2019),

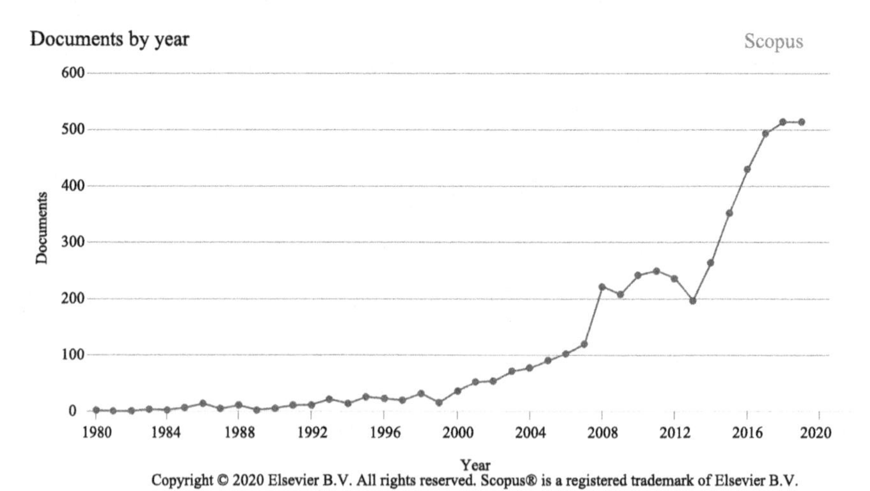

Fig. 1 Data on medical robotics literature from Scopus in February 2020. Query: KEY (medical AND robot) AND (LIMIT-TO (LANGUAGE, "English")).

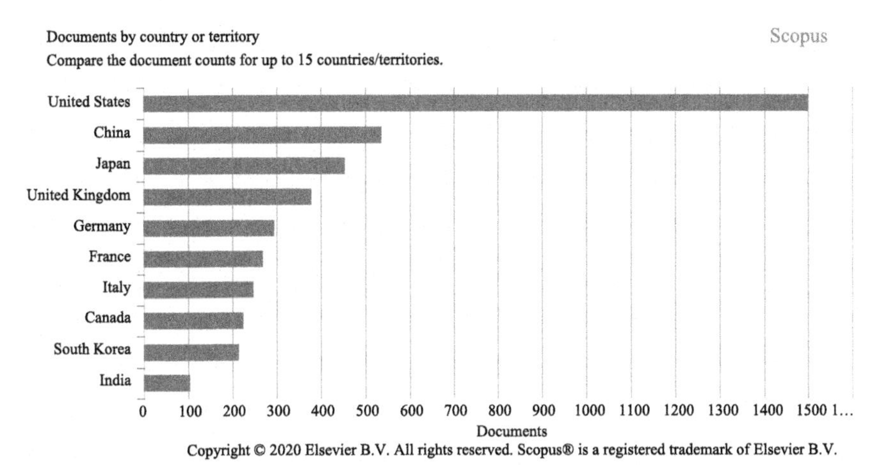

Fig. 2 Data on medical robotics literature by country from Scopus in February 2020. Query: KEY (medical AND robot) AND (LIMIT-TO (LANGUAGE, "English")).

a number of books (Rosen et al., 2011; Gomes, 2012; Troccaz, 2013; Kim, 2014; Bai et al., 2018), and several survey papers (Preising et al., 1991; Dario et al., 1996; Taylor, 2006; Okamura et al., 2010; Bogue, 2011; Ferrigno et al., 2011; Beasley, 2012; Poorten et al., 2012; Ni, 2015; Cianchetti et al., 2018; Troccaz et al., 2019).

In particular, most of the survey papers are devoted to surgical robotics (Howe and Matsuoka, 1999; Cleary and Nguyen, 2001; Jaikumar et al., 2002; Lanfranco et al., 2004; Puangmali et al., 2008; Taylor, 2008; van der Meijden and Schijven, 2009; Kalan et al., 2010; Trejos et al., 2010; Dogangil et al., 2010; Gomes, 2011; Mattei et al., 2014; Burgner-Kahrs et al., 2015; Davies, 2015; Leal Ghezzi and Campos Corleta, 2016; Enayati et al., 2016; Shi et al., 2017; Díaz et al., 2017; Peters et al., 2018; Simaan et al., 2018; Gifari et al., 2019).

Several survey papers are also devoted to rehabilitation and assistive robotics (Miller, 1998; Feil-Seifer and Mataric, 2005; Tapus et al., 2007; Johnson et al., 2008; Broadbent et al., 2009; Chen et al., 2013a,b; Yan et al., 2015; Liu et al., 2016; Aliman et al., 2017; Meng et al., 2017; Rupal et al., 2017).

In this framework, we have also reported many interesting survey book chapters (Davies, 1996; Hillman, 2006; Speich and Rosen, 2008; Taylor et al., 2008, 2016; Walker and Green, 2009; Okamura et al., 2011; Martins et al., 2012; Matarić and Scassellati, 2016; Frisoli, 2018; Moreno et al., 2018; Ruiz-Olaya et al., 2019; Witte and Collins, 2020).

Survey papers devoted to medical robotics and related subjects are evaluated in Scopus Database at 243 (see Fig. 3), revealing the development of this field.

The most widespread surgical robot is the da Vinci Intuitive Surgical System (http://www.intuitivesurgical.com/) shown in Fig. 4. The robotic system comprises three components: (1) a surgeon's console, (2) a patient-side robotic cart with four arms manipulated by the surgeon (one to control the camera and three to manipulate instruments), and (3) a high-definition

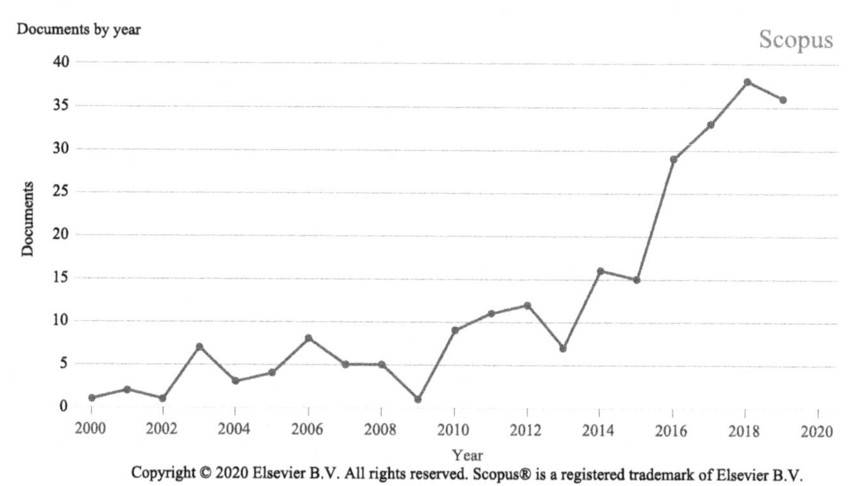

Fig. 3 Data on medical robotics from Scopus in February 2020. Query: KEY (medical AND robot) AND (LIMIT TO (LANGUAGE, "English")) AND (LIMIT-TO (DOCTYPE, "re")).

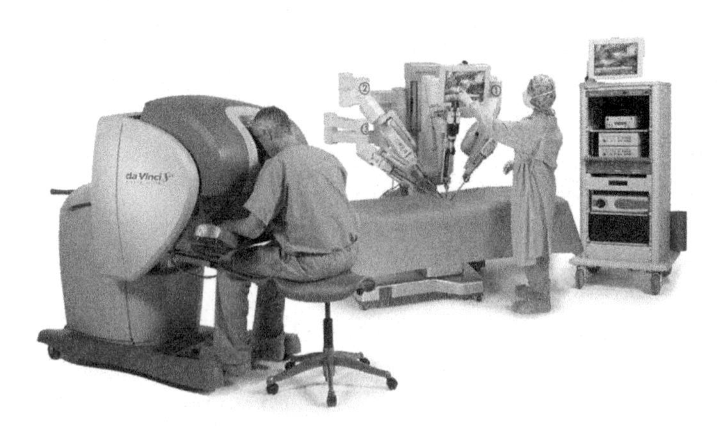

Fig. 4 The intuitive surgical Da Vinci system by ker—Norberak egina, used under CC BY-SA 4.0, https://commons.wikimedia.org/w/index.php?curid=84236736.

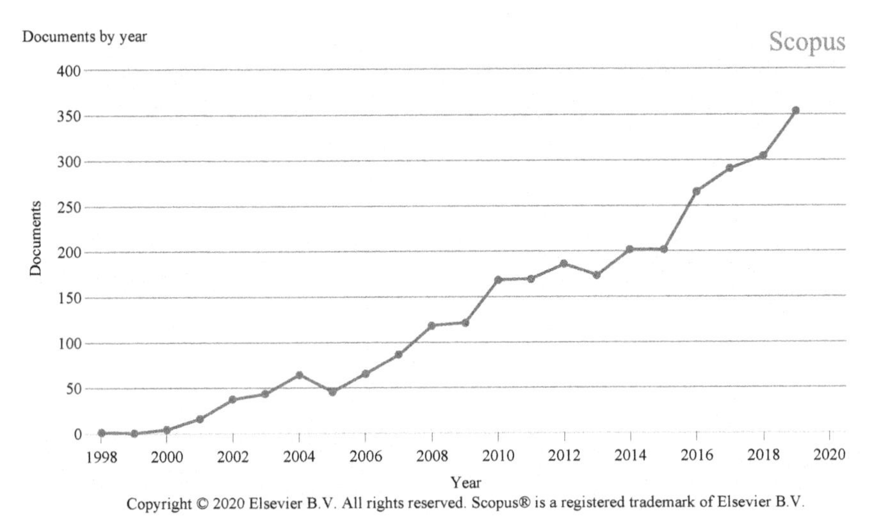

Fig. 5 Data on da Vinci robot publications from Scopus in February 2020. Query: TITLE-ABS-KEY (Da AND Vinci AND robot).

3D vision system. Articulated surgical instruments are mounted on the robotic arms, which are introduced into the body through cannulas. The original robotic system that the da Vinci was created on was designed at Stanford Research Institute International in Menlo Park with funding support from the Defense Advanced Research Projects Agency (DARPA) and the National Aeronautics and Space Administration (NASA). In 2000, the da Vinci System was cleared by the United States' Food and Drug Administration (FDA). As shown in Fig. 5, the da Vinci System has been referenced in more than 2909 peer-reviewed research works with most publications published by the United States with 1136 articles (39.05%), Italy with 264 (9.07%), Corea with 239 (08.21%), France with 185 (06.35%), and China with 166 (05.70%). According to the Intuitive company (https://www.intuitive.com/en-us/about-us/company) founded in 1995, since its first robotic-assisted systems were cleared by the FDA for general laparoscopic surgery in June 2000, 5582 da Vinci Systems are being used in 67 countries and more than 7.2 million da Vinci procedures have been completed worldwide through 2019 (Fig. 6).

3 Classification of medical robotics

Medical robots can be classified and studied using various taxonomies (Daneshmand et al., 2017). As shown in Fig. 7, medical robots can be

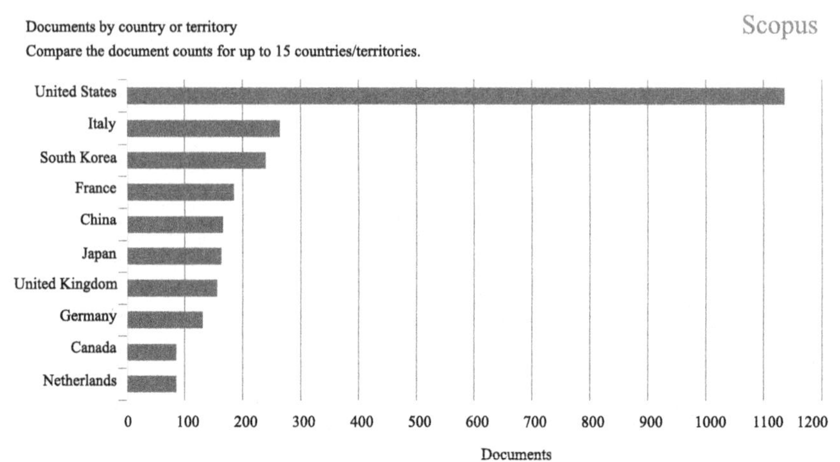

Fig. 6 Data on da Vinci robots by country from Scopus in February 2020. Query: TITLE-ABS-KEY (Da AND Vinci AND robot).

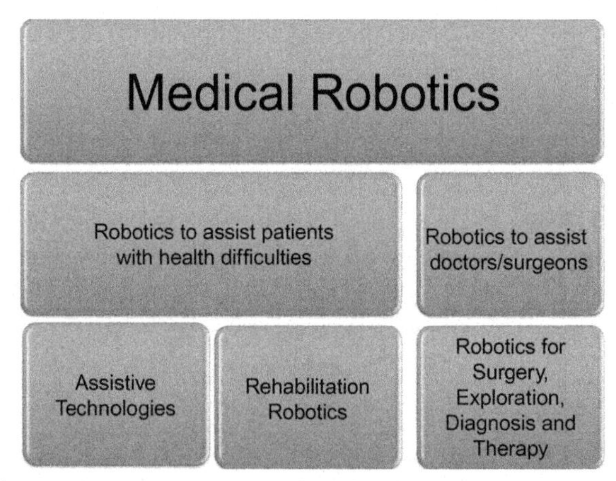

Fig. 7 Classification-based applications of medical robotics. *Adapted from Dombre, E., 2003. Introduction to Medical Robotics, Summer European University on Surgical Robotics, Montpellier, https://www.lirmm.fr/UEE07/presentations/lecturers/Dombre.pdf.*

classified according to their applications into two main groups (Dombre, 2003): robotics to assist doctors/surgeons and robotics to assist patients with health difficulties. The first class includes robotics for surgery, exploration, diagnosis, and therapy. The second class incorporates on one hand all assistive-technologies including robots that improve the quality of life of disabled and elderly people by increasing personal independence, and on the

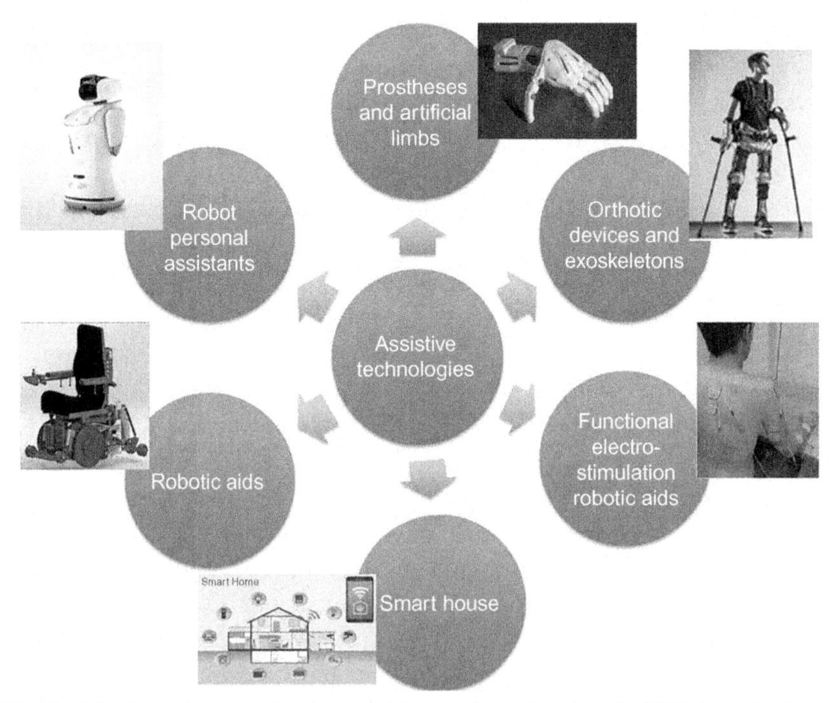

Fig. 8 Robotic assistive technologies. *Adapted from Dombre, E., 2003. Introduction to Medical Robotics, Summer European University on Surgical Robotics, Montpellier, https://www.lirmm.fr/UEE07/presentations/lecturers/Dombre.pdf.*

other hand robotics assisted-therapy tools used temporarily by people with motor disorders caused by stroke or spinal cord diseases (Dombre, 2003).

As shown in Fig. 8, assistive technologies include (Garcia-Aracil et al., 2017; Mohammed et al., 2017; Shishehgar et al., 2018):

- Prosthetic devices and artificial limbs (Marks and Michael, 2001),
- Orthotic devices and exoskeletons (Segil, 2019; Moreno et al., 2018; Pons, 2008),
- Functional electrical stimulation (FES) systems (Popović, 2014),
- Smart houses (Stefanov et al., 2004),
- Robotic aids (Faria et al., 2014),
- Robot personal assistants (Graf et al., 2009; Clotet et al., 2016; Costa et al., 2018; Martinez-Martin and del Pobil, 2018).

In Dario et al. (1996), medical robots are classified according to their scale on three sets: (1) macro-robotics, (2) micro-robotics, and (3) bio-robotics. Macro-robotics include manipulators for surgery and rehabilitation and wheelchairs. Micro-robotics include robotics devices for minimally invasive surgery (MIS) as well miniaturized mechatronics tools for conventional surgery. Bio-robotics deal with simulating biological systems in order to provide a better understanding of human physiology.

The classification scheme developed by Taylor (1997) categorizes surgical robots by the role they play in medical applications as tools that can work cooperatively with physicians to carry out surgical interventions. In this framework, five classes of systems are identified: (1) intern replacements, (2) telesurgical systems, (3) navigational aids, (4) precise positioning systems, and (5) precise path systems.

In Cleary and Nguyen (2001), the authors focus on the role of surgical robots within the context of Computer Integrated Systems (CIS). They classify the systems into two broad families: (1) surgical CAD/CAM and (2) surgical assistants.

In Camarillo et al. (2004), the authors define a role-based classification of medical robots: (1) passive role where the role of the robot is limited in scope, or its involvement is largely low risk, (2) restricted role where the robot is responsible for more invasive tasks with higher risk, but is still restricted from essential portions of the procedure, and (3) active role where the robot is intimately involved in the procedure and carries high responsibility and risk.

In Hockstein et al. (2007), surgical robots are classified as (1) automated arms, (2) mobile devices, (3) mills, or (4) telerobotic devices. Additionally, they are also classified as (1) active, (2) semi-active, or (3) passive devices. Active devices are totally programmable and carry out tasks independently. Semi-active devices and passive robotic devices translate movements from an operator's or surgeon's hands into powered or unpowered movements of the robot's end-effector arms. Surgical robots can include both active mills and semi-active telerobotic devices.

In Dogangil et al. (2010), medical robots are classified according to the following surgical specialties: neurosurgery, eye surgery and ear, nose, and throat (ENT) surgery; general, thoracic, and cardiac surgery; gastrointestinal and colorectal surgery; and urologic surgery. In the same work, the systems are also cross-classified according to their engineering design and robotics technology.

In Lee et al. (2017), the authors classify soft robotics according to the following criteria: actuation, sensing, structure, control and electronics, materials, fabrication and system, and applications.

In Yang et al. (2017a), medical robots (in commercial use as well as in research) are classified according to robot autonomy. The spectrum of automation in surgical robotics relates to the level of dependence on the human surgeon to guide the procedure, as compared to the robot guiding itself.

These levels are categorized as (1) direct control, (2) shared control, (3) supervised autonomy, and (4) full autonomy. They are discussed according to ethics and legality including ensuring patient safety, culpability, and approval from the FDA prior to marketing in the United States.

In Cianchetti et al. (2018), the authors classify soft medical robotics in three classes: (1) medical devices including surgery robotic devices and drug-delivery robotic devices, (2) assistive robotics including wearable robots and rehabilitation devices, and (3) robots mimicking the human body including prostheses, artificial organs, and body-part simulators.

In George Thuruthel et al. (2018), the article attempts to classify continuum robots for medical applications according to their control strategies and provide an insight into various controllers developed for continuum/soft robots. The classification schema include (1) controller's design based on modeling approaches (model-based controllers, model-free controllers, and hybrid controllers combining model-based and model-free approaches), (2) controllers design based on actuators (actuation technologies, actuator arrangement, actuator shape, actuator number, material), and (3) controllers design based on feedback loops (operating spaces, required sensors, performances).

In Chikhaoui and Burgner-Kahrs (2018), the authors classified continuum robots for medical applications according to control strategies and focused on open-loop control, closed-loop control, or the combination of both.

In Yang et al. (2018), the survey systemically summarizes the state-of-the-art force control technologies for robot-assisted needle insertion, such as force modeling, measurement, factors that influence the interaction force, parameter identification, and force control algorithms.

In Runciman et al. (2019), the authors classify soft robotics in MIS according to six criteria: (1) the working principles, (2) materials, (3) manufacturing, (4) actuation, (5) variable stiffness, and (6) locomotion and sensing methods.

In Gifari et al. (2019), the authors classify soft surgical robots for endoscopic applications according to (1) actuation principles, (2) sensing, (3) control, and (4) stiffness adjustability.

In summary, medical robots can be classified from many viewpoints, including:

- The application, the targeted anatomy, or technique (Preising et al., 1991; Dombre, 2003; Dogangil et al., 2010; Cianchetti et al., 2018).
- The robotic scale (Dario et al., 1996).
- The role (Taylor, 1997; Camarillo et al., 2004)
- The structural or mechanical design (Burgner-Kahrs et al., 2015; Taylor et al., 2016); (Runciman et al., 2019; Gifari et al., 2019).

- The actuation strategy (Cianchetti et al., 2018; Runciman et al., 2019; Gifari et al., 2019; Le et al., 2016).
- The force sensing and its application in MIS and therapy (Trejos et al., 2010).
- The haptics technologies (Enayati et al., 2016).
- The level of autonomy (Payne and Yang, 2014; Yang et al., 2017a).
- The control strategies (George Thuruthel et al., 2018; Chikhaoui and Burgner-Kahrs, 2018).
- The possible applications of participatory and opportunistic mobile sensing (Daneshmand et al., 2017).
- The variable stiffness and the locomotion and sensing methods (Runciman et al., 2019; Gifari et al., 2019).

4 Advantages and fundamental requirements

To date, fields such as surgical robotics, rehabilitation robotics, and assistive robotics have been generally considered as separate fields in spite of the fact that they offer the same advantages and need the same requirements. We highlight these items in the sections that follow.

4.1 Advantages

Despite their high cost and limitations in haptic sensing and judgment in complex situations, medical robots have to supply measurable and complementary abilities compared to those of human ones to be accepted and extensively used in healthcare applications (Taylor et al., 2008). Due to their accuracy, repeatability, and indefatigability, robotic technology is increasingly affecting the entire healthcare sector through advances in diagnosis, preoperative planning, surgery, postoperative evaluation, acute rehabilitation, and chronic assistive devices. Particularly, medical robots have proved their abilities in (Taylor, 1997; Yang et al., 2017a):

- improving doctor's technical qualifications characterized by lack of precision, fatigue, tremor, and inattention either by making existing procedures more accurate, faster, or less invasive;
- making it possible to perform infeasible interventions by humans in hazardous environments;
- advancing surgical safety by enhancing technical performance of difficult procedures in dangerous proximity to delicate anatomical structures by using online monitoring and information supports for surgical procedures;

- recording and understanding data as valuable tools for training, skill assessment, and certification for surgeons;
- keeping immune to infection;
- being potentially unaffected by radiation; and
- ability to incorporate many sensors (force, acoustic, etc.).

4.2 Fundamental requirements

As reported in Cianchetti et al. (2018), biocompatibility and biomimicry are key considerations for medical robots. It is crucial to remember that in surgery and endoscopy, robots operate inside a human body, that rehabilitation robots are in physical interaction with the patient, that artificial limbs and organs replace human limbs, and drug-delivery robotic devices can be integrated inside or on the skin of patients. Medical robots need to be compatible with the human body and tissues to guarantee system functionality and body acceptability to avoid allergies and contact reactions, immediate immune responses, and rejection. The materials also need to match the mechanical properties of human tissues to a certain degree. Fig. 9 presents a comparison

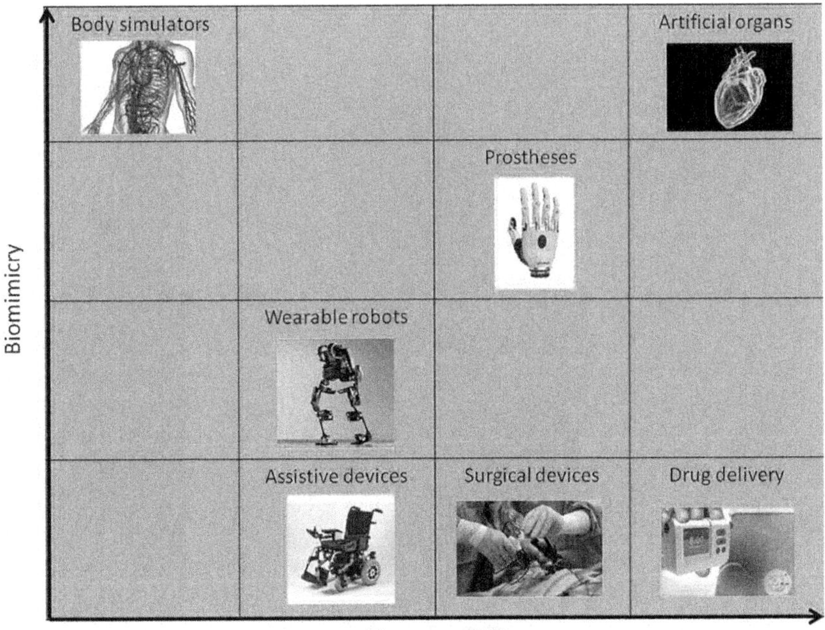

Fig. 9 Levels of biomimicry and biocompatibility comparison of biomedical soft robots. *Adapted from Cianchetti, M., et al., 2018. Biomedical applications of soft robotics, Nat. Rev. Mater. 3(6), 143–153. doi: 10.1038/s41578-018-0022-y.*

between levels of biomimicry and biocompatibility related to different classes of biomedical soft robots.

In Wolf and Shoham (2009), the authors stressed four requirements for medical robots: (1) simple operation, (2) safety, (3) easy sterilization, and (4) compact size and light weight. These requirements are shown in Fig. 10 in which safety considerations are highlighted and further explained.

It is obvious that all parties in contact with the patient or surgeon must be easily sterilized. A surgery robotic system should be sterile using one of the following three methods (Troccaz, 2012):

- Sterilization according to different processes such as autoclave or the Sterrad process,
- Disposable sterile packaging,
- Reusable sterile packaging.

Fig. 11 shows a sterilizable robot vs. a robot in sterile packaging.

In Daneshmand et al. (2017), the authors highlighted six requirements of utmost importance for medical robotics: (1) safety, (2) sterilization, (3) precision in the positioning of the instrument, (4) efficiency, (5) user-friendliness, and (6) cost.

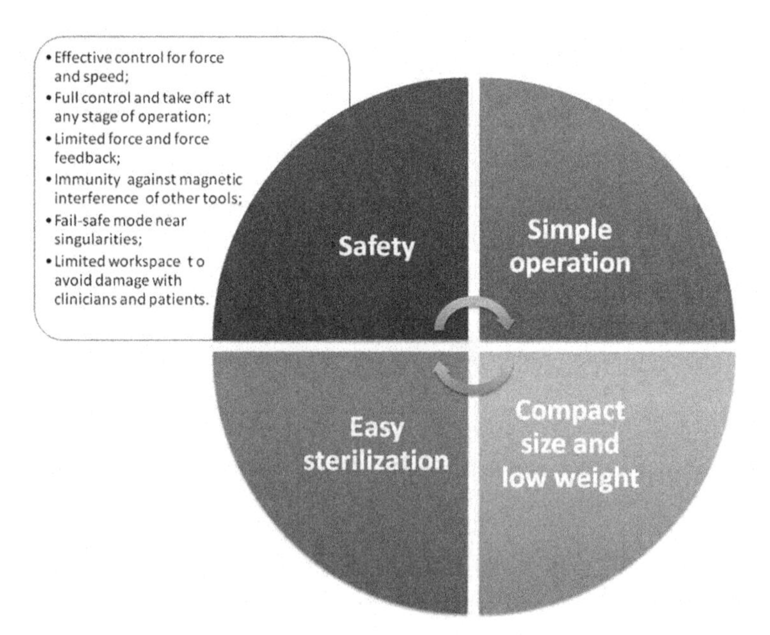

Fig. 10 Main requirements for medical robots. *Adapted from Wolf, A., Shoham, M., 2009. Medical automation and robotics. In Springer Handbook of Automation. Springer, Berlin, Heidelberg, pp. 1397–1407. doi: 10.1007/978-3-540-78831-7_78.*

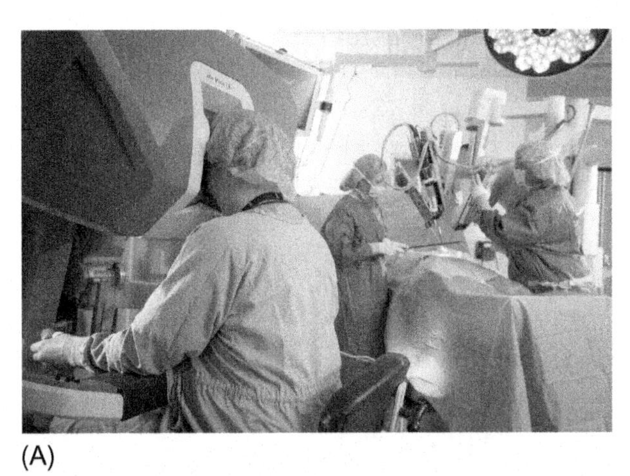

(A)

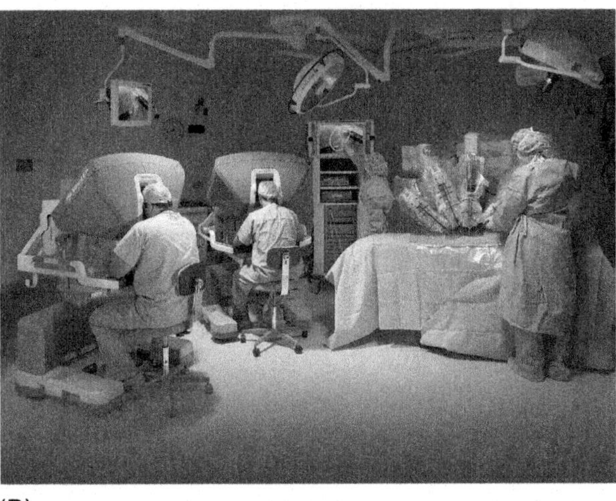

(B)

Fig. 11 Sterilizable robot vs. robot in sterile packaging. (A) By Fatma Darsti, used under CC-BY-SA-4.0, https://upload.wikimedia.org/wikipedia/commons/3/34/Robot_assisted_surgery.jpg. (B) By Trend Kim used under CC-BY-SA-3.0, https://upload.wikimedia.org/wikipedia/commons/7/7c/Da_Vinci_action_023874_10x7_150dpi.jpg.

Indeed, the interaction between humans and robots requires safety considerations to prevent doing harms to the human component, the environment, and the robot itself. Two levels of safety can be introduced: software safety and hardware safety (Daneshmand et al., 2017).

For precision considerations constrained to imaging systems distortions, the positioning tolerance is fixed to 1–3 mm (Daneshmand et al., 2017). Precise control of robotic devices configured for stable, safe, and compliant motion in a changing environment can be realized with several strategies and schemes for position and force control (Hogan, 1985a,b,c; Pierrot et al., 1999; Duchemin et al., 2005; Cortesão et al., 2006; Zarrad et al., 2007a,b; Mehdi and Boubaker, 2010, 2011a, b, 2012a, b, c, 2013, 2015, 2016; Amirkhani et al., 2019). Different methods are also presented to overcome the physical limitations of instrument sensorization to meet safety requirements (Katsura et al., 2005; Trejos et al., 2010; Haidegger et al., 2009; Joubair et al., 2016; Enayati et al., 2016),(Yang et al., 2018; Guo et al., 2019; Mobayen et al., 2019).

As important as the sterilization of the parts that are in direct contact with the patient is, cost is also an important factor that should be given careful consideration (Daneshmand et al., 2017).

5 Robot-assisted surgery

5.1 History

Robotic surgery has undergone a profound revolution in the past 35 years and the history of its evolution is described in many articles (Cleary and Nguyen, 2001; Taylor and Stoianovici, 2003; Hockstein et al., 2007; Kalan et al., 2010; Dogangil et al., 2010; Pugin et al., 2011; Valero et al., 2011; Lee-Kong and Feingold, 2013; Hussain et al., 2014; Shah et al., 2015; Leal Ghezzi and Campos Corleta, 2016; Lane, 2018; Marino et al., 2018; George et al., 2018).

Actually, the general idea of telerobotic healthcare was born in the early 1970s. It was proposed by NASA to provide surgical care for astronauts. However, the first prototypes were built 15 years later, in the late 1980s, when multiple academic centers started developing new prototypes. According to (Bogue, 2011), the world's first surgical robot was the "Arthrobot" developed in Vancouver, Canada in 1983. It was able to position a patient's leg based on a voice command. In 1984, it was used in an orthopedic surgical procedure. Since then, numerous products have been developed and the best known today is the da Vinci Surgical System produced by Intuitive Surgical, Inc.

As reported by Hockstein et al. (2007), the first surgical application was described in 1985 when an industrial robotic arm, the PUMA 200 produced by Unimation Limited, was modified to perform a stereotactic brain biopsy with 0.05 mm accuracy. Among the Programmable Universal Machines for Assembly, the PUMA 560 was used then for neuro-stereoctatic surgery and was the first robot-assisted surgical procedure reported in the literature

(Kwoh et al., 1988). This has served as the prototype for neuromate (Integrated Surgical Systems, Sacramento, CA, USA), which received FDA approval in 1999. Thus, in spite of the encouraging preliminary results, the work on the PUMA robot was ceased. In 1991, Imperial College in London developed a small-sized robot called the ProBot. This later was used to perform a prostatic surgery. In 1992, the Robodoc (Integrated Surgical Systems) was introduced for use in hip replacement surgery. In 1996, similar devices were designed for use in knee replacement and temporal bone surgery, notably the Acrobot (The Acrobot Company, Ltd., London, UK) and the RX-130 robot (Staubli Unimation Inc., Faverges, France), respectively. Neither device has yet completed clinical testing nor received FDA approval.

Since then, surgical robotics was deeply marked by MIS. MIS involves the use of long rigid or flexible surgical instruments that are inserted into the body through small incisions or natural orifices, in contrast to open surgery where large incisions are used to access the target organ directly. The goal of MIS is to complete a surgical procedure as safely and quickly as possible, while minimizing damage to peripheral tissue. MIS is being used with increasing frequency as an alternative to open surgery because of the improvements it can bring to patient safety, cosmetics, recovery time, hospital stays, postoperative complications, and pain. The development of a tele-operated surgical robot is motivated by the desire to enhance the effectiveness of the surgical procedure and provide a comfortable operating environment for the surgeon. In this framework, Automated Endoscopic System for Optimal Positioning (AESOP), produced by Computer Motion Company and initially funded by NASA, was introduced in 1994 as the first laparoscopic camera holder to be approved by the FDA. Voice control was then added in 1996 for position control of the camera. Seven degrees of freedom were added in 1998 to mimic a human hand. The endoscope can be inserted into the patient through a smaller incision.

In 1998, ZEUS was introduced commercially by the American robotics company Computer Motion with the original idea of telepresence surgery. For such a system, the surgeon operates on the patient at a distance from a robot console. ZEUS's major success was in cardiac surgery. ZEUS had three robotic arms remotely controlled by the surgeon. The first arm, AESOP, the voice-activated endoscope, allowed the surgeon to see inside the patient's body. The other two robotic arms mimicked the surgeon's movements to make precise incisions and extractions.

In 1999, several projects on robotic tele-ecography were launched (Gonzales et al., 2001). Such robotic systems, like the SYRTECH system

(Gourdon et al., 1999a,b), Salcudean System (Salcudean et al., 1999), Hippocrate (Pierrot et al., 1999), and others (Masuda et al., 2001), are very useful diagnostic tools because they are noninvasive, generally nonexpensive, and use highly portable methods that do not use ionizing radiation. However, generating and interpreting ultrasound images are highly operator-dependent. As a result, performance and interpretation of these examinations have traditionally been limited to medical specialists (Ferreira et al., 2015).

In 2003, ZEUS was discontinued, following the merger of Computer Motion with its rival Intuitive Surgical, maker of the da Vinci Surgical System described previously. The da Vinci system was the first surgical system to address adequately all laparoscopic surgery's previous problems. This is due to the reduction of the range of motion, 3D imaging, and the surgical arm unit that positions and maneuvers detachable surgical instruments. The main technological advantages of this system are stereovision, dexterity, realistic 3D imaging, motion-scaling, and tremor filtration. Since then, more precise and accurate endoscopic surgery is guaranteed (Bodner et al., 2005). Such telepresence surgery overcomes two main problems of classical laparoscopic surgery, i.e., the limitation to only four degrees of freedom and the lack of stereovision (Rassweiler et al., 2001). Since then, the da Vinci System has been used in many surgical procedures where the first intervention was a laparoscopic radical prostatectomy (Abbou et al., 2001). Other applications in laparoscopic tele-surgery soon followed.

Since 2004, the field of surgical robotics is marked by a new generation of small patient-mounted robots (Walsh et al., 2008) and surgical robots for dedicated applications (Morgia and De Renzis, 2009; Barrett et al., 2007), transluminal endoscopy interventions as flexible endoscopy (Dogangil et al., 2010) and intra-body continuum robots (Burgner-Kahrs et al., 2015; Runciman et al., 2019). A review of emerging surgical robotic technology is described in Peters et al. (2018). Several applications of the new generation of medical robots will be described in the following section. The relatively short history of surgical robotics can be then depicted in three stages as described by Fig. 12.

5.2 Applications

Robotic MIS has undergone a profound revolution in the past 35 years. MIS is performed by introducing two to three long and rigid tools into the patient's abdomen through small incisions (about 10–15 mm in diameter) for endoscopic vision. For single-port laparoscopic surgery, surgeons create

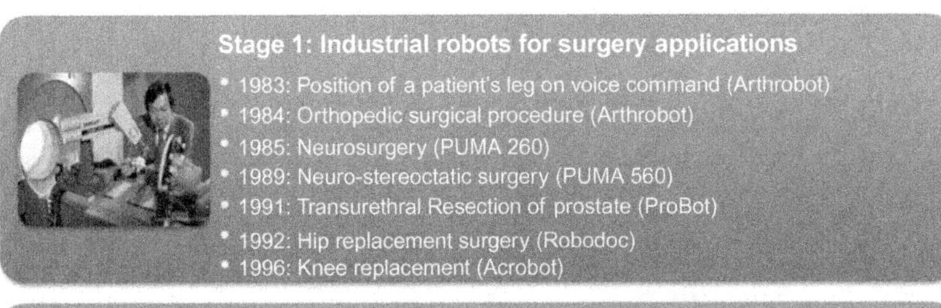

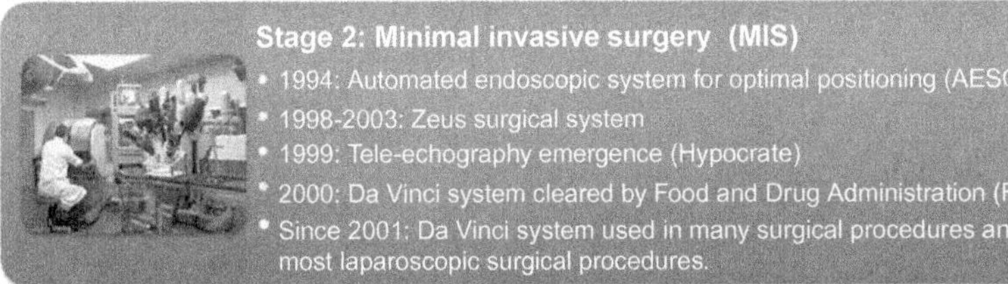

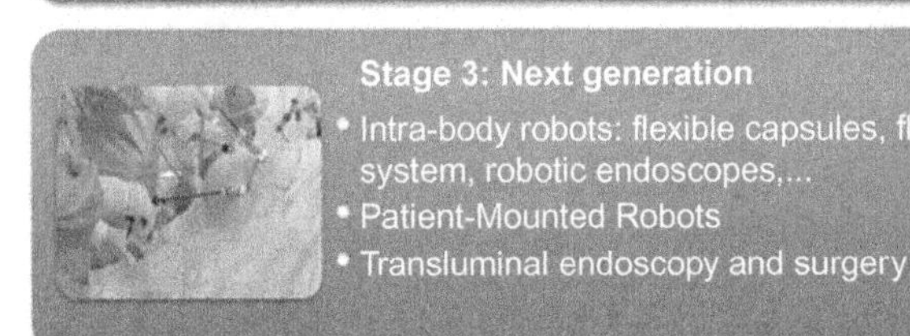

Fig. 12 Timetable of surgical robotics.

one larger incision (about 20–40 mm) at the patient's navel and introduce several semi- flexible instruments through the incision. This approach has the advantage to reduce the number of ports, but it can lead to tool encumbrance. MIS can also be performed through natural orifices such as the mouth, vagina, or anus using a flexible endoscope. This procedure is called natural orifice transluminal endoscopic surgery (NOTES) (Cianchetti et al., 2018). In MIS, single-port surgery and NOTES, reachability of the target organ can be a crucial issue. Indeed, the instruments clash during the operation, which increases the overall complexity of the procedure. Redesigning the instruments is an imperative measure in order to make this emerging operative method safe and reproducible. Fig. 13 describes the latest surgical approaches.

Nowadays robot-assisted surgery is used in many clinical sub-domains of surgery, for example, neurosurgery, orthopedic surgery, throat surgery, abdominal surgery/laparoscopy, radio surgery, and so on. Some of these sub-domains are shown in Fig. 14. The main clinical applications of surgical robots and related clinical needs, technical requirements, and normativity in

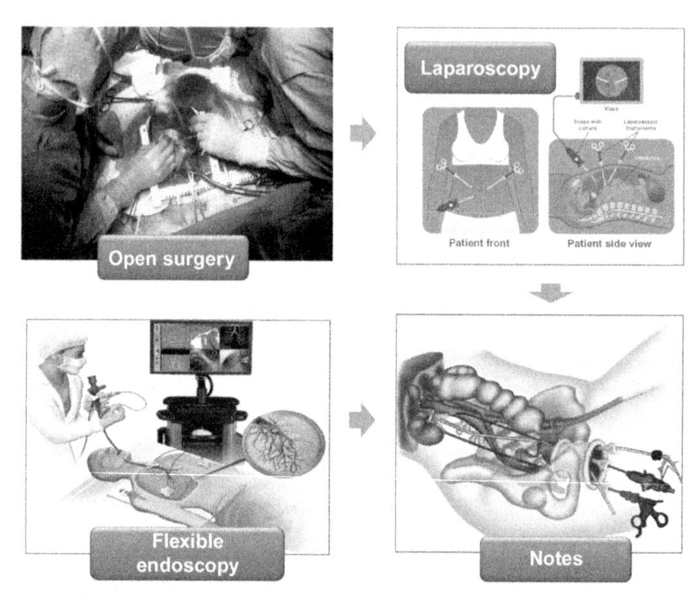

Fig. 13 Surgical approaches. *Adapted from Wang, X., Meng, M. Q.-H., 2012. Robotics for natural orifice transluminal endoscopic surgery: a review, J. Robot. doi: 10.1155/2012/512616.*

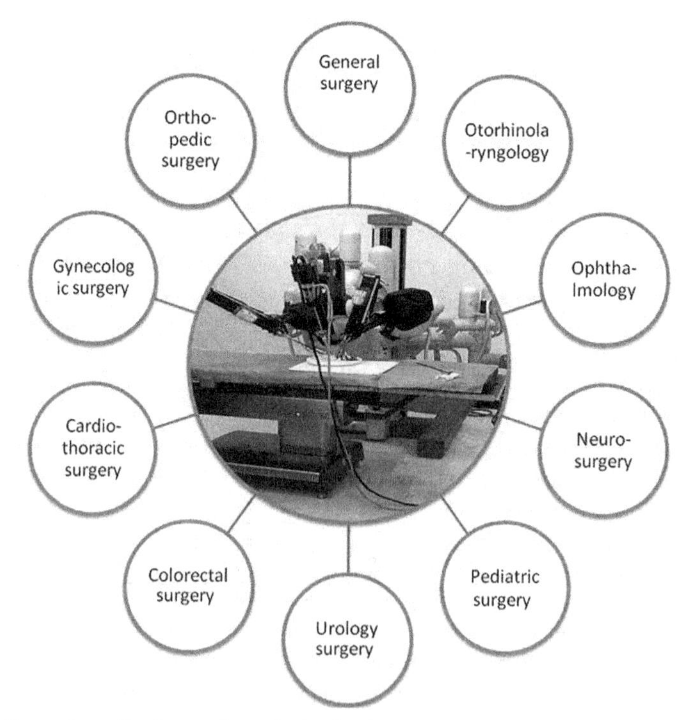

Fig. 14 Several applications of surgical robotics.

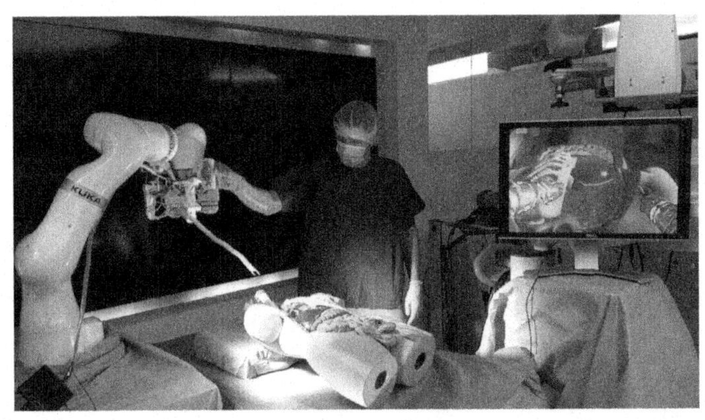

Fig. 15 The i^2Snake Robot (Berthet-Rayne et al., 2018), used under CC-BY-4.0, https://upload.wikimedia.org/wikipedia/commons/6/6e/The_i%C2%B2Snake_Robot.png.

design are described in Díaz et al. (2017). Commercially available robotics systems dedicated for these applications can be found in Beasley (2012) and Peters et al. (2018).

Fig. 15 shows an endoscopic application using the i^2Snake robot (Berthet-Rayne et al., 2018). The i^2Snake robot is a prototype of an articulated endoscopic robot with two small arms equipped with a camera, light, and a suction/irrigation channel allowing the surgeon to perform surgery remotely. It can navigate through the mouth and reach deep-seated lesions.

Another example of robotic surgical application is shown in Fig. 16, which shows images of the neuromate robot and its surgical workflow in positioning frameless stereotactic biopsies (Yasin et al., 2019).

Another application is shown in Fig. 17 for a robotic single-port surgery performed using a fourth-generation model of the da Vinci Surgical System, da Vinci SP, for a benign gynecologic disease at Ewha Womans University Medical Center, Seoul, Republic of Korea (Shin et al., 2020). The patient cart of the da Vinci SP surgical system includes an articulating camera and three robotic instruments able to be positioned simultaneously through a 25-mm SP multichannel single port. As reported in Shin et al. (2020), this surgical system may replace conventional laparoscopic surgery or robotic multiport laparoscopy even in complicated and severe adhesion cases. Other advantages are reported (Shin et al., 2020) compared to the previous versions of the da Vinci Surgical System, the Si and Xi systems. This new mechanical system includes various kinds of articulating instruments and is designed to

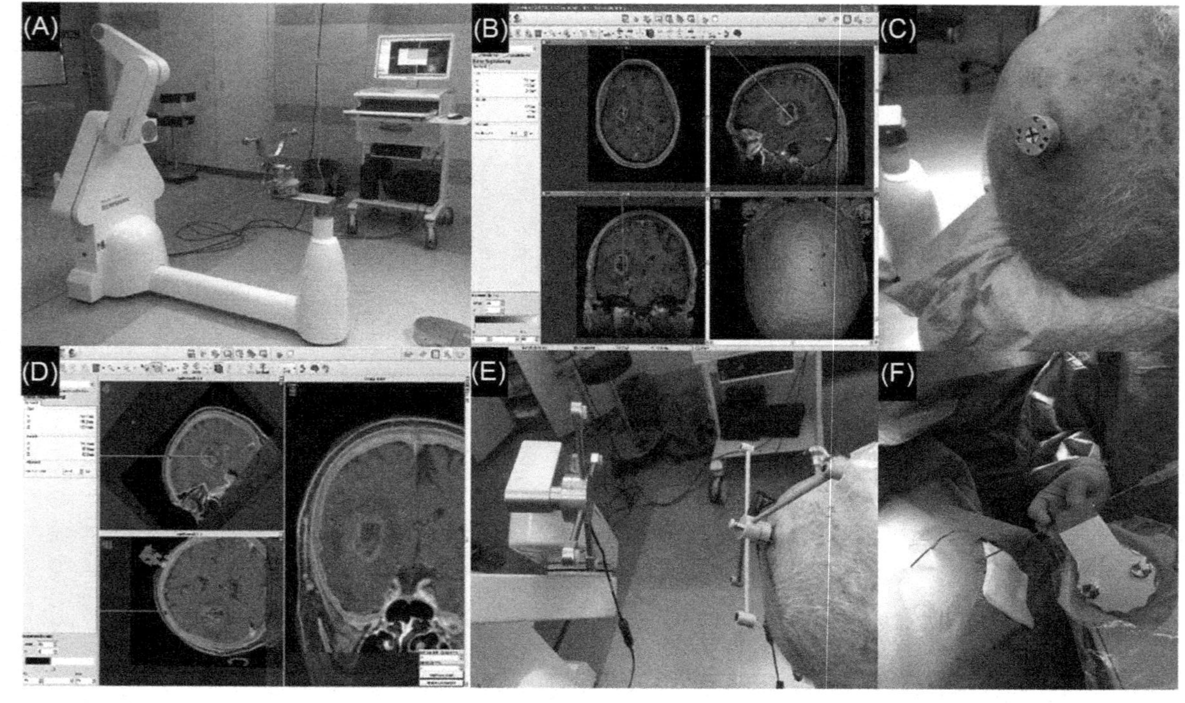

Fig. 16 Images of the Neuromate robot and surgical workflow. (A) Robot and planning station. (B) Trajectory plan. (C) The base of the localizing device is mounted to the skull. (D) The preoperative magnetic resonance imaging including the trajectory is fused to the computed tomography reference scan. (E) Patient registration using the ultrasonic tracking system. (F) Positioning the biopsy needle through the instrument holder along the trajectory. *Reproduced with permission from Yasin, H., et al., 2019. Experience with 102 frameless stereotactic biopsies using the neuromate robotic device, World Neurosurg. 123, e450–e456. doi: 10.1016/j.wneu.2018.11.187,* ***,*** *Copyright Elsevier, 2020.*

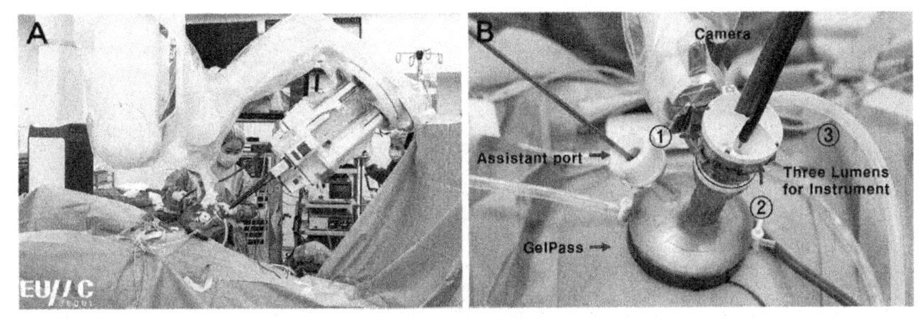

Fig. 17 The da Vinci SP surgical system performing a single-port surgery for a benign gynecologic disease. (A) Operating room setup. (B) The da Vinci SP cannula can be inserted through the Gel Pass and one additional trocar is inserted 1.5 cm away from the cannula. *Reproduced with permission from Shin, H. J., et al., 2020. Robotic single-port surgery using the da Vinci SP® surgical system for benign gynecologic disease: a preliminary report, Taiwan. J. Obstet. Gynecol. 59(2), 243–247. doi: 10.1016/j.tjog.2020.01.012, Copyright Elsevier, 2020.*

be more ergonomic (Fig. 18). The EndoWrist SP instruments and camera have joggle joints in combination with snake-like wrist joints that mimic the human wrist, shoulder, and elbow. Also, the wrist joint allows for seven degrees of freedom and the elbow joint maintains intracorporeal triangulation. A more remarkable feature is the instrument guidance system, which tracks the locations of the robot camera, port, and instruments within the operational field and simultaneously repositions the instruments. With this system, it is possible to operate efficiently in a narrow space without clashing the instrument.

Fig. 19 shows a new conceptual design for microrobots to be developed for NOTES. For these robots, the tools are attached directly to the magnetic joint, which is coupled to the external magnetic handle. The advantage of the design lies in its simple structure and compact size. Each modular robot has an ingestible size and can be ingested/inserted into the lumen through the natural orifices/endoscope channels. After reaching the working space, the robots start the self-assembly process by magnetic force. An external controller is operated by the surgeon for wireless control of the robotic configuration, the assembly, and the surgical task. After completion of the surgical procedure, the robots either reassemble into a snake shape or disassemble into individual modules and excrete naturally. The self-assembly design not only reduces the size of each module but also enables convenient addition and replacement of different interventional functions/modules even during the operation (Nakadate and Hashizume, 2018).

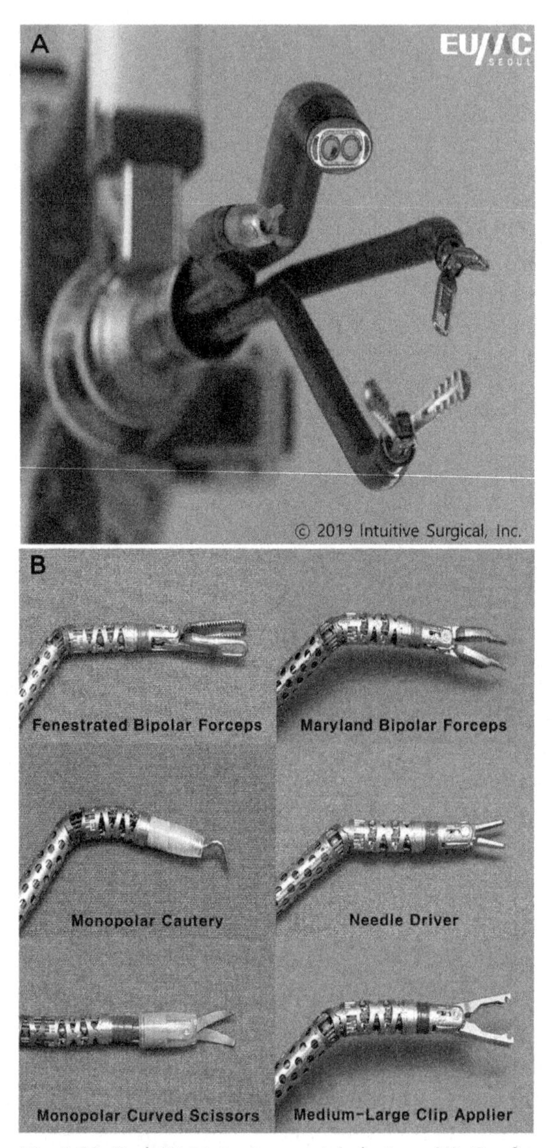

Fig. 18 The da Vinci SP EndoWrist instruments' design. (A) Configuration of three articulating instruments and camera. (B) Variable instruments. *Obtained with permission from Shin, H. J., et al., 2020. Robotic single-port surgery using the da Vinci SP® surgical system for benign gynecologic disease: a preliminary report, Taiwan. J. Obstet. Gynecol. 59(2), 243–247. doi: 10.1016/j.tjog.2020.01.012, Copyright Elsevier, 2020.*

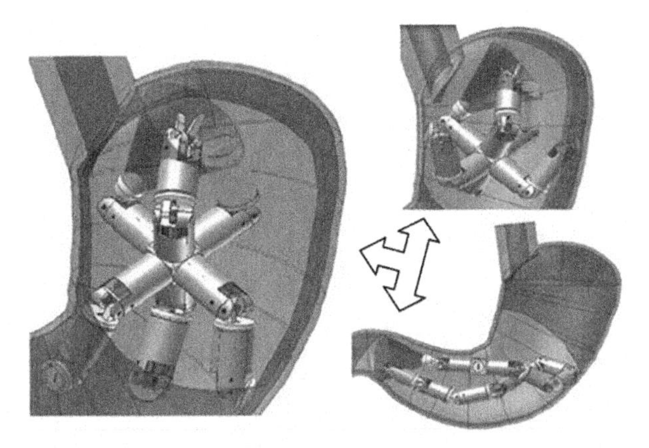

Fig. 19 Conceptual representation of configurable modular robots inside the stomach (Nakadate and Hashizume, 2018), used under CC BY 3.0.

5.3 Commercially available/FDA-approved robotic devices and platforms

Nowadays, a number of FDA-approved devices and platforms for robotic surgery are commercially available (see Table 1), including the da Vinci Surgical System described previously, CyberKnife M6 System (Morgia and De Renzis, 2009), Sensei X Robotic Catheter System (Dello Russo et al., 2016), FreeHand 1.2, invendoscopy E200 system (Peters et al., 2018), Flex Robotic System (Mattheis et al., 2017), Senhance Sugical System (Spinelli et al., 2018), ARES system (Caversaccio et al., 2008), ViaCath System (Abbott et al., 2007), ROSA Spine (Lefranc and Peltier, 2016), and ROSA Brain (De Benedictis et al., 2017). Most of the latter commercial devices are described in (Peters et al., 2018). A list of commercial stereotactic neurosurgery devices can be found in Faria et al. (2015).

Figs. 20–24 present photos related to the CyberKnife M6 System, Sensei X robotic catheter system, Senhance Surgical System, Flex Robotic System, and ROSA ONE's arm, respectively.

In addition to FDA-approved surgical robotic systems, several platforms are available that have not yet obtained FDA approval. These include MirοSurge, SPORT Surgical System, SurgiBot, Versius Robotic System, Master and Slave Transluminal Endoscopic Robot, Verb Surgical, Miniature In Vivo Robot, and the Einstein Surgical Robot (Peters et al., 2018).

Table 1 Commercially available/FDA-approved surgical robots.

Robotic system	Company	Country	DOF	Branch of surgery
da Vinci surgical system	Intuitive Surgical Inc.	US	3+7/wrist	Urologic Laparoscopic Gynecologic General non-cardiovascular Thoracoscopic Cardiotomy
CyberKnife M6 system	Accuray Incorporated	US	6	Stereotactic radiosurgery Radiation therapy Treating cancer and benign tumors
Sensei X robotic catheter system	Hansen Medical Inc	US	3	Cardiac catheter insertion
FreeHand v1.2	Freehand 2010 Ltd.	UK	3	Laparoscopy
Invendoscopy E200 system	Invendo Medical GmbH	Denmark	180°	Colonoscopy
Flex robotic system	Medrobotics Corp	US	180°	Transoral: oropharynx, hypopharynx, larynx
Senhance	TransEnterix/US	US	7	Gynecologic Laparoscopy
ARES	Auris Surgical Robotics	US	6	Bronchoscopy
ViaCath System	BIOTRONIC	Germany	9	Ureteroscopy Endovascular procedures
ROSA Spine	Zimmer Biomet Medtech	France	6	Brain and spinal surgeries
ROSA Brain	Zimmer Biomet Medtech	France	6	Minimally invasive brain surgeries
SurgiScope	ISIS Robotics	France	7	Neurological MIS and endoscopy
AURORA Surgiscope System	NDI	Canada	–	Maxillofacial surgery, ENT surgery, spine surgery, neurosurgery

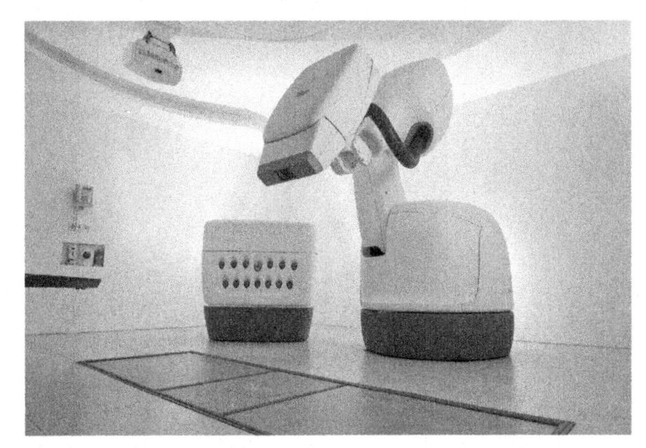

Fig. 20 A CyberKnife M6 System at the European CyberKnife Center Munich-Großhadern, Germany, by Textefuermedizin, used under CC BY-SA 4.0, https://en.wikipedia.org/wiki/Cyberknife#/media/File:Cyberknife_M6_und_MLC.jpg.

Fig. 21 Sensei X by Hansen Medical (Beasley, 2012), used under CC BY 3.0.

6 Rehabilitation robotics and assistive technologies

6.1 Motivations

According to the United Nations report on World Population Aging (United Nations and Department of Economic and Social Affiars Population Division, 2017), older persons aged 60 years or older numbered 962 million in 2017. This number is expected to double again by 2050. As the average

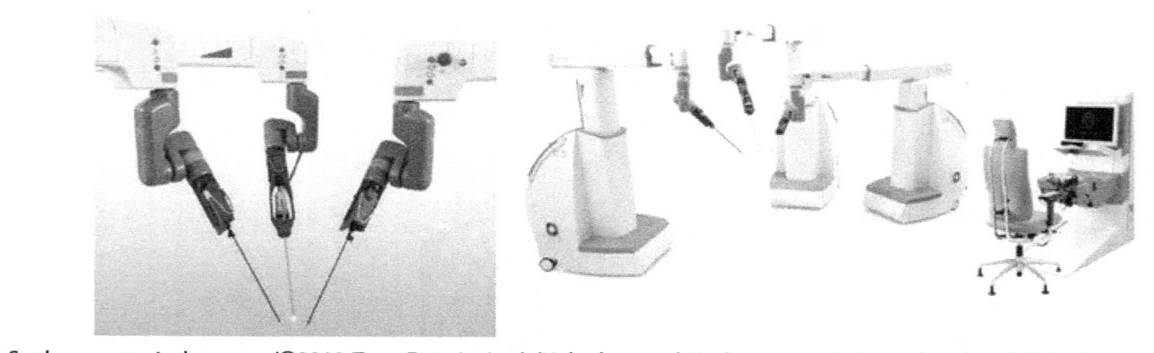

Fig. 22 Senhance surgical system (©2018 TransEnterix, Inc.) (Nakadate and Hashizume, 2018), used under CC BY 3.0.

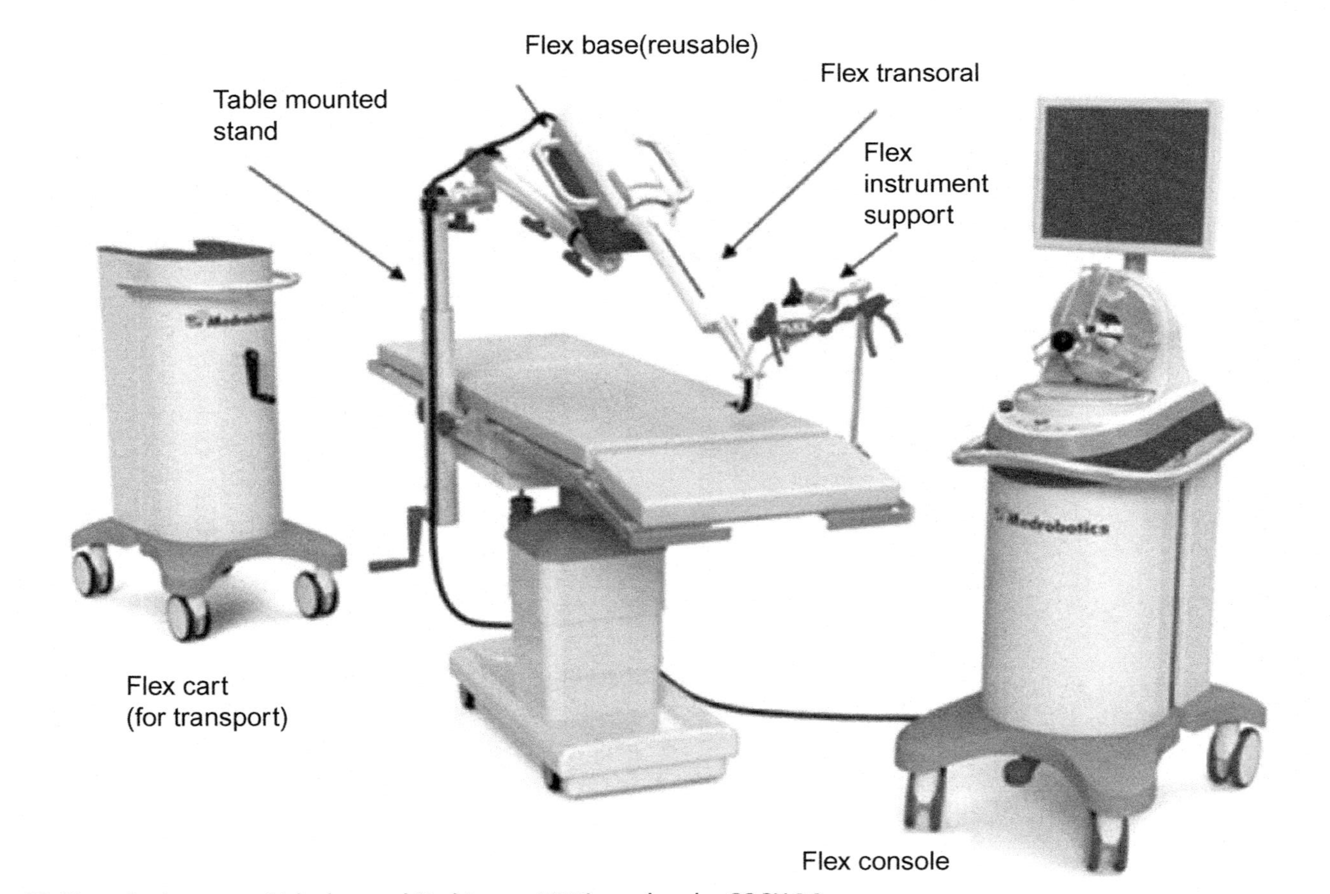

Fig. 23 Flex robotic system (Nakadate and Hashizume, 2018), used under CC BY 3.0.

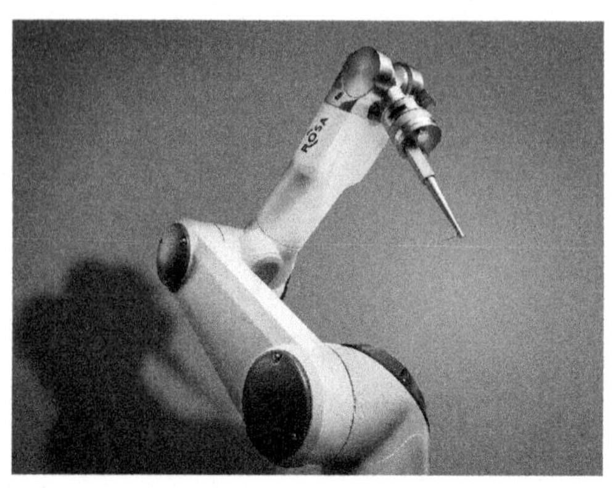

Fig. 24 ROSA ONE's arm, by LucileBssg, used under CC BY-SA 4.0, https://en.wikipedia.org/wiki/ROSA#/media/File:ROSA_One%C2%AE_Robot_.jpg.

age of populations continues to rise, policies should be implemented to address the needs and interests of older persons, including healthcare. Furthermore, as neurological conditions, especially stroke, are a major cause of disability among older people (Krebs and Volpe, 2013), this situation creates an urgent need for new approaches to improve the effectiveness and efficiency of rehabilitation. It also creates an unprecedented opportunity to deploy technologies such as robotics to assist in the recovery process (Krebs and Volpe, 2013). Movement disorders significantly reduce a patient's quality of life and limit the independence of affected subjects. Fortunately, there are various approaches to restore functionality like ortheses, functional electrical stimulation, and physical therapy (Maciejasz et al., 2014). Rehabilitation and assistive robotics have the potential to change older people lives, improving their recovery and/or supporting them to perform everyday tasks (Garcia-Aracil et al., 2017). Some robotic devices like wheelchairs not only provide support to stakeholders but also to caregivers (Shishehgar et al., 2018).

Among assistive devices, the most popular are rehabilitation robotic devices (Colombo and Sanguineti, 2018), exoskeletons (Aliman et al., 2017; Bai et al., 2018), and prosthetic (Tucker et al., 2015) and orthotic devices (Dzahir and Yamamoto, 2014). Rehabilitation robotics are machines used in order to actively assist and/or resist the motion of the stroke patient, increase efficiency, and reduce cost by decreasing the amount of one-on-one time that a therapist must spend with a patient. Therapeutic robots can also continuously collect data that can be used to quantitatively measure the patient's progress throughout the recovery process, enabling therapists to optimize treatment techniques. Robotic orthotics are mechanisms used to assist or support weak or ineffective joints, muscles, or limbs. Orthotics often take the form of an exoskeleton (Najarian et al., 2011). An exoskeleton is a powered anthropomorphic suit that is worn by the patient having links and joints that correspond to those of the human and actuators that assist the patient with moving his or her limb or lifting external loads. A prosthetic is a robotic device that substitutes a missing part of the human body providing mobility or manipulation abilities when a limb is lost.

6.2 Literature review

The field of rehabilitation and assistive robotics is distinguished by a comprehensive bibliography marked a number of books (Pons, 2008; Xie, 2016; Bai et al., 2018) and a large number of survey papers (Krebs et al., 1998; Tejima, 2001; Dellon and Matsuoka, 2007; Pignolo, 2009; Pons, 2010; Díaz et al., 2011; Heo et al., 2012; Poli et al., 2013; Masiero et al., 2014; Dzahir and Yamamoto, 2014; Maciejasz et al., 2014; Yang et al., 2017b; Qian and Bi, 2015; Gassert and Dietz, 2018; Shishehgar et al., 2018; Aggogeri et al., 2019; Shi et al., 2019; Pamungkas et al., 2019).

A number of book chapters (Van der Loos and Reinkensmeyer, 2008; Zhang et al., 2017; Krebs and Volpe, 2013; Van der Loos et al., 2016) is also available.

Fig. 25 shows the number of academic papers written in English found using a typical search engine in Scopus using the keywords "rehabilitation robot." The number of papers illustrates an increase since 2006.

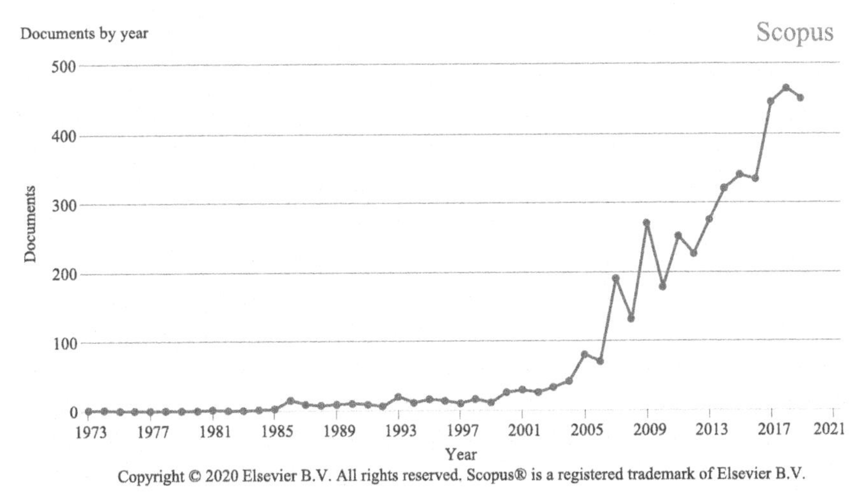

Fig. 25 Data on rehabilitation robots from Scopus in February 2020. Query: KEY (rehabilitation AND robot) AND (LIMIT-TO (LANGUAGE, "English")).

6.3 A brief history

In 1910, Theodor Büdingen proposed a patent for the first rehabilitation machine driven by an electric motor to guide and maintain stepping movements in patients with heart disease (Gassert and Dietz, 2018). In the 1930s, Richard Scherb developed a cable-driven apparatus to move joints for orthopedics therapy supporting multiple interaction modes (passive, active- and active-resisted movements). The first powered exoskeletons were proposed in the 1970s (Gassert and Dietz, 2018). The first robot deployed in clinical trials delivering rehabilitation therapy was the MIT-Manus developed in 1989, which uses a 6-DOF PUMA 560 robot to interact with the impaired patient. This system can operate in three unilateral modes and one bilateral mode (Krebs and Volpe, 2013). The unilateral modes are passive, in which the patient remains passive while the robot moves the arm along a preprogrammed path; active-assisted, in which the patient initiates movement and the robot assists and guides the motion along the desired path; and active-constrained, in which the robot resists motion along the path and provides a restoring force in all other directions. The Mirror Image Motion Enabler (MIME) was the first system to explore bilateral training in robotic applications for stroke rehabilitation (Burgar et al., 2000).

Development of rehabilitation robots for the lower extremitis began in 1994, with the design of the Lokomat (Gassert and Dietz, 2018). The most widely commercially available robotic systems for assisting individuals with disabilities are the Handy 1 (Rehab Robotics Limited, UK) developed in 1987 by Mike Topping, which enables people with little or no hand function to independently complete everyday functions such as eating, drinking, washing, shaving, and teeth cleaning (Topping and Smith, 1998); the wheelchair-mounted MANUS (Exact Dynamics, Netherlands) with two-fingered gripper (Driessen et al., 2001); and the wheelchair-mounted Raptor (Applied Resources Corporation, U.S.A.) that allows individuals with disabilities to feed themselves and reach objects on the floor, on a table, or above their heads (Speich and Rosen, 2008).

6.4 Classification and related devices

Rehabilitation robots can be classified and studied using various taxonomies. The most natural classification is based on the anatomy of the two extremities of the human body (Hesse et al., 2003; Pitkin, 2015) by considering upper extremity rehabilitation devices (Butcher and Meals, 2002; Balasubramanian et al., 2010; Brochard et al., 2010; Gopura et al., 2011;

Lo and Xie, 2012; Rodríguez-Prunotto et al., 2014; Wang et al., 2014; Jobbágy et al., 2014; Pitkin, 2015; Platz et al., 2017; Wang et al., 2017; Hakim et al., 2017; Ergasheva, 2017; Nadas et al., 2017; Rupal et al., 2017; Moreno et al., 2018; Huo et al., 2016) and lower body extremity rehabilitation devices (Dollar and Herr, 2008; Mohammed and Amirat, 2009; Mohammed et al., 2012; Carpino et al., 2013; Li et al., 2013; Tucker et al., 2015; Huo et al., 2016; Chen et al., 2016; Rupal et al., 2017; Yang et al., 2017b; Pamungkas et al., 2019). In reality, the development of bipedal robots has promoted the development of lower extremity rehabilitation devices as bipedal robots and related technologies, such as stability control and motion planning, can be applied in human lower limb rehabilitation training and assisted walking (Yang et al., 2017b). The model-based gait planning method mainly includes the multi-link models (see (Aloulou and Boubaker, 2010, 2011, 2012, 2013a,b, 2015, 2016 and related references) and the inverted pendulum model (Benrejeb and Boubaker, 2012; Boubaker, 2012, 2013, 2017; Boubaker and Iriarte, 2017).

Some research papers focus on hand rehabilitation devices (Butcher and Meals, 2002; Balasubramanian et al., 2010; Heo et al., 2012; Sale et al., 2012; Zuo and Olson, 2014; Meng et al., 2017; Yue et al., 2017; Chu and Patterson, 2018; Aggogeri et al., 2019).

In Brewer et al. (2007), the authors classify rehabilitation robots by rehabilitation applications: gross motor movements (e.g., reaching), bilateral training, fine motor movements (e.g., grasping), and motor or visual feedback distortion to induce after-effects, telerehabilitation, and assessment.

In Gassert and Dietz (2018), rehabilitation robots are categorized into grounded exoskeletons, grounded end-effectors devices, and wearable exoskeletons (See Fig. 26). The design approaches depend on whether the extremity of the limb is trained or not. Grounded end-effectors devices will typically achieve higher motion dynamics and allow the rendering of a wider range of impedances than exoskeleton devices with a serial kinematic structure, where proximal joints need to move distal joints (Gassert and Dietz, 2018).

A coarser classification than that of the previous one is proposed by Aggogeri et al. (2019), who claimed that a primary categorization of rehabilitation robotic technologies is based on the design concepts of the device: end-effectors or exoskeleton. An exoskeleton is a wearable robot attached to the user's limbs, in order to enhance the user's movements. It focuses on the anatomy of the subject's hand following the limb segments; each degree of freedom is aligned with the corresponding human joint. Fig. 27 illustrates a

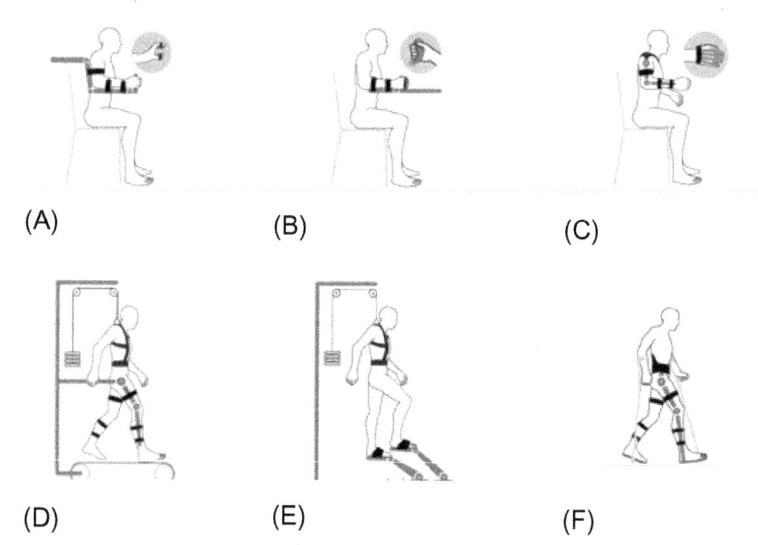

Fig. 26 Classifications of rehabilitation robots. (1) A-B-C: upper body rehabilitation devices, D-E-F: lower body rehabilitation devices; (2) A-D: grounded rehabilitation devices, B-E: end-effectors devices and C-F: wearable exoskeletons. *Modified from Gassert, R., Dietz, V., 2018. Rehabilitation robots for the treatment of sensorimotor deficits: a neurophysiological perspective, J. NeuroEng. Rehabil. doi: 10.1186/s12984-018-0383-x.*

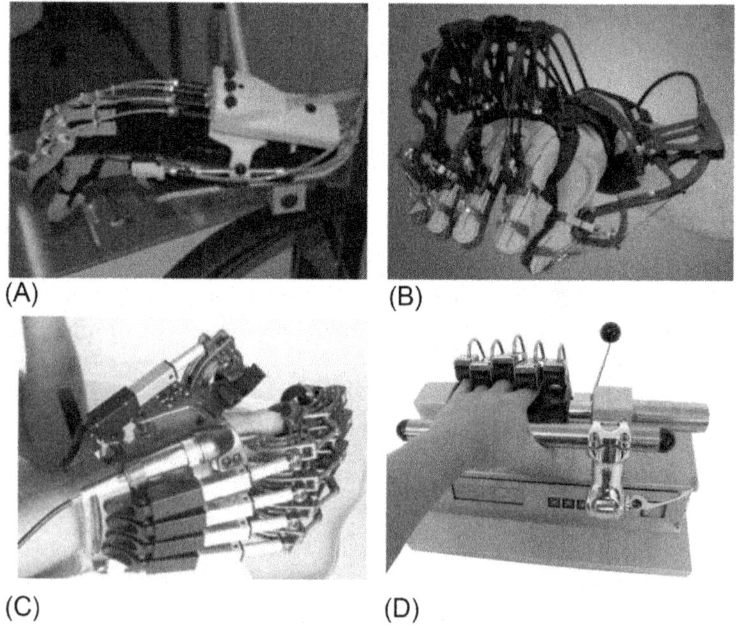

Fig. 27 Examples of hand exoskeletons: (A) Gloreha (Borboni et al., 2016), (B) CyberGrasp, (C) Hand of Hope, and (D) Reha-Digit (Aggogeri et al., 2019), used under CC BY 4.0.

number of examples. The end-effectors robots are more flexible than exoskeleton devices in fitting the different sizes of hands, reducing the setup time, and increasing the usability for new patients. They suffer from the control of distal joints and haptic aspects of object manipulation. Nevertheless their design is complex and a deep investigation of hand biomechanics and physical human-robot interaction is required. Fig. 28 illustrates a number of examples for upper lim rehabilitation devices. Exoskeletons, on the contrary, need to be adjusted to fit different sizes of hand due to their complexity, and as a consequence, the geometric parameters also need to be updated in the controller to guarantee the accuracy of the kinematic models. Fig. 29 illustrates an FDA-approved robotic exoskeleton (EksoGT, Ekso Bionics, Inc. Richmond, CA, USA).

Table 2 summarizes popular commercial rehabilitation robotic devices, whereas Table 3 presents commercial Exoskeleton devices. Fig. 30 shows

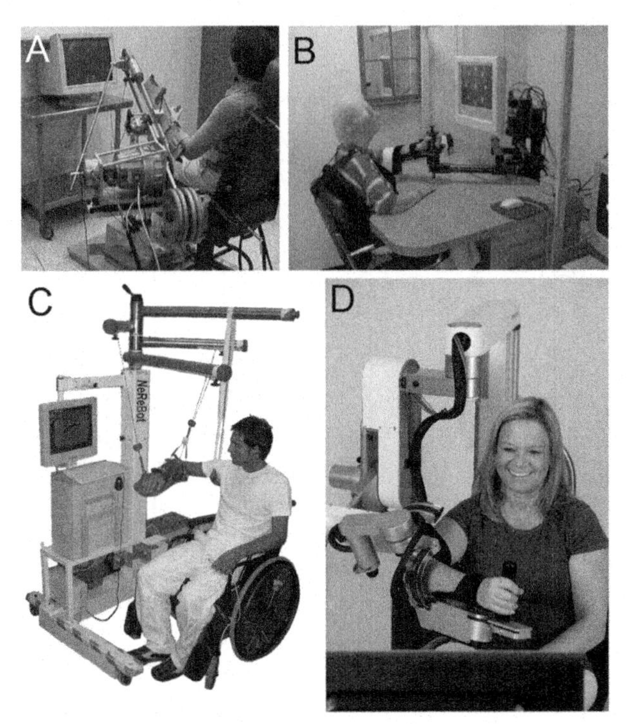

Fig. 28 Examples of rehabilitation devices for upper limb rehabilitation. (A) ARM Guide; (B) InMotion ARM; (C) NeReBot; (D) ArmeoPower (Maciejasz et al., 2014); licensee BioMed Central Ltd.

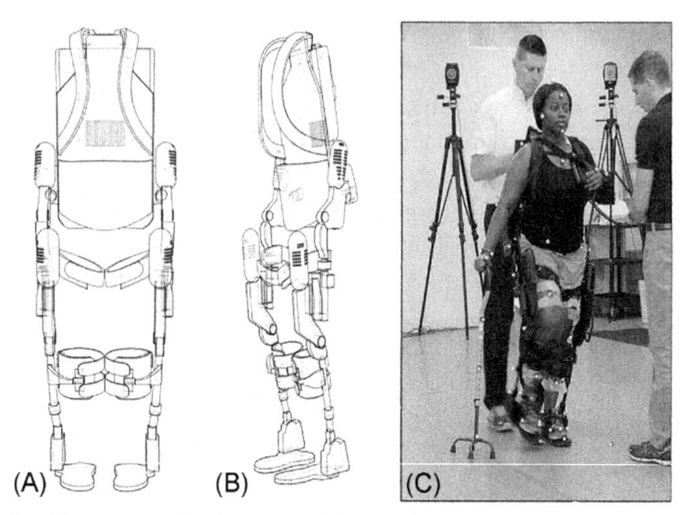

Fig. 29 The FDA-approved robotic exoskeleton EksoGT (Ekso Bionics, Inc. Richmond, CA, USA). (A) Frontal view of the EksoGT, (B) Oblique view of EksoGT, (C) One representative participant in the commercially available RE (Androwis et al., 2018), used under CC BY 4.0.

Table 2 Commercial rehabilitation robotic devices.

Name	Company	Country	DOF	Training furthers
Bi-Manu-Track	Reha-Stim Medtec	Switzerland	1	Active hand rehabilitation (wrist and forearm)
Reha-Digit	Reha-Stim Medtec	Switzerland	4	Passive finger rehabilitation
Amadeo	Ectron	UK	5	Passive finger rehabilitation
LOKOMAT	Médimex	France	–	Lower extremity rehabilitation
Armeo Power	Hocoma	Switzerland	6	Lower extremity rehabilitation
Welwalk-1000	Toyota Motor Corporation	Japan	–	Support for individuals with lower limb paralysis to assist in knee bending and stretching movements

Table 3 Commercial exoskeleton devices.

Name	Company	Country	DOF	Training furthers
CyberGrasp	CyberGlove Systems Inc.	US	4	Extension and flexion of fingers
Hand of Hope	Rehab-Robotics Company	China	5	Move each finger separately
Gloreha-Hand	Gloreha	Italy	5	Hand continuous passive motion training
Hand Mentor	Motus Nova	US	1	Hand passive/assistive motion training
ReoGo	Motorika USA Inc.	US	4	Shoulder, elbow and also wrist training or fingers if special handle is used.
Ekso	Ekso Bionics	US	6	Medical-Paraplegics
Rex	Rex Bionics Pty Ltd.	Australia	–	Medical-Paraplegics
Phoenix	SuitX (formerly US Bionics)	US	–	Medical-Compensation
Walking Assist	Honda	Japan	2	Walking Assist
Roki	Roki Robotics	Mexico	4	Walking Assist
HAL	Cyberdyne	Japan	–	Full-body exoskeleton for the arms, legs, and torso controlled by human minds.

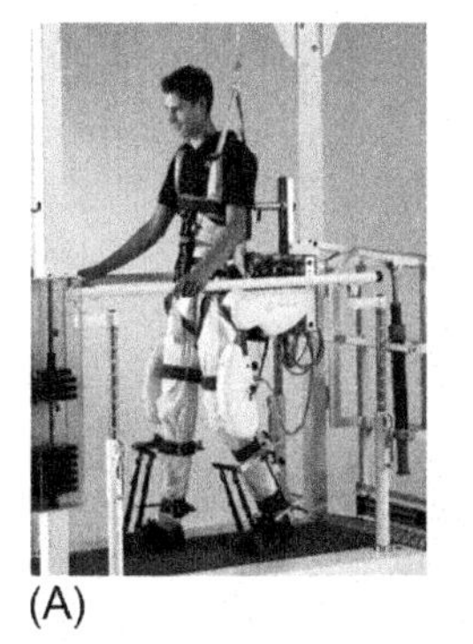
(A)

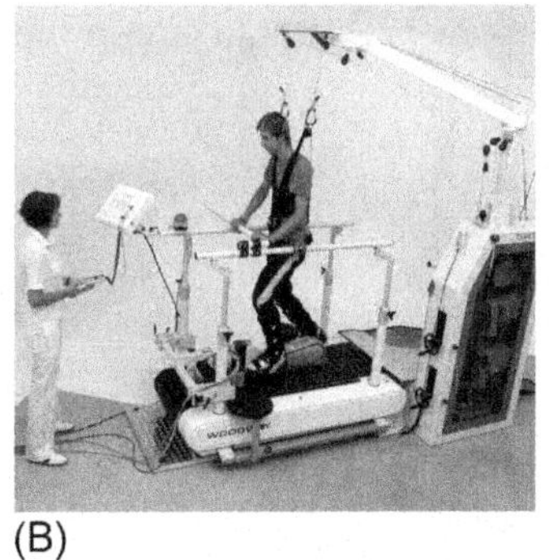
(B)

(C)

Fig. 30 Examples of treadmill gait trainers for lower-limb commercial orthotic systems (A) LOKOMAT; (B) LokoHelp and (C) ReoAmbulator (Dzahir and Yamamoto, 2014), used under CC BY license.

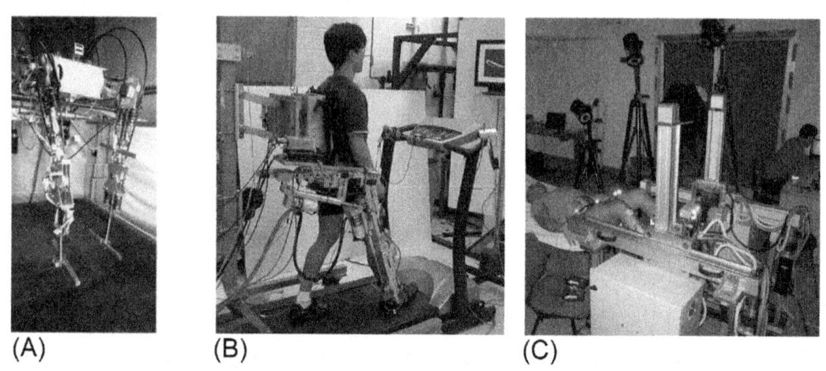

Fig. 31 Examples of lower-limb exoskeleton systems (A) LOPES; (B) active leg exoskeleton (ALEX); and (C) NEUROBike (Dzahir and Yamamoto, 2014), used under CC BY license.

three commercial treadmill walking trainers for lower limb orthopedic systems while Fig. 31 presents three examples of exoskeletons.

Robotic assistive technologies can be oriented for rehabilitation of patients with special needs, such as children and the eldery. Indeed, in addition to their educational roles, robots have been also introduced to help children with autism spectrum disorder, a neurodevelopment disorder characterized by impairments in social-communication abilities and restricted and repetitive behaviors (see (Zhang et al., 2019) and related references). Fig. 32 shows a scenario of rehabilitation assistance for a child with autism.

In the same framework, in Shishehgar et al. (2018), the authors categorize robots for addressing aged care problems in nine groups: (1) companion robots,

Fig. 32 Social robots for children rehabilitation with autism spectrum disorder (Zhang et al., 2019), used under CC BY 4.0.

(2) manipulators service, (3) telepresence robots, (4) rehabilitation robots, (5) health monitoring robots, (6) reminders, (7) entertainment robots, (8) domestic robots, and (9) fall detection/prevention robots. All these robotic types have been applied to eight key problem areas in aged care, namely social isolation, dependent living, physical or cognitive impairment, mobility problems, poor health monitoring, lack of recreation, memory problems, and fall problems. Fig. 33 shows a collection of companion robots used to solve previous healthcare problems as they have a social effect, such as activating communication. Most of them are designed in the shape of an animal (cat, dog, rabbit, seal, etc.), but some are humanoid robots. They have the ability to listen, respond to speech, talk in a few circumstances, recognize touch, and detect sound and light. Companion robots are widely used in social isolation problems, but not so much in the physical/mental problem area. In a few cases they

Fig. 33 Pictures of companion robots. *Reproduced with permission from Shishehgar, M., Kerr, D., Blake, J., 2018. A systematic review of research into how robotic technology can help older people, Smart Health doi: 10.1016/j.smhl.2018.03.002, Copyright Elsevier, 2020.*

are helpful in resolving a lack of recreation. Telepresence robots are basically used to facilitate two-way communication with older adults; through their visual content they can offer a richer communication experience than voice alone. One of the most important aspects of communication is a sense of connectedness, which helps individuals not to feel lonely. These robots could also have benefits for people with cognitive impairment problems and help them to stay in their homes. Manipulator service robots are robots with an arm-shaped attachment(s) to handle and carry objects upon request for the purpose of supporting the basic tasks of independent living.

7 Robots in medical training as body-part simulators

Motivated by the need of standardized medical procedures and in order to reduce animal and patient tests, several commercial realistic body-part simulators have been developed for the training of medical students. Detailed history of the development of mannequins and some screen simulators are given in Cooper (2004), Cooper and Taqueti (2008), and Chiniara and Crelinsten (2019). The combination of soft materials, contractile actuators, and flexible sensors have opened many opportunities for the generation of body-part simulators for the investigation of physiology and to generate realistic healthcare scenarios (Cianchetti et al., 2018).

Harvey, currently sold by the Laerdal Corporation (https://www.laerdal.com/doc/172/Next-Generation-Harvey-The-Cardiopulmonary-Patient-Simulator), was one of the earliest medical simulators available for training of healthcare professionals (Cooper and Taqueti, 2008). Harvey was created in 1968 by Dr. Michael Gordon at the University of Miami. Next Generation Harvey provides cardiopulmonary training for all healthcare providers. This full-size manikin realistically simulates nearly any cardiac disease at the touch of a button. It includes 50 patient cases, 10 comprehensive standardized patient curriculums, and enhanced physical features.

Among birthing simulators reproducing a mother in labor (Kim et al., 2005; Leigh and Hurst, 2008; Smith et al., 2014), Victoria, produced by Gaumard Scientific (https://www.gaumard.com/), simulates events seen during a full-term or premature delivery. These include any complications

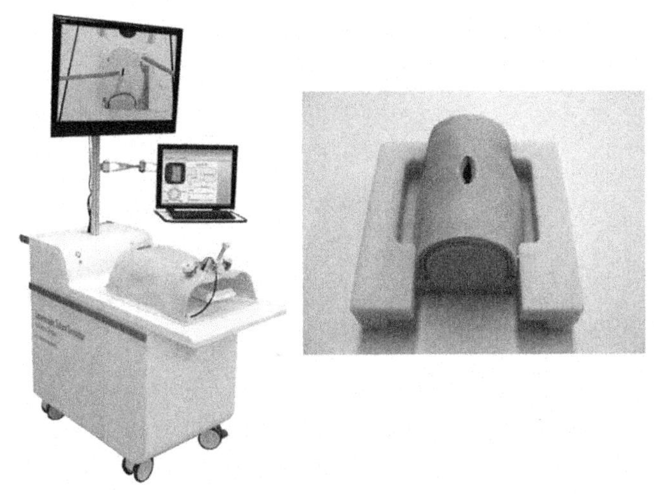

Fig. 34 Laparoscopic training simulator (Nakadate and Hashizume, 2018). Used under CC BY 3.0C.

like breech presentation, shoulder dystocia, maternal bleeding, or umbilical cord prolapse.

Nowadays, there are also some surgical simulators using computer graphics, rubber phantom, and harvested animal organs used for training young, less experienced surgeons. Fig. 34 shows an example. The phantom mimicks the small intestine. The task is anastomosis of the defect on the intestine by three interrupted sutures. The result is evaluated by five category skills: volume of air pressure leakage, number of full-thickness sutures, suture tension, wound area, and performance time (Takeoka et al., 2017).

Other types of simulators are also available, like the Advanced VR-based Cardiac Life Support (ACLS) simulator (Khanal et al., 2014) and bronchoscopy simulators (Baker et al., 2016).

Another type of simulators for lung physiology and mechanical ventilation of intensive care patients are also commercially available. A typical example is shown in Fig. 25, which describes a baby lung simulator produced by the Swiss Company neosim (https://www.neosim.ch/Welcome/). Robotic simulation-based training in neonatal resuscitation offers many benefits not inherent in traditional paradigms of medical education (Halamek et al., 2000) (Fig. 35).

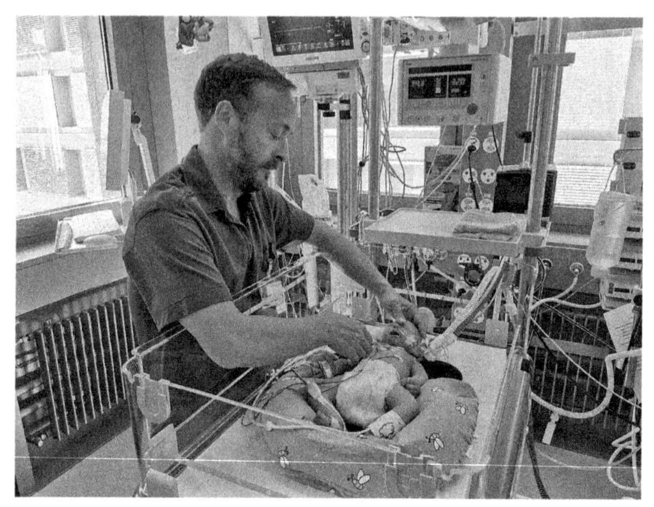

Fig. 35 Ventilation training on a baby lung simulator (LuSi, neosim AG) at the Cantonal Hospital in Chur, Switzerland, by Josef X Brunner, used under CC BY-SA 4.0, https://de.wikipedia.org/wiki/Medizinische_Simulation#/media/Datei:Beatmungstraining_Neonatologie_KSGR_Chur,_Schweiz.jpg.

8 Conclusion

Medical robotics has reached a state of maturity with which it will be possible to improve our healthcare system and our society. In the field of surgery, the operating room is currently a sophisticated mix of micro-robots, telepresence workstations, computer-integrated surgery, and advanced imaging systems. Robots have improved the dexterity of surgeons, offering the least invasive techniques and widening the reach to target organs. They have also improved hospital efficiency in moving materials and drugs. Many more are yet to be invented and will provide a value proposition that has not been thought of yet. The continued growth of this industry is amplified by the fact that our society is fastly aging and will continue to want the best quality care we can afford. Robotic technology has successfully produced valuable tools for therapy, rehabilitaiton, children and elderly care settings, and medical training. Medical robotics, which brings forth a new set of capabilities, will surely continue to find an expanding role in the healthcare area. The need and appetite for innovation will undoubtedly continue.

References

Abbott, D.J., et al., 2007. Design of an endoluminal NOTES robotic system. In: 2007 IEEE/RSJ International Conference on Intelligent Robots and Systems. IEEE, pp. 410–416. https://doi.org/10.1109/IROS.2007.4399536.

Abbou, C.C., et al., 2001. Laparoscopic radical prostatectomy with a remote controlled robot. J. Urol. 165, 1964–1966.

Aggogeri, F., Mikolajczyk, T., O'Kane, J., 2019. Robotics for rehabilitation of hand movement in stroke survivors. Adv. Mech. Eng. https://doi.org/10.1177/1687814019841921.

Aliman, N., Ramli, R., Haris, S.M., 2017. Design and development of lower limb exoskeletons: a survey. Robot. Auton. Syst. 95, 102–116. https://doi.org/10.1016/j.robot.2017.05.013.

Aloulou, A., Boubaker, O., 2010. Control of a step walking combined to arms swinging for a three dimensional humanoid prototype. J. Comput. Sci. 6 (8), 886–895. https://doi.org/10.3844/jcssp.2010.886.895.

Aloulou, A., Boubaker, O., 2011. Minimum jerk-based control for a three dimensional bipedal robot. In: Lecture Notes in Computer Science (including subseries Lecture Notes in Artificial Intelligence and Lecture Notes in Bioinformatics). pp. 251–262. https://doi.org/10.1007/978-3-642-25489-5_25.

Aloulou, A., Boubaker, O., 2012. A relevant reduction method for dynamic modeling of a seven-linked humanoid robot in the three-dimensional space. Procedia Eng. 41, 1277–1284. https://doi.org/10.1016/j.proeng.2012.07.311.

Aloulou, A., Boubaker, O., 2013a. Model validation of a humanoid robot via standard scenarios. In: 14th International Conference on Sciences and Techniques of Automatic Control & Computer Engineering—STA'2013. IEEE, pp. 288–293. https://doi.org/10.1109/STA.2013.6783145.

Aloulou, A., Boubaker, O., 2013b. On the dynamic modeling of an upper-body humanoid robot in the three-dimensional space. In: 10th International Multi-Conferences on Systems, Signals & Devices 2013 (SSD13). IEEE, pp. 1–6. https://doi.org/10.1109/SSD.2013.6564097.

Aloulou, A., Boubaker, O., 2015. A minimum jerk-impedance controller for planning stable and safe walking patterns of biped robots. In: Mechanisms and Machine Science. pp. 385–415. https://doi.org/10.1007/978-3-319-14705-5_13.

Aloulou, A., Boubaker, O., 2016. An optimal jerk-stiffness controller for gait pattern generation in rough terrain. ROBOMECH J. 3 (1), 13. https://doi.org/10.1186/s40648-016-0052-4.

Amirkhani, S., et al., 2019. Fast terminal sliding mode tracking control of nonlinear uncertain mass–spring system with experimental verifications. Int. J. Adv. Robot. Syst. 16(1), 172988141982817. https://doi.org/10.1177/1729881419828176.

Androwis, G.J., et al., 2018. Electromyography assessment during gait in a robotic exoskeleton for acute stroke. Front. Neurol. 9. https://doi.org/10.3389/fneur.2018.00630.

Bai, S., Virk, G.S., Sugar, T., 2018. Wearable Exoskeleton Systems: Design, Control and Applications. In: Bai, S., Virk, G.S., Sugar, T.G. (Eds.), Institution of Engineering and Technology. https://doi.org/10.1049/PBCE108E.

Baker, P.A., et al., 2016. 'Evaluating the ORSIM ® simulator for assessment of anaesthetists' skills in flexible bronchoscopy: aspects of validity and reliability. Br. J. Anaesth. 117, i87–i91. https://doi.org/10.1093/bja/aew059.

Balasubramanian, S., Klein, J., Burdet, E., 2010. Robot-assisted rehabilitation of hand function. Curr. Opin. Neurol. 23 (6), 661–670. https://doi.org/10.1097/WCO.0b013e32833e99a4.

Barrett, A.R.W., et al., 2007. Computer-assisted hip resurfacing surgery using the Acrobot® navigation system. Proc. Inst. Mech. Eng. H J. Eng. Med. https://doi.org/10.1243/09544119JEIM283.

Beasley, R.A., 2012. Medical robots: current systems and research directions. J. Robot. 2012, 1–14. https://doi.org/10.1155/2012/401613.

Benrejeb, W., Boubaker, O., 2012. FPGA modelling and real-time embedded control design via labview software: application for swinging-up a pendulum. Int. J. Smart Sens. Intell. Syst. 5 (3), 576–591. https://doi.org/10.21307/ijssis-2017-496.

Berthet-Rayne, P., et al., 2018. The i2Snake robotic platform for endoscopic surgery. Ann. Biomed. Eng. 46 (10), 1663–1675. https://doi.org/10.1007/s10439-018-2066-y.

Bodner, J., et al., 2005. The da Vinci robotic system for general surgical applications: a critical interim appraisal. Swiss Med. Wkly. doi: 2005/45/smw-11022.

Bogue, R., 2011. Robots in healthcare. Ind. Robot. 38 (3), 218–223. https://doi.org/10.1108/01439911111122699.

Borboni, A., Mor, M., Faglia, R., 2016. Gloreha-hand robotic rehabilitation: design, mechanical model, and experiments. J. Dyn. Syst. Trans. ASME. https://doi.org/10.1115/1.4033831.

Boubaker, O., 2012. The inverted pendulum: a fundamental benchmark in control theory and robotics. In: 2012 International Conference on Education and e-Learning Innovations, ICEELI 2012. https://doi.org/10.1109/ICEELI.2012.6360606.

Boubaker, O., 2013. The inverted pendulum benchmark in nonlinear control theory: a survey. Int. J. Adv. Robot. Syst. https://doi.org/10.5772/55058.

Boubaker, O., 2017. The inverted pendulum: history and survey of open and current problems in control theory and robotics. In: The Inverted Pendulum in Control Theory and Robotics: From Theory to New Innovations. https://doi.org/10.1049/pbce111e_ch1.

Boubaker, O., Iriarte, R. (Eds.), (2017). The Inverted Pendulum in Control Theory and Robotics: From theory to new innovations. Institution of Engineering and Technology. https://doi.org/10.1049/PBCE111E.

Brewer, B.R., McDowell, S.K., Worthen-Chaudhari, L.C., 2007. Poststroke upper extremity rehabilitation: a review of robotic systems and clinical results. Top. Stroke Rehabil. 14 (6), 22–44. https://doi.org/10.1310/tsr1406-22.

Broadbent, E., Stafford, R., MacDonald, B., 2009. Acceptance of healthcare robots for the older population: review and future directions. Int. J. Soc. Robot. 1 (4), 319–330. https://doi.org/10.1007/s12369-009-0030-6.

Brochard, S., et al., 2010. What's new in new technologies for upper extremity rehabilitation? Curr. Opin. Neurol. 23 (6), 683–687. https://doi.org/10.1097/WCO.0b013e32833f61ce.

Burgar, C.G., et al., 2000. Development of robots for rehabilitation therapy: the Palo Alto VA/Stanford experience. J. Rehabil. Res. Dev. 37 (6), 663–673.

Burgner-Kahrs, J., Rucker, D.C., Choset, H., 2015. Continuum robots for medical applications: a survey. IEEE Trans. Robot. 31 (6), 1261–1280. https://doi.org/10.1109/TRO.2015.2489500.

Butcher, C., Meals, R.A., 2002. Rehabilitation of the hand and upper extremity. J. Hand. Surg. [Am.] 27 (5), 919. https://doi.org/10.1053/jhsu.2002.35304.

Camarillo, D.B., Krummel, T.M., Salisbury, J.K., 2004. Robotic technology in surgery: past, present, and future. Am. J. Surg. 188 (4), 2–15. https://doi.org/10.1016/j.amjsurg.2004.08.025.

Carpino, G., et al., 2013. Lower limb wearable robots for physiological gait restoration: state of the art and motivations. Medic 21 (2), 72–80.

Caversaccio, M., et al., 2008. Augmented reality endoscopic system (ARES): preliminary results. Rhinology 46, 156–158.

Chen, G., et al., 2013a. A review of lower extremity assistive robotic exoskeletons in rehabilitation therapy. Crit. Rev. Biomed. Eng. 41 (4–5), 343–363. https://doi.org/10.1615/CritRevBiomedEng.2014010453.
Chen, T.L., et al., 2013b. Robots for humanity: using assistive robotics to empower people with disabilities. IEEE Robot. Autom. Mag. 20 (1), 30–39. https://doi.org/10.1109/MRA.2012.2229950.
Chen, B., et al., 2016. Recent developments and challenges of lower extremity exoskeletons. J. Orthop. Transl. https://doi.org/10.1016/j.jot.2015.09.007.
Chikhaoui, M.T., Burgner-Kahrs, J., 2018. Control of continuum robots for medical applications: state of the art. In: International Conference on New Actuators.
Chiniara, G., Crelinsten, L., 2019. A brief history of clinical simulation: how did we get here? In: Clinical Simulation. Elsevier, pp. 3–16. https://doi.org/10.1016/B978-0-12-815657-5.00001-2.
Chu, C.Y., Patterson, R.M., 2018. Soft robotic devices for hand rehabilitation and assistance: a narrative review. J. NeuroEng. Rehabil. https://doi.org/10.1186/s12984-018-0350-6.
Cianchetti, M., et al., 2018. Biomedical applications of soft robotics. Nat. Rev. Mater. 3 (6), 143–153. https://doi.org/10.1038/s41578-018-0022-y.
Cleary, K., Nguyen, C., 2001. State of the art in surgical robotics: clinical applications and technology challenges. Comput. Aided Surg. 6 (6), 312–328. https://doi.org/10.3109/10929080109146301.
Clotet, E., et al., 2016. Assistant personal robot (APR): conception and application of a teleoperated assisted living robot. Sensors 16 (5), 610. https://doi.org/10.3390/s16050610.
Colombo, R., Sanguineti, V., 2018. Rehabilitation robotics: technology and applications. In: Rehabilitation Robotics. https://doi.org/10.1016/b978-0-12-811995-2.09991-4.
Cooper, J.B., 2004. A brief history of the development of mannequin simulators for clinical education and training. Qual. Saf. Health Care. https://doi.org/10.1136/qhc.13.suppl_1.i11.
Cooper, J.B., Taqueti, V.R., 2008. A brief history of the development of mannequin simulators for clinical education and training. Postgrad. Med. J. 84 (997), 563–570. https://doi.org/10.1136/qshc.2004.009886.
Cortesão, R., et al., 2006. Haptic control design for robotic-assisted minimally invasive surgery. In: IEEE International Conference on Intelligent Robots and Systems. https://doi.org/10.1109/IROS.2006.282168.
Costa, A., Novais, P., Julian, V., 2018. A survey of cognitive assistants. In: Intelligent Systems Reference Library. https://doi.org/10.1007/978-3-319-62530-0_1.
Daneshmand, M., et al., 2017. Medical robots with potential applications in participatory and opportunistic remote sensing: a review. Robot. Auton. Syst. https://doi.org/10.1016/j.robot.2017.06.009.
Dario, P., et al., 1996. Robotics for medical applications. IEEE Robot. Autom. Mag. 3 (3), 44–56. https://doi.org/10.1109/100.540149.
Davies, B.L., 1996. A discussion of safety issues for medical robots. In: Computer-Integrated Surgery: Technology and Clinical Applications. p. 756.
Davies, B., 2015. Robotic surgery—a personal view of the past, present and future. Int. J. Adv. Robot. Syst. 12 (5), 54. https://doi.org/10.5772/60118.
De Benedictis, A., et al., 2017. Robot-assisted procedures in pediatric neurosurgery. Neurosurg. Focus. https://doi.org/10.3171/2017.2.FOCUS16579.
Dello Russo, A., et al., 2016. Analysis of catheter contact force during atrial fibrillation ablation using the robotic navigation system: results from a randomized study. J. Interv. Card. Electrophysiol. https://doi.org/10.1007/s10840-016-0102-0.
Dellon, B., Matsuoka, Y., 2007. Prosthetics, exoskeletons, and rehabilitation [grand challenges of robotics]. IEEE Robot. Autom. Mag. 14 (1), 30–34. https://doi.org/10.1109/MRA.2007.339622.

Díaz, I., Gil, J.J., Sánchez, E., 2011. Lower-limb robotic rehabilitation: literature review and challenges. J. Robot. https://doi.org/10.1155/2011/759764.
Díaz, C.E., et al., 2017. A research review on clinical needs, technical requirements, and normativity in the design of surgical robots. Int. J. Med. Robot. 13(4), e1801. https://doi.org/10.1002/rcs.1801.
Dogangil, G., Davies, B.L., Rodriguez y Baena, F., 2010. A review of medical robotics for minimally invasive soft tissue surgery. Proc. Inst. Mech. Eng. H J. Eng. Med. 224 (5), 653–679. https://doi.org/10.1243/09544119JEIM591.
Dollar, A.M., Herr, H., 2008. Lower extremity exoskeletons and active orthoses: challenges and state-of-the-art. IEEE Trans. Robot. 24 (1), 144–158. https://doi.org/10.1109/TRO.2008.915453.
Dombre, E., 2003. Introduction to Medical Robotics. Summer European University on Surgical Robotics, Montpellier. http://www.lirmm.fr/uee07/presentations/lecturers/Dombre.pdf.
Driessen, B.J.F., Evers, H.G., van Woerden, J.A., 2001. MANUS—a wheelchair-mounted rehabilitation robot. Proc. Inst. Mech. Eng. H J. Eng. Med. 215 (3), 285–290. https://doi.org/10.1243/0954411011535876.
Duchemin, G., et al., 2005. A hybrid position/force control approach for identification of deformation models of skin and underlying tissues. IEEE Trans. Biomed. Eng. 52 (2), 160–170. https://doi.org/10.1109/TBME.2004.840505.
Dzahir, M.A.M., Yamamoto, S.I., 2014. Recent trends in lower-limb robotic rehabilitation orthosis: control scheme and strategy for pneumatic muscle actuated gait trainers. Robotics. https://doi.org/10.3390/robotics3020120.
Enayati, N., De Momi, E., Ferrigno, G., 2016. Haptics in robot-assisted surgery: challenges and benefits. IEEE Rev. Biomed. Eng. 9, 49–65. https://doi.org/10.1109/RBME.2016.2538080.
Ergasheva, B.I., 2017. Lower limb exoskeletons: brief review. Sci. Tech. J. Inform. Technol. Mech. Opt. 17 (6), 1153–1158. https://doi.org/10.17586/2226-1494-2017-17-6-1153-1158.
Faria, B.M., Reis, L.P., Lau, N., 2014. A survey on intelligent wheelchair prototypes and simulators. In: Advances in Intelligent Systems and Computing. https://doi.org/10.1007/978-3-319-05951-8_52.
Faria, C., et al., 2015. Review of robotic technology for stereotactic neurosurgery. IEEE Rev. Biomed. Eng. 8, 125–137. https://doi.org/10.1109/RBME.2015.2428305.
Fei, B., et al., 2001. The safety issues of medical robotics. Reliab. Eng. Syst. Saf. 73 (2), 183–192. https://doi.org/10.1016/S0951-8320(01)00037-0.
Feil-Seifer, D., Mataric, M.J., 2005. Socially assistive robotics. In: 9th International Conference on Rehabilitation Robotics, 2005 (ICORR 2005). IEEE, pp. 465–468. https://doi.org/10.1109/ICORR.2005.1501143.
Ferreira, A.C., et al., 2015. Teleultrasound: historical perspective and clinical application. Int. J. Telemed. Appl. https://doi.org/10.1155/2015/306259.
Ferrigno, G., et al., 2011. Medical robotics. IEEE Pulse 2 (3), 55–61. https://doi.org/10.1109/MPUL.2011.941523.
Frisoli, A., 2018. Exoskeletons for upper limb rehabilitation. In: Rehabilitation Robotics. Elsevier, pp. 75–87. https://doi.org/10.1016/B978-0-12-811995-2.00006-0.
Garcia-Aracil, N., Casals, A., Garcia, E., 2017. Rehabilitation and assistive robotics. Adv. Mech. Eng. https://doi.org/10.1177/1687814017699338.
Gassert, R., Dietz, V., 2018. Rehabilitation robots for the treatment of sensorimotor deficits: a neurophysiological perspective. J. NeuroEng. Rehabil. https://doi.org/10.1186/s12984-018-0383-x.
George Thuruthel, T., et al., 2018. Control strategies for soft robotic manipulators: a survey. Soft Rob. 5 (2), 149–163. https://doi.org/10.1089/soro.2017.0007.

George, E.I., et al., 2018. Origins of robotic surgery: from skepticism to standard of care. J. Soc. Laparoend. Surg. https://doi.org/10.4293/JSLS.2018.00039.

Gifari, M.W., et al., 2019. A review on recent advances in soft surgical robots for endoscopic applications. Int. J. Med. Robot. 15(5). https://doi.org/10.1002/rcs.2010.

Gomes, P., 2011. Surgical robotics: reviewing the past, analysing the present, imagining the future. Robot. Comput. Integr. Manuf. 27 (2), 261–266. https://doi.org/10.1016/j.rcim.2010.06.009.

Gomes, P., 2012. Medical Robotics, Medical Robotics: Minimally Invasive Surgery. Woodhead Publishing Limited. https://doi.org/10.1533/9780857097392.

Gonzales, A.V., et al., 2001. TER: a system for robotic tele-echography. In: Lecture Notes in Computer Science (including Subseries Lecture Notes in Artificial Intelligence and Lecture Notes in Bioinformatics). https://doi.org/10.1007/3-540-45468-3_39.

Gopura, R.A.R.C., Kiguchi, K., Bandara, D.S.V., 2011. A brief review on upper extremity robotic exoskeleton systems. In: 2011 6th International Conference on Industrial and Information Systems. IEEE, pp. 346–351. https://doi.org/10.1109/ICIINFS.2011.6038092.

Gourdon, A., et al., 1999a. A tele-scanning robotic system using satellite communication. In: Proceedings European Medical & Biological Engineering Conference "EMBEC," Vienna.

Gourdon, A., et al., 1999b. New robotic mechanism for medical application. In: IEEE/ASME International Conference on Advanced Intelligent Mechatronics, AIM. https://doi.org/10.1109/aim.1999.803139.

Graf, B., Parlitz, C., Hägele, M., 2009. Robotic home assistant care-o-bot® 3 product vision and innovation platform. In: Lecture Notes in Computer Science (including Subseries Lecture Notes in Artificial Intelligence and Lecture Notes in Bioinformatics). https://doi.org/10.1007/978-3-642-02577-8_34.

Guo, S., et al., 2019. A novel robot-assisted endovascular catheterization system with haptic force feedback. IEEE Trans. Robot. 35 (3), 685–696. https://doi.org/10.1109/TRO.2019.2896763.

Haidegger, T., et al., 2009. Force sensing and force control for surgical robots. In: IFAC Proceedings Volumes (IFAC-PapersOnline). https://doi.org/10.3182/20090812-3-DK-2006.0035.

Hakim, R.M., Tunis, B.G., Ross, M.D., 2017. Rehabilitation robotics for the upper extremity: review with new directions for orthopaedic disorders. Disabil. Rehabil. Assist. Technol. 12 (8), 765–771. https://doi.org/10.1080/17483107.2016.1269211.

Halamek, L.P., et al., 2000. Time for a new paradigm in pediatric medical education: teaching neonatal resuscitation in a simulated delivery room environment. Pediatrics 106 (4), E45.

Heo, P., et al., 2012. Current hand exoskeleton technologies for rehabilitation and assistive engineering. Int. J. Precis. Eng. Manuf. 13 (5), 807–824. https://doi.org/10.1007/s12541-012-0107-2.

Hesse, S., et al., 2003. Upper and lower extremity robotic devices for rehabilitation and for studying motor control. Curr. Opin. Neurol. 16 (6), 705–710. https://doi.org/10.1097/00019052-200312000-00010.

Hillman, M.R., 2006. Assistive robotics. In: Wiley Encyclopedia of Biomedical Engineering. John Wiley & Sons, Inc., Hoboken, NJ, USA. https://doi.org/10.1002/9780471740360.ebs1029

Hockstein, N.G., et al., 2007. A history of robots: from science fiction to surgical robotics. J. Robot. Surg. https://doi.org/10.1007/s11701-007-0021-2.

Hogan, N., 1985a. Impedance control—an approach to manipulation. I—theory. II—implementation. III—applications. ASME Trans. J. Dyn. Syst. Meas. Control B 107, 304–313.

Hogan, N., 1985b. Impedance control: an approach to manipulation: part II—implementation. J. Dyn. Syst. Meas. Control. 107 (1), 8–16. https://doi.org/10.1115/1.3140713.

Hogan, N., 1985c. Impedance control: an approach to manipulation: part III—applications. J. Dyn. Syst. Meas. Control. 107 (1), 17–24. https://doi.org/10.1115/1.3140701.

Howe, R.D., Matsuoka, Y., 1999. Robotics for surgery. Annu. Rev. Biomed. Eng. 1 (1), 211–240. https://doi.org/10.1146/annurev.bioeng.1.1.211.

Huo, W., et al., 2016. Lower limb wearable robots for assistance and rehabilitation: a state of the art. IEEE Syst. J. 1068–1081. https://doi.org/10.1109/JSYST.2014.2351491.

Hussain, A., et al., 2014. The use of robotics in surgery: a review. Int. J. Clin. Pract. https://doi.org/10.1111/ijcp.12492.

Jaikumar, S., Kim, D.H., Kam, A.C., 2002. History of minimally invasive spine surgery. Neurosurgery 51 (Suppl_2), S2-1–S2-14. https://doi.org/10.1097/00006123-200211002-00003.

Jobbágy, B., et al., 2014. Robotic exoskeleton for rehabilitation of the upper limb. Am. J. Mech. Eng. 2 (7), 299–302. https://doi.org/10.12691/ajme-2-7-27.

Johnson, M.J., et al., 2008. Rehabilitation and assistive robotics [TC spotlight]. IEEE Robot. Autom. Mag. 15 (3), 16–110. https://doi.org/10.1109/MRA.2008.928304.

Joubair, A., et al., 2016. Use of a force-torque sensor for self-calibration of a 6-DOF medical robot. Sensors 16 (6), 798. https://doi.org/10.3390/s16060798.

Kalan, S., et al., 2010. History of robotic surgery. J. Robot. Surg. 4 (3), 141–147. https://doi.org/10.1007/s11701-010-0202-2.

Katsura, S., Iida, W., Ohnishi, K., 2005. Medical mechatronics—an application to haptic forceps. Annu. Rev. Control. 29 (2), 237–245. https://doi.org/10.1016/j.arcontrol.2005.05.003.

Khanal, P., et al., 2014. Collaborative virtual reality based advanced cardiac life support training simulator using virtual reality principles. J. Biomed. Inform. 51, 49–59. https://doi.org/10.1016/j.jbi.2014.04.005.

Kim, K.C., 2014. Kim, K.C. (Ed.), Robotics in General Surgery. Springer, New York, NY. https://doi.org/10.1007/978-1-4614-8739-5.

Kim, E.J., et al., 2005. A biofidelic birthing simulator. IEEE Eng. Med. Biol. Mag. https://doi.org/10.1109/memb.2005.1549728.

Kramme, R., Hoffmann, K.P., Pozos, R.S., 2011. Springer Handbook of Medical Technology. In: Kramme, R., Hoffmann, K.-P., Pozos, R.S. (Eds.), Springer, Berlin, Heidelberg. https://doi.org/10.1007/978-3-540-74658-4.

Krebs, H.I., Volpe, B.T., 2013. Rehabilitation robotics. In: Handbook of Clinical Neurology. pp. 283–294. https://doi.org/10.1016/B978-0-444-52901-5.00023-X.

Krebs, H.I., et al., 1998. Robot-aided neurorehabilitation. IEEE Trans. Rehabil. Eng. 6 (1), 75–87. https://doi.org/10.1109/86.662623.

Kwoh, Y.S., et al., 1988. A robot with improved absolute positioning accuracy for CT guided stereotactic brain surgery. IEEE Trans. Biomed. Eng. https://doi.org/10.1109/10.1354.

Lane, T., 2018. A short history of robotic surgery. Ann. R. Coll. Surg. Engl. https://doi.org/10.1308/rcsann.supp1.5.

Lanfranco, A.R., et al., 2004. Robotic surgery. Ann. Surg. 239 (1), 14–21. https://doi.org/10.1097/01.sla.0000103020.19595.7d.

Le, H.M., Do, T.N., Phee, S.J., 2016. A survey on actuators-driven surgical robots. Sensors Actuators A Phys. 247, 323–354. https://doi.org/10.1016/j.sna.2016.06.010.

Leal Ghezzi, T., Campos Corleta, O., 2016. 30 years of robotic surgery. World J. Surg. 40 (10), 2550–2557. https://doi.org/10.1007/s00268-016-3543-9.

Lee, C., et al., 2017. Soft robot review. Int. J. Control. Autom. Syst. https://doi.org/10.1007/s12555-016-0462-3.

Lee-Kong, S., Feingold, D.L., 2013. The history of minimally invasive surgery. Semin. Colon Rectal Surg. https://doi.org/10.1053/j.scrs.2012.10.003.

Lefranc, M., Peltier, J., 2016. Evaluation of the ROSA™ Spine robot for minimally invasive surgical procedures. Expert Rev. Med. Devices 13 (10), 899–906. https://doi.org/10.1080/17434440.2016.1236680.

Leigh, G., Hurst, H., 2008. We have a high-fidelity simulator, now what? Making the most of simulators. Int. J. Nurs. Educ. Scholarsh. 5 (1), 1–9. https://doi.org/10.2202/1548-923X.1561.

Li, C., Sui, J., Ji, L., 2013. Lower limb rehabilitation robots: a review. In: IFMBE Proceedings, pp. 2042–2045. https://doi.org/10.1007/978-3-642-29305-4_536.

Liu, W., Yin, B., Yan, B., 2016. A survey on the exoskeleton rehabilitation robot for the lower limbs. In: 2016 2nd International Conference on Control, Automation and Robotics (ICCAR). IEEE, pp. 90–94. https://doi.org/10.1109/ICCAR.2016.7486705.

Lo, H.S., Xie, S.Q., 2012. Exoskeleton robots for upper-limb rehabilitation: state of the art and future prospects. Med. Eng. Phys. 34 (3), 261–268. https://doi.org/10.1016/j.medengphy.2011.10.004.

Maciejasz, P., et al., 2014. A survey on robotic devices for upper limb rehabilitation. J. NeuroEng. Rehabil. 11 (1), 3. https://doi.org/10.1186/1743-0003-11-3.

Marino, M.V., et al., 2018. From illusion to reality: a brief history of robotic surgery. Surg. Innov. https://doi.org/10.1177/1553350618771417.

Marks, L.J., Michael, J.W., 2001. Science, medicine, and the future: artificial limbs. BMJ 323 (7315), 732–735. https://doi.org/10.1136/bmj.323.7315.732.

Martinez-Martin, E., del Pobil, A.P., 2018. Personal robot assistants for elderly care: an overview. In: Intelligent Systems Reference Library. pp. 77–91. https://doi.org/10.1007/978-3-319-62530-0_5.

Martins, M.M., et al., 2012. Assistive mobility devices focusing on smart walkers: classification and review. Robot. Auton. Syst. 60 (4), 548–562. https://doi.org/10.1016/j.robot.2011.11.015.

Masiero, S., et al., 2014. The value of robotic systems in stroke rehabilitation. Expert Rev. Med. Devices. https://doi.org/10.1586/17434440.2014.882766.

Masuda, K., et al., 2001. Three dimensional motion mechanism of ultrasound probe and its application for tele-echography system. In: Proceedings 2001 IEEE/RSJ International Conference on Intelligent Robots and Systems. Expanding the Societal Role of Robotics in the the Next Millennium (Cat. No.01CH37180). IEEE, pp. 1112–1116. https://doi.org/10.1109/IROS.2001.976317.

Matarić, M.J., Scassellati, B., 2016. Socially assistive robotics. In: Springer Handbook of Robotics. Springer International Publishing, Cham, pp. 1973–1994. https://doi.org/10.1007/978-3-319-32552-1_73.

Mattei, T.A., et al., 2014. Current state-of-the-art and future perspectives of robotic technology in neurosurgery. Neurosurg. Rev. 37 (3), 357–366. https://doi.org/10.1007/s10143-014-0540-z.

Mattheis, S., et al., 2017. Flex robotic system in transoral robotic surgery: the first 40 patients. Head Neck 39 (3), 471–475. https://doi.org/10.1002/hed.24611.

Mehdi, H., Boubaker, O., 2010. Rehabilitation of a human arm supported by a robotic manipulator: a position/force cooperative control. J. Comput. Sci. 6 (8), 912–919. https://doi.org/10.3844/jcssp.2010.912.919.

Mehdi, H., Boubaker, O., 2011a. Impedance controller tuned by particle swarm optimization for robotic arms. Int. J. Adv. Robot. Syst. 8 (5), 57. https://doi.org/10.5772/45692.

Mehdi, H., Boubaker, O., 2011b. Position/force control optimized by particle swarm intelligence for constrained robotic manipulators. In: International Conference on

Intelligent Systems Design and Applications, ISDA. https://doi.org/10.1109/ISDA.2011.6121653.
Mehdi, H., Boubaker, O., 2012a. Robot-assisted therapy: design, control and optimization. Int. J. Smart Sens. Intell. Syst. https://doi.org/10.21307/ijssis-2017-522.
Mehdi, H., Boubaker, O., 2012b. Robust tracking control for constrained robots. Procedia Eng. https://doi.org/10.1016/j.proeng.2012.07.313.
Mehdi, H., Boubaker, O., 2012c. Stiffness and impedance control using Lyapunov theory for robot-aided rehabilitation. Int. J. Soc. Robot. 4 (S1), 107–119. https://doi.org/10.1007/s12369-011-0128-5.
Mehdi, H., Boubaker, O., 2013. Robust stiffness control for constrained robots under model uncertainties. In: 2013 International Conference on Electrical Engineering and Software Applications, ICEESA 2013. https://doi.org/10.1109/ICEESA.2013.6578388.
Mehdi, H., Boubaker, O., 2015. Robust impedance control-based Lyapunov-Hamiltonian approach for constrained robots. Int. J. Adv. Robot. Syst. 1. https://doi.org/10.5772/61992.
Mehdi, H., Boubaker, O., 2016. PSO-Lyapunov motion/force control of robot arms with model uncertainties. Robotica 34 (3), 634–651. https://doi.org/10.1017/S0263574714001775.
Meng, Q., et al., 2017. A survey on sEMG control strategies of wearable hand exoskeleton for rehabilitation. In: 2017 2nd Asia-Pacific Conference on Intelligent Robot Systems (ACIRS). IEEE, pp. 165–169. https://doi.org/10.1109/ACIRS.2017.7986086.
Miller, D.P., 1998. Assistive robotics: an overview. In: Lecture Notes in Computer Science (Including Subseries Lecture Notes in Artificial Intelligence and Lecture Notes in Bioinformatics). https://doi.org/10.1007/bfb0055975.
Mobayen, S., Boubaker, O., Fekih, A., 2019. Design of observer-based tracking controller for robotic manipulators. In: New Trends in Observer-Based Control. https://doi.org/10.1016/b978-0-12-817034-2.00020-4.
Mohammed, S., Amirat, Y., 2009. Towards intelligent lower limb wearable robots: challenges and perspectives—state of the art. In: 2008 IEEE International Conference on. Robotics and Biomimetics, ROBIO 2008, pp. 312–317. https://doi.org/10.1109/ROBIO.2009.4913022.
Mohammed, S., Amirat, Y., Rifai, H., 2012. Lower-limb movement assistance through wearable robots: state of the art and challenges. Adv. Robot. 26, 1–22. https://doi.org/10.1163/016918611X607356.
Mohammed, S., et al., 2017. Special issue on assistive and rehabilitation robotics. Auton. Robot. 41 (3), 513–517. https://doi.org/10.1007/s10514-017-9627-z.
Moreno, J.C., Figueiredo, J., Pons, J.L., 2018. Exoskeletons for lower-limb rehabilitation. In: Rehabilitation Robotics. pp. 89–99. https://doi.org/10.1016/B978-0-12-811995-2.00008-4.
Morgia, G., De Renzis, C., 2009. CyberKnife in the treatment of prostate cancer: a revolutionary system. Eur. Urol. 56 (1), 40–42. https://doi.org/10.1016/j.eururo.2009.02.020.
Nadas, I., et al., 2017. Considerations for designing robotic upper limb rehabilitation devices. In: AIP Conference Proceedings. p. 030005. https://doi.org/10.1063/1.5018278.
Najarian, S., Fallahnezhad, M., Afshari, E., 2011. Advances in medical robotic systems with specific applications in surgery—a review. J. Med. Eng. Technol. https://doi.org/10.3109/03091902.2010.535593.
Nakadate, R., Hashizume, M., 2018. Intelligent information-guided robotic surgery. In: Recent Advances in Laparoscopic Surgery. Intech Open https://doi.org/10.5772/intechopen.82191 Working Title.
Ni, Z., 2015. Survey on medical robotics. J. Mech. Eng. 51 (13), 45. https://doi.org/10.3901/JME.2015.13.045.

Okamura, A., Mataric, M., Christensen, H., 2010. Medical and health-care robotics. IEEE Robot. Autom. Mag. 17 (3), 26–37. https://doi.org/10.1109/MRA.2010.937861.

Okamura, A.M., et al., 2011. Force feedback and sensory substitution for robot-assisted surgery. In: Surgical Robotics. Boston, MA, Springer US, pp. 419–448. https://doi.org/10.1007/978-1-4419-1126-1_18.

Pamungkas, D.S., et al., 2019. Overview: types of lower limb exoskeletons. Electronics (Switzerland). https://doi.org/10.3390/electronics8111283.

Payne, C.J., Yang, G.-Z., 2014. Hand-held medical robots. Ann. Biomed. Eng. 42 (8), 1594–1605. https://doi.org/10.1007/s10439-014-1042-4.

Peters, B.S., et al., 2018. Review of emerging surgical robotic technology. Surg. Endosc. 32 (4), 1636–1655. https://doi.org/10.1007/s00464-018-6079-2.

Pierrot, F., et al., 1999. Hippocrate: a safe robot arm for medical applications with force feedback. Med. Image Anal. 3 (3), 285–300. https://doi.org/10.1016/S1361-8415(99)80025-5.

Pignolo, L., 2009. Robotics in neuro-rehabilitation. J. Rehabil. Med. https://doi.org/10.2340/16501977-0434.

Pitkin, M., 2015. Prosthetic restoration and rehabilitation of the upper and lower extremity. Prosthetics Orthot. Int. https://doi.org/10.1177/0309364614537109.

Platz, T., et al., 2017. A survey on robotic devices for upper limb rehabilitation. Nervenarzt 74 (10), 841–849. https://doi.org/10.1007/s00115-003-1549-7.

Poli, P., et al., 2013. Robotic technologies and rehabilitation: new tools for stroke patients therapy. Biomed. Res. Int. 2013, 1–8. https://doi.org/10.1155/2013/153872.

Pons, J.L., 2008. Pons, J.L. (Ed.), Wearable Robots: Biomechatronic Exoskeletons. John Wiley & Sons, Ltd., Chichester, UK. https://doi.org/10.1002/9780470987667

Pons, J.L., 2010. Rehabilitation exoskeletal robotics. IEEE Eng. Med. Biol. Mag. https://doi.org/10.1109/MEMB.2010.936548.

Poorten, E., Vander, B., Demeester, E., Lammertse, P., 2012. Haptic feedback for medical applications, a survey. In: Actuator.

Popović, D.B., 2014. Advances in functional electrical stimulation (FES). J. Electromyogr. Kinesiol. 24 (6), 795–802. https://doi.org/10.1016/j.jelekin.2014.09.008.

Preising, B., Hsia, T.C., Mittelstadt, B., 1991. A literature review: robots in medicine. IEEE Eng. Med. Biol. Mag. 10 (2), 13–22. https://doi.org/10.1109/51.82001.

Puangmali, P., et al., 2008. State-of-the-art in force and tactile sensing for minimally invasive surgery. IEEE Sensors J. 8 (4), 371–381. https://doi.org/10.1109/JSEN.2008.917481.

Pugin, F., Bucher, P., Morel, P., 2011. History of robotic surgery: from AESOP® and ZEUS® to da Vinci®. J. Visc. Surg. https://doi.org/10.1016/j.jviscsurg.2011.04.007.

Qian, Z., Bi, Z., 2015. Recent development of rehabilitation robots. Adv. Mech. Eng. 7 (2), 563062. https://doi.org/10.1155/2014/563062.

Rassweiler, J., et al., 2001. Telesurgical laparoscopic radical prostatectomy: initial experience. Eur. Urol. https://doi.org/10.1159/000049752.

Rodríguez-Prunotto, L., et al., 2014. Upper limb robotic devices in rehabilitation for neurological disease. Rehabilitacion 48 (2), 104–128. https://doi.org/10.1016/j.rh.2014.01.001.

Rosen, J., Hannaford, B., Satava, R.M., 2011. Surgical Robotics: Systems Applications and Visions. In: Rosen, J., Hannaford, B., Satava, R.M. (Eds.), Springer US, Boston, MA. https://doi.org/10.1007/978-1-4419-1126-1.

Ruiz-Olaya, A.F., Lopez-Delis, A., da Rocha, A.F., 2019. Upper and lower extremity exoskeletons. In: Handbook of Biomechatronics. Elsevier, pp. 283–317. https://doi.org/10.1016/B978-0-12-812539-7.00011-8.

Runciman, M., Darzi, A., Mylonas, G.P., 2019. Soft robotics in minimally invasive surgery. Soft Rob. 6 (4), 423–443. https://doi.org/10.1089/soro.2018.0136.

Rupal, B.S., et al., 2017. Lower-limb exoskeletons: research trends and regulatory guidelines in medical and non-medical applications. Int. J. Adv. Robot. Syst. https://doi.org/10.1177/1729881417743554.

Salcudean, S.E., et al., 1999. Robot-assisted diagnostic ultrasound—design and feasibility experiments. In: Lecture Notes in Computer Science (Including Subseries Lecture Notes in Artificial Intelligence and Lecture Notes in Bioinformatics). https://doi.org/10.1007/10704282_115.

Sale, P., Lombardi, V., Franceschini, M., 2012. Hand robotics rehabilitation: feasibility and preliminary results of a robotic treatment in patients with hemiparesis. Stroke Res. Treat. https://doi.org/10.1155/2012/820931.

Schweikard, A., Ernst, F., 2015. Medical Robotics. Springer International Publishing, Cham. https://doi.org/10.1007/978-3-319-22891-4.

Segil, J., 2019. Handbook of Biomechatronics. Elsevier. https://doi.org/10.1016/C2016-0-02397-5.

Shah, J., Vyas, A., Vyas, D., 2015. The history of robotics in surgical specialties. Am. J. Robot. Surg.. https://doi.org/10.1166/ajrs.2014.1006.

Shi, C., et al., 2017. Shape sensing techniques for continuum robots in minimally invasive surgery: a survey. IEEE Trans. Biomed. Eng. 64 (8), 1665–1678. https://doi.org/10.1109/TBME.2016.2622361.

Shi, D., et al., 2019. A review on lower limb rehabilitation exoskeleton robots. Chin. J. Mech. Eng. 32 (1), 74. https://doi.org/10.1186/s10033-019-0389-8.

Shin, H.J., et al., 2020. Robotic single-port surgery using the da Vinci SP® surgical system for benign gynecologic disease: a preliminary report. Taiwan. J. Obstet. Gynecol. 59 (2), 243–247. https://doi.org/10.1016/j.tjog.2020.01.012.

Shishehgar, M., Kerr, D., Blake, J., 2018. A systematic review of research into how robotic technology can help older people. Smart Health. https://doi.org/10.1016/j.smhl.2018.03.002.

Simaan, N., Yasin, R.M., Wang, L., 2018. Medical technologies and challenges of robot-assisted minimally invasive intervention and diagnostics. Annu. Rev. Control Robot. Auton. Syst. 1 (1), 465–490. https://doi.org/10.1146/annurev-control-060117-104956.

Smith, S., et al., 2014. A robotic system to simulate child birth design and development of the pneumatic artificial muscle (PAM) birthing simulator. In: 2014 13th International Conference on Control Automation Robotics and Vision, ICARCV 2014. https://doi.org/10.1109/ICARCV.2014.7064499.

Speich, J., Rosen, J., 2008. Medical robotics. In: Encyclopedia of Biomaterials and Biomedical Engineering. second ed. Four Volume Set. CRC Press, pp. 1804–1815. https://doi.org/10.1201/b18990-174.

Spinelli, A., et al., 2018. First experience in colorectal surgery with a new robotic platform with haptic feedback. Color. Dis. 20 (3), 228–235. https://doi.org/10.1111/codi.13882.

Stefanov, D.H., Bien, Z., Bang, W.-C., 2004. The smart house for older persons and persons with physical disabilities: structure, technology arrangements, and perspectives. IEEE Trans. Neural Syst. Rehabil. Eng. 12 (2), 228–250. https://doi.org/10.1109/TNSRE.2004.828423.

Takeoka, T., et al., 2017. Assessment potential of a new suture simulator in laparoscopic surgical skills training. Minim. Invasive Ther. Allied Technol. 26 (6), 338–345. https://doi.org/10.1080/13645706.2017.1312456.

Tapus, A., Mataric, M., Scassellati, B., 2007. Socially assistive robotics [grand challenges of robotics]. IEEE Robot. Autom. Mag. 14 (1), 35–42. https://doi.org/10.1109/MRA.2007.339605.

Taylor, R.H., 1997. Robots as surgical assistants: where we are, wither we are tending, and how to get there. In: Lecture Notes in Computer Science (Including Subseries Lecture

Notes in Artificial Intelligence and Lecture Notes in Bioinformatics). https://doi.org/10.1007/BFb0029430.
Taylor, R.H., 2006. A perspective on medical robotics. Proc. IEEE 94 (9), 1652–1664. https://doi.org/10.1109/JPROC.2006.880669.
Taylor, R.H., 2008. Medical robotics and computer-integrated surgery. In: 2008 32nd Annual IEEE International Computer Software and Applications Conference. IEEE, p. 1. https://doi.org/10.1109/COMPSAC.2008.234.
Taylor, R.H., Stoianovici, D., 2003. Medical robotics in computer-integrated surgery. IEEE Trans. Robot. Automat. 19 (5), 765–781. https://doi.org/10.1109/TRA.2003.817058.
Taylor, R.H., et al., 2008. Medical robotics and computer-integrated surgery. In: Springer Handbook of Robotics. Springer, Berlin, Heidelberg, pp. 1199–1222. https://doi.org/10.1007/978-3-540-30301-5_53.
Taylor, R.H., et al., 2016. Medical robotics and computer-integrated surgery. In: Springer Handbook of Robotics. pp. 1657–1683. https://doi.org/10.1007/978-3-319-32552-1_63.
Tejima, N., 2001. Rehabilitation robotics: a review. Adv. Robot. 14 (7), 551–564. https://doi.org/10.1163/156855301742003.
Topping, M., Smith, J., 1998. The development of Handy 1, a rehabilitation robotic system to assist the severely disabled. Ind. Robot. 25 (5), 316–320. https://doi.org/10.1108/01439919810232459.
Trejos, A.L., Patel, R.V., Naish, M.D., 2010. Force sensing and its application in minimally invasive surgery and therapy: a survey. Proc. Inst. Mech. Eng. C J. Mech. Eng. Sci. 224 (7), 1435–1454. https://doi.org/10.1243/09544062JMES1917.
Troccaz, J., 2012. Robotique médicale: sous la direction de Jocelyne Troccaz, Robotique médicale. Hermes Science Publications.
Troccaz, J., 2013. Troccaz, J. (Ed.), Medical Robotics. John Wiley & Sons, Inc., Hoboken, NJ, USA. https://doi.org/10.1002/9781118562147.
Troccaz, J., Delnondedieu, Y., Poyet, A., 1995. Safety issues in surgical robotics. IFAC Proc. Vol. 28 (20), 19–26. https://doi.org/10.1016/S1474-6670(17)45019-1.
Troccaz, J., Dagnino, G., Yang, G.-Z., 2019. Frontiers of medical robotics: from concept to systems to clinical translation. Annu. Rev. Biomed. Eng. 21 (1), 193–218. https://doi.org/10.1146/annurev-bioeng-060418-052502.
Tucker, M.R., et al., 2015. Control strategies for active lower extremity prosthetics and orthotics: a review. J. NeuroEng. Rehabil. https://doi.org/10.1186/1743-0003-12-1.
United Nations and Department of Economic and Social Affiars Population Division, 2017. World Population Ageing 2017—Highlights. United Nations doi: ST/ESA/SER.A/348.
Valero, R., et al., 2011. Robotic surgery: history and teaching impact. Acta. Urol. Esp. https://doi.org/10.1016/j.acuroe.2011.12.004.
Van der Loos, H.F.M., Reinkensmeyer, D.J., 2008. Rehabilitation and health care robotics. In: Springer Handbook of Robotics. https://doi.org/10.1007/978-3-540-30301-5_54.
Van der Loos, H.F.M., Reinkensmeyer, D.J., Guglielmelli, E., 2016. Rehabilitation and health care robotics. In: Springer Handbook of Robotics. Springer International Publishing, Cham, pp. 1685–1728. https://doi.org/10.1007/978-3-319-32552-1_64.
van der Meijden, O.A.J., Schijven, M.P., 2009. The value of haptic feedback in conventional and robot-assisted minimal invasive surgery and virtual reality training: a current review. Surg. Endosc. 23 (6), 1180–1190. https://doi.org/10.1007/s00464-008-0298-x.
Walker, I.D., Green, K.E., 2009. Continuum robots. In: Encyclopedia of Complexity and Systems Science. Springer, New York, NY, pp. 1475–1485. https://doi.org/10.1007/978-0-387-30440-3_96.

Walsh, C.J., et al., 2008. A patient-mounted, telerobotic tool for CT-guided percutaneous interventions. J. Med. Devices 2(1). https://doi.org/10.1115/1.2902854.

Wang, Q., Chen, W., Markopoulos, P., 2014. Literature review on wearable systems in upper extremity rehabilitation. In: 2014 IEEE-EMBS International Conference on Biomedical and Health Informatics, BHI. https://doi.org/10.1109/BHI.2014.6864424.

Wang, Q., et al., 2017. Interactive wearable systems for upper body rehabilitation: a systematic review. J. NeuroEng. Rehabil. https://doi.org/10.1186/s12984-017-0229-y.

Witte, K.A., Collins, S.H., 2020. Design of lower-limb exoskeletons and emulator systems. In: Wearable Robotics. Elsevier, pp. 251–274. https://doi.org/10.1016/B978-0-12-814659-0.00013-8.

Wolf, A., Shoham, M., 2009. Medical automation and robotics. In: Springer Handbook of Automation. Springer, Berlin, Heidelberg, pp. 1397–1407. https://doi.org/10.1007/978-3-540-78831-7_78.

Xie, S., 2016. Advanced robotics for medical rehabilitation. In: Advanced Robotics for Medical Rehabilitation. Springer International Publishing, Cham. https://doi.org/10.1007/978-3-319-19896-5 Springer Tracts in Advanced Robotics.

Yan, T., et al., 2015. Review of assistive strategies in powered lower-limb orthoses and exoskeletons. Robot. Auton. Syst. https://doi.org/10.1016/j.robot.2014.09.032.

Yang, G.-Z., et al., 2017a. Medical robotics—regulatory, ethical, and legal considerations for increasing levels of autonomy. Sci. Robot. 2(4), eaam8638. https://doi.org/10.1126/scirobotics.aam8638.

Yang, X., et al., 2017b. State of the art: bipedal robots for lower limb rehabilitation. Appl. Sci. 7 (11), 1182. https://doi.org/10.3390/app7111182.

Yang, C., et al., 2018. 'Force modeling, identification, and feedback control of robot-assisted needle insertion: a survey of the literature. Sensors 18 (2), 561. https://doi.org/10.3390/s18020561.

Yasin, H., et al., 2019. Experience with 102 frameless stereotactic biopsies using the neuromate robotic device. World Neurosurg. 123, e450–e456. https://doi.org/10.1016/j.wneu.2018.11.187.

Yue, Z., Zhang, X., Wang, J., 2017. Hand rehabilitation robotics on poststroke motor recovery. Behav. Neurol. https://doi.org/10.1155/2017/3908135.

Zarrad, W., et al., 2007a. Stability and transparency analysis of a haptic feedback controller for medical applications. In: Proceedings of the IEEE Conference on Decision and Control. https://doi.org/10.1109/CDC.2007.4434677.

Zarrad, W., et al., 2007b. Towards teleoperated needle insertion with haptie feedback controller. In: IEEE International Conference on Intelligent Robots and Systems. https://doi.org/10.1109/IROS.2007.4399085.

Zhang, X., Yue, Z., Wang, J., 2017. Robotics in lower-limb rehabilitation after stroke. Behav. Neurol. https://doi.org/10.1155/2017/3731802.

Zhang, Y., et al., 2019. Could social robots facilitate children with autism spectrum disorders in learning distrust and deception? Comput. Hum. Behav. 98, 140–149. https://doi.org/10.1016/j.chb.2019.04.008.

Zuo, K.J., Olson, J.L., 2014. The evolution of functional hand replacement: from iron prostheses to hand transplantation. Can. J. Plast. Surg. https://doi.org/10.1177/229255031402200111.

CHAPTER 8

Wearable mechatronic devices for upper-limb amputees

Juan J. Huaroto[a,b], Etsel Suárez[a,b], Emir A. Vela[c]
[a]Department of Engineering, Faculty of Science and Philosophy, Universidad Peruana Cayetano Heredia, Lima, Peru
[b]Department of Mechanical Engineering, Universidad Nacional de Ingenieria, Lima, Peru
[c]Department of Energy and Mechanical Engineering, Universidad de Ingenieria y Tecnologia-UTEC, Barranco, Peru

1 Introduction

The first upper-limb prosthetic device was reported in early AD. This is evidence that humans strived to design and provide wearable prosthetic devices to replace lost body limbs, whether they be upper or lower limbs. When a person loses a limb many issues occur, from physical disability to psychological problems such as loss of self-confidence and even depression (Kristjansdottir et al., 2019; Roche et al., 2019). Physical disability in amputees causes serious difficulties in realizing activities of daily life, entering the labor market, and simply leading normal lives, especially in developing countries (Resnik et al., 2019; Daniels et al., 2019). It is in this context that prosthetic devices appeared. At the beginning, upper-limb prostheses were purely mechanical devices with very limited use for grasping objects, or designed solely for aesthetics. With the advance of technology, they began to evolve increasing their range of tasks, degrees of freedom (DOFs), and interactions with the user (Huaroto et al., 2018; Marasco et al., 2018; Castellini, 2020; Motti, 2020).

As first reported, wearable devices (WDs) started off as purely mechanical systems. However, with the creation of electrical and electronic components, WDs are now mostly known as purely electronic devices. Current WDs are composed of a set of sensors for acquiring and measuring body information and signals for health monitoring and diagnosis (Motti, 2020). There are also many WDs that are composed of both actuators and sensors for augmenting human capacities, rehabilitation, providing assistance, and replacing body limbs lost due to amputation or congenital diseases. These are referred as mechatronic and robotic WDs (Pons, 2008; Resnik et al., 2012). Examples include robotic exoskeletons that a user wears

Control Theory in Biomedical Engineering
https://doi.org/10.1016/B978-0-12-821350-6.00008-1

for augmenting strength to lift heavy objects or enhance gait or running (also known as extenders) (Maciejasz et al., 2014; Jarrassé et al., 2014; Shen and Rosen, 2020); orthotic robots for rehabilitation and assistance after a disease such as stroke; and prosthetic robots for replacing a lost limb, either upper or lower, to recover lost function such as grasping or walking (Nemah et al., 2019; Voloshina and Collins, 2020; Castellini, 2020). All these WDs rely on human anatomy to enhance their use in a more natural way. Fig. 1 illustrates the classification of mechatronic wearable devices presented in this chapter.

Nowadays, more people are losing limbs due to job accidents and diseases that may lead to amputation, such as diabetes (Gopura et al., 2016; Islam et al., 2017; Nemah et al., 2019; Geiss et al., 2019; Shen and Rosen, 2020). Prosthetic WDs are being developed by a significant number of

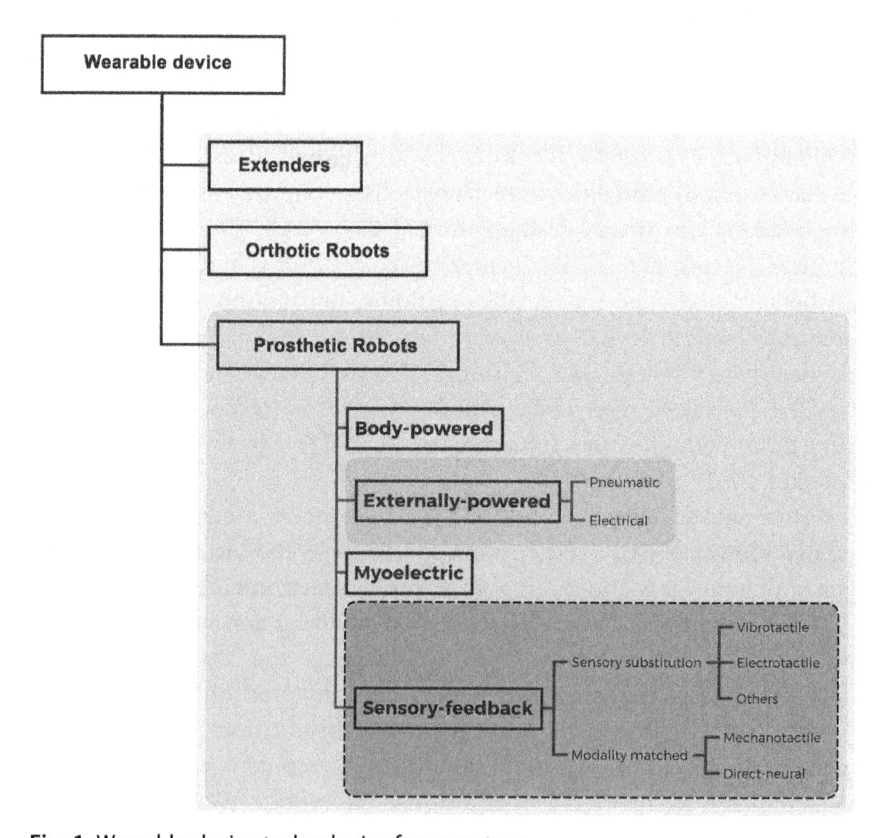

Fig. 1 Wearable device technologies for amputees.

research groups and companies worldwide (Maciejasz et al., 2014; Jarrassé et al., 2014; Castellini, 2020). Despite these efforts, WDs are not yet widely used, as there are still many challenges to overcome (Armstrong et al., 2019; Kristjansdottir et al., 2019). WD users would like a cost-friendly device that is similar in shape and function to their lost limb; low-cost WDs are still very limited (Biddiss and Chau, 2007). Another big challenge is that amputees do not feel as though the WD is a part of their body due to the lack of bilateral communication and sensory feedback between the device and the amputee (Kuiken, 2006; Marasco et al., 2018).

In this chapter, we review wearable mechatronic devices for upper limb amputees. First, we examine the human sensory feedback and physiology of the skin, a topic we consider very important to improve bilateral communication between amputees and their WD. Second, we provide some preliminary concepts about WDs and classify them in order to better understand the existing types (see Fig. 1). Then, we delve deeper into prosthetic WDs by describing body-powered and externally powered prosthetic WDs, putting emphasis on myoelectric WDs. Finally, we explain the modalities of sensory feedback prostheses as a way of integrating the bilateral communication between user and WD in a manner that significantly improves user adherence.

2 Human sensory feedback and physiology of the human skin

Every WD is in physical contact with the human skin, and communicates different sensations to the human user through sensory feedback. This human sensory feedback generally includes hearing, sight, taste, smell, and sense of touch. Sense of touch is described by the term "haptics," which is the science that studies the sense of touch on human skin. Sense of touch allows us to manipulate objects and perceive sensations around us, such as textures, temperature, pain, and so on. In recent years, haptics has been used in several applications such as telerobotics, virtual reality, video games, and others for simulating physical experiences perceived by the other human senses. Unlike the other human sensations, sense of touch is distributed throughout the human skin, muscles, and tendons (Culbertson et al., 2018). Hence, the human sensory feedback of the sense of touch is considered haptic feedback and divided into two modalities: tactile and kinesthetic feedback, depending on the sensation experienced by the individual.

2.1 Tactile feedback

Tactile feedback such as pressure, stretch, vibration, and combined stimuli is recorded by the human body through sensory organs called mechanoreceptors, which are located in our whole skin, being in greater density on the fingers and palm of our hands (Johnson, 2001). These mechanoreceptors are the Meissner's corpuscles, Merkel's disks, Ruffini endings and Pacinian corpuscles (PCs), which are connected to the peripheral nervous system by peripheral nerve bundles along with the human dermis.

Mechanoreceptors have temporal and spatial resolution ranges. Hence they are categorized by types and adaptation due to a stimulus. Mechanoreceptors type I have small receptive fields, while those of type II have large receptive fields. In addition, mechanoreceptors can be PC rapid adaptors (RAs), whose response ranges cannot register static stimuli signals, or they can be slow adaptors (SAs), which can register transient stimuli. The two types of SAs are type I and type II.

Type I SAs are responsible for roughness and shape recording, while type II SAs sense skin stretches with a large receptive field. Otherwise, RAs sense the slip and flutter on the skin with a small receptive field. PCs, unlike RAs, respond to a dynamic stimulus but have large receptive fields (see the characteristics of skin mechanoreceptors in Caldwell et al. (1997) for further insight). In effect, the mechanoreceptors' features must be considered carefully when developing wearable haptic devices, as they are key for stimulating amputees' residual limbs to generate the illusion of movement as well as to improve the adherence of prostheses.

2.2 Kinesthetic feedback

Kinesthetic feedback is characterized by the sense of movement and strength in our limbs. The receiving organs that allow this sensation are the muscular spindles and Golgi tendon organs (Proske and Gandevia, 2012). When these are stimulated, the individual can perceive an illusion of movement (Culbertson et al., 2018). The Golgi tendon organ is a capsule in the connection between tendon and muscle fibers. In this capsule, there are several sensory nerve fibers called Ib afferents that transport movement sensation to the nervous system. Some experiments of tactile stimulation were reported in the 1970s, for instance, one study induced vibration at 100 Hz on the muscles and tendons that resulted in an illusion of movement or a change in limb position (Goodwin et al., 1972; Eklund, 1972).

3 Wearable device: Preliminary concepts

3.1 Definitions

Wearable device

In the last few decades, technology has evolved such that WD systems now include on-body sensors for detecting physiological signals without discomfort, and are able to capture and continuously record data in real time (Bonfiglio and De Rossi, 2011). In this sense, WDs have the potential to help in both diagnosis and ongoing treatment of a vast number of individuals with neurological, cardiovascular, and pulmonary diseases (Patel et al., 2012). Since the development of transistors, WDs have become much smaller, and with the advent of the Internet of Things (IoT) WDs are becoming smart, portable, multi-functional, and able to connect to the Internet and run on batteries (Hung et al., 2004). These devices have been used for many applications, including communication, sports, optics, Virtual Reality (VR), and Artificial Intelligence (AI). In the field of mechatronics, WDs are proposed as wearable mechatronic devices or wearable robots to be worn by humans either to augment the function of a limb or to replace it completely (Pons, 2008). WDs are used externally to the body, either attached as an accessory or embedded in clothes (Raskovic et al., 2004; Yang and Sahabi, 2016; Motti, 2020). To physically assist weakened and/or disabled individuals with impaired upper limb function, extensive research has been carried out in many branches of robotics, particularly wearable robots (e.g., exoskeletons, powered orthotic devices, etc.) and end-effector-based robotic devices (i.e., devices that do not actively support or hold the subject's arm, but rather connect with the subject's hand or forearm) (Rahman et al., 2014).

According to Pons (2008), a wearable robot can be seen as a technology that extends, complements, substitutes, or enhances human function and capability or empowers or replaces (a part of) the human limb where it is worn. This definition is considered in this chapter. In the next section, we present a classification of wearable robots.

Empowering robotic exoskeletons (extenders)

Extenders were defined for the first time in the 1990s as a class of robotic manipulators that extend the strength of the human arm while maintaining human control of the task. The defining characteristic of an extender is the "transmission of power and information signals." The individual wears the extender and the physical contact between the extender and the user allows direct transfer of mechanical power and information signals (Kazerooni, 1989). The difference between master-slave and extenders is that the human operator is either at a remote location or close to the slave manipulator, but is not in direct physical contact with the slave in the sense of transfer of power as in the case of extenders.

Orthotic robots

The term orthotic is commonly used to define a wearable structure assembled on the human anatomy for restoring a human function previously lost (Perry et al., 2007; Polygerinos et al., 2014; Merchant et al., 2018; Radder et al., 2018; Shen and Rosen, 2020; Armeo Power, 2020; among others). Although orthotics restore the capabilities to natural levels, exoskeletons can even increase them. For instance, The Man Amplifier (Clark et al., 1962), conceived by Cornell Aeronautical Laboratory, Inc. (CAL), is one of the first developed exoskeletons. It was created by employing powered joints that are worn by a man to augment or amplify his muscular strength and increases his endurance by performing tasks requiring large amounts of physical exertion.

Prosthetic robots

A modern definition of prosthetics includes the terms robotics, control, signal processing, and human–machine interfaces. A prosthesis is a wearable structure that replaces one missing part of the human body (often as a part of the upper or lower limb). For many years, prostheses were considered as solely mechanical parts; however, nowadays it is possible to connect directly a prosthesis through the human nervous system to achieve more natural movements.

In this chapter, we survey prosthetic robot technologies for upper-limb devices. However, the technical aspects presented in this chapter are also applicable to lower-limb prosthetic technologies (Voloshina and Collins, 2020).

3.2 Features

One of the most important requirements of any device interacting with humans is safety. As exoskeleton devices move under close contact conditions with the wearer, any malfunction can be seriously harmful to the user. Mechanical designs should therefore consider the possibilities of unpredicted erroneous operation of the device controller when it is actively actuated. Limits to the range of motion can be set using a mechanical stopper or corresponding structural designs so that the exoskeleton cannot force the wearer's body to move in an excessive range of motion (Heo and Kim, 2014). Designing a comfortable mechanical interface between the device and its wearer is a critical yet underrated issue in wearable robots (Xiloyannis et al., 2019). Wearable robots provide a physical interface between their human wearers and the wearers' environment, enhancing the wearers' interaction with the world. This is beneficial because these devices can assist in rehabilitation, improve the independence of disabled people, and even provide ergonomic and safe support of industrial workers. Currently, great efforts are being made to create actuators and sensors with inherent physical softness for ergonomic and safe interaction (Veale et al., 2018).

4 Upper-limb prosthetic technologies

4.1 Overview

Losing an upper limb is a major cause of decreased bodily function because the hand is an integral component of the human body that performs a wide range of tasks, including grasping and manipulating the environment as well as communicating nonverbally. Thus, upper-limb loss is a shocking experience that needs psychological and physical treatment. To mitigate the effects of limb loss, artificial WDs (prostheses) were developed that could help people gain and recover functionalities in the amputated regions. Upper-limb prostheses (Cordella et al., 2016) can be classified into two main categories based on their function: passive prostheses (which in turn are divided into cosmetic and functional) and active prostheses (which include body-powered and externally powered devices). The first mechanisms were denominated passive devices because no moving parts were included in the prostheses in order to move them; external and direct force interaction was needed to move and operate them. One of the first documented prosthetic works is described in 77 CE in Naturalis Historia (Zuo and Olson, 2014), where in the Second Punic War (218–201 BCE) a Roman general received a prosthesis that enabled him to successfully return to battle. The concept of an "automatic" body-powered upper-limb prosthesis was pioneered by the German dentist Peter Baliff in 1818 (Childress, 1985; Meier, 2004). Using transmission of tension through leather straps, Baliff's device enabled the intact muscles of the trunk and shoulder girdle to elicit motion in a terminal device attached to the amputated stump. In 1948, the Bowden cable body-powered prosthesis was introduced replacing bulky straps with a sleek, sturdy cable. Despite new materials and improved craftsmanship, today's body-powered prostheses are essentially adaptations of the Bowden design (Zuo and Olson, 2014). Durable, portable, and relatively affordable, body-powered prostheses allow the user an impressive range of motion, speed, and force in operating a terminal device by changing the tension in a cable via preserved shoulder and body movements. Although prolonged wearing can be uncomfortable, dexterous motor tasks are limited, and appearance is not human-like, body-powered prostheses are widely used (Ostlie et al., 2012).

With the design and development of mechanisms and technologies, more sophisticated and active upper-limb prostheses appeared in which moving parts were key elements. In this sense, body-powered prostheses received an upgrade and thus gave more capabilities to amputees. For automation and better control of tasks, electrical-powered prostheses were proposed. This technology adds electrical power sources to utilize electric motors with batteries and some variations with different mechanisms (e.g., gears and pulleys). In the late 1950s, the use of external power was introduced in the United States to help high-level bilateral arm amputees who were victims of war or accidents (Muilenburg and LeBlanc, 1989). Another kind of active mechanism is related to fluid systems (pneumatic and hydraulic), where the working principle is based on pressurized fluids to move upper-limb prosthetic elements (Marquardt, 1965). The first myoelectric prosthesis with clinical trials was reported by Alexander Kobrinski in 1960. It used transistors (Scott, 1992) such that identification, acquisition, and interpretation of myoelectric signals of the patient could be better controlled.

Another interesting concept is the use of "underactuated mechanisms." These systems have an input vector of a smaller dimension than that of the output vector (Birglen et al., 2008). In this sense, underactuated mechanisms for prosthetic devices are used to increase energy efficiency and optimize its kinematics mimicking biomechanics (Massa et al., 2002; Cipriani et al., 2006).

At present, the main drawback of using a prosthesis is user rejection. Surveys on use and satisfaction in using prosthetic hands showed that 30% to 50% of upper-limb amputees do not use their prosthesis for activities of daily living (Atkins et al., 1996; Schulz et al., 2001). Rejection is mainly caused by non-intuitive control, lack of sufficient feedback, and insufficient functionality.

4.2 Body-powered prosthesis

Body-powered prostheses were the initial design of prostheses. The first historical reference for these devices came in 500 BCE when Hegesistratus, imprisoned in chains, cut off his foot to escape and subsequently made himself a wooden foot to replace it. The first pictured arm prosthesis dates to ca. 1400 AC. The Alt-Ruppin hand had artificial digits and small holes to allow the air to cool the residual limb. A picture of another prosthesis, Ballif's arm (1812), shows the fixing belts and the system for moving the artificial limb. This prosthesis introduced for the first time the principle of hand operation by the shoulder and arm movement, which is currently the standard for body-powered prostheses. This allows for grasping tasks using cable-based mechanisms that amplify the movement of other body parts, commonly the shoulder and arm (Jacobsen et al., 1999). Currently, the cost of a commercial body-powered prosthetic hand ranges from $4000 to $10,000 (Resnik et al., 2012). However, due to the revolution of advanced manufacturing and 3D printing, many systems have been developed as body-powered prostheses (e.g., Zero Point Frontiers prosthetic device (2013), Talon Flextensor 1.0 by profbink (2014), Hollies Hand Version 4 by Anthromod (2015), and Victoria Hand (2019)), thus reducing the fabrication time and cost to about $500 (Ten Kate et al., 2017). An important technical point is that although body-powered prostheses allowed a greater range of elbow flexion, shoulder flexion is also required for completing a continuous grasp, which nowadays is addressed using externally powered prostheses.

4.3 Externally powered prosthesis

Historically, a crucial turning point in changing from body-powered to externally powered prostheses was World War II due to the increase in the number of amputees. Since then, the technology has changed from body-powered prostheses that use external energy (externally powered prostheses), such as pneumatic or electrical. The first externally powered prosthesis was patented in Germany in 1915 (Dahlheim, 1915). It used a pneumatic-powered hand to deliver grasp movements. Four years later, also in Germany, the first publication related to an electrical-powered prosthesis was released (Schlesinger, 1919). Since then, many studies have compared user adherence of body-powered vs. electrical-powered prostheses. For instance, the work of Muilenburg and LeBlanc (1989) provides a significant discussion about the inherent proprioceptive feedback using body-powered prostheses relating to low cost, reliability, and functionality apropos of externally powered prostheses. In contrast, Heger et al. (1985) provide statistics of prosthesis fitting in a residual limb, resulting in 80% of patients having complete or useful acceptance of an electrically powered prosthesis. These patients experimented with a satisfactory combination of comfort, aesthetics, and functional parameters. As a result, it was evident that this technology might be better than body-powered prostheses.

In terms of costs, externally powered prostheses can cost $25,000 to $75,000 (Resnik et al., 2012), which is 7.5 times more than a body-powered prosthesis. For this reason, 3D-printed electrical-powered prostheses controlled by electromyography (myoelectrical prostheses), in some cases by individual voice, or by electroencephalography (EEG) have been developed to reduce the cost of hardware and prototyping time.

4.4 Myoelectric prostheses

Before the development of externally powered prostheses, devices were commanded solely by buttons or switches in an open-loop control manner. However, it was not until the myoelectric prosthesis appeared that this approach changed. Myoelectric prostheses belong to the electrical-powered active prostheses spectrum with capability of reading the electromyographic (EMG) activity from the voluntary movements of muscles (Popov, 1965). This technology was developed by Reiter in the period 1944–1948 (Weihe, 1998). He used vacuum tubes to create a myoelectric prosthesis but with huge dimensions. Consequently, his system did not gain clinical or commercial acceptance. It was until late 1960s that researchers around the world "re-invented" the technology with the help of the transistor and its application as an amplifier. The electromyography upper-limb wearable device for amputees consisted of several elements to detect the user's intention, trying to mimic the natural human movement, from electrodes for EMG detection, to signal processors and controllers, ending with actuators for prosthesis movements. Sensors established communication between human control signals and the upper-limb prostheses. Then, modern prosthetic hands incorporated surface electrodes located on the skin surface to acquired myoelectric signals. These electrodes are preferred because noninvasive techniques are much easier to access. In conventional myoprostheses (prostheses with electromyography techniques), two bipolar EMG-signals are placed on antagonistic muscle groups, such as the wrist extensors and flexors, and are used to control the velocity of one DOF (Mazumdar, 2004). However, there exist techniques to collect intramuscular signals (Weir et al., 2009; Al-Ajam et al., 2013) with implantable myoelectric sensors. More recently, a surgical procedure known as Targeted

Muscle Reinnervation consists of biological signal amplification by means of innervation of electrical nerves into new groups of surface muscles so that surface electrodes can acquire and record the user movement intention.

4.4.1 EMG control strategies

Because a myoelectric prosthesis is considered a wearable robotic device, control techniques for acquiring and processing EMG signals should be taken into account.

To control upper-limb prostheses by means of the acquisition of EMG signals, there are seven known control schemes well summarized by Geethanjali (2016):

- ON-OFF control: As its name suggests, this is a control mode where the electric motor in the terminal device is either ON or OFF. This action is achieved by setting a threshold value, for which the processed information from the EMG signal is usually compared with a Mean Absolute Value (MAV) or root-square mean value.
- Proportional control: This scheme focuses on controlling the velocity of the actuator, that is, motor velocity, as a function of the amplitude or mean value of the acquired EMG signal. In other words, the speed of the terminal device becomes proportional to the levels of EMG signals (Mazumdar, 2004; Bottomley et al., 1963).
- Direct control: This belongs to the proportional scheme where a direct control (Hahne et al., 2014) and communication between the incomplete electrical nerve and muscle control the exact part of the limb that was amputated, for example, each finger has its terminal nerve where the EMG is extracted and processed to be used as input signal to control it.
- Finite-State Machine control: This is a mode where some states are predefined and programmed for some finite positions (Dosen et al., 2010).
- Pattern Recognition-based control, Regression and Posture control schemes: These are modern techniques where signal classification, regression analysis with a pre-processing of feature extraction, and estimation using adaptive approaches are used (i.e., machine learning). (Fougner et al., 2012; Muceli and Farina, 2011).

The technology used to develop affordable upper-limb prostheses is based on the approaches previously discussed, from ON-OFF to myoelectric control strategies.

The functionality requirements of the prosthesis increase with the level of amputation, which leads to a paradox seen in myoelectric control. The functionality and therefore the control site requirements (i.e., EMG sensors' location) increase with the level of amputation while the number of sites available for EMG adquisition decreases (the muscle normally involved in the function of the limb that has been lost is used as the control source). In this sense, medical pre-procedures to relocate and amplify terminal nerves and muscle signals that were lost after amputation are a good option to increase the functionality of prostheses where high-level amputations are identified.

4.4.2 Targeted muscle reinnervation

Although a limb is lost with an amputation, the control signals remain in the residual peripheral nerves of the amputated limb. Thus, instead of surface EMG detection, another method exploiting the residual nerves was the development of intramuscular direct connections of a device to remaining terminal nerves in order to increase the robustness of the control of upper-limb prostheses (Dewald et al., 2019). This work is encouraging, but several inherent problems appear when it comes to high-level amputations. The neuroelectric signal is very small (microvolts), difficult to record in a long term, and difficult to separate from the EMG signals of surrounding muscle (which have similar frequency content). Nerve atrophy (Upshaw and Sinkjaer, 1998), electrical components' durability, and good electrical signal wire transmission should be maintained. Kuiken (2006) proposed a method to avoid these disadvantages by means of targeted reinnervation. "Targeted motor reinnervation enabled to denervate expendable regions of muscle in or near an amputated limb and transfer residual arm peripheral nerve endings to these muscles." A basic scheme of targeted reinnervation of a trans-humeral amputee is shown in Fig. 2.

The extension of Kuiken's work (2007) was presented in a case study of a woman with a left arm amputation at the humeral neck, with great outcomes (better intuitive prosthesis control and maneuverability) using box-and-blocks tests (Mathiowetz et al., 1985). In other studies (Miller et al., 2008) box-and-blocks tests and clothespin tests show from two to six times better outcomes in high-level amputees (i.e., trans-humeral, shoulder amputations), and then proposed techniques for targeted muscle reinnervation presented in (Cheesborough et al., 2015; Kuiken et al., 2017). In addition to the greater outcomes found with Targeted Muscle Reinnervation in the sense of dexterity and intuitive prostheses control, other promising results were found recently in the field of Neuroma and Physical Phantom, presenting reduction in residual limb pain and phantom limb pain (Dumanian et al., 2019).

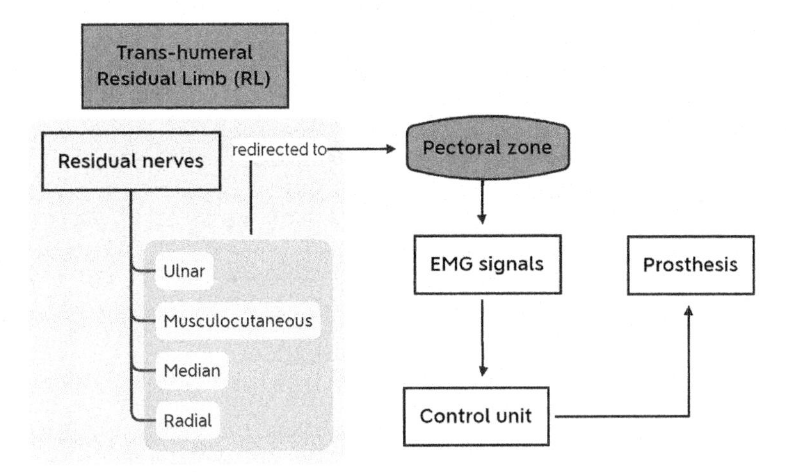

Fig. 2 The overall scheme of targeted muscle reinnervation in a trans-humeral amputee.

4.5 Sensory feedback prosthesis

As discussed in Section 2, human sensory feedback concerns all human sensations from external to internal sensation (exteroception and proprioception feedback). All individuals perceive their surroundings thanks to their sensory feedback; however, when a traumatic or congenital amputation is presented, inherent sensory feedback is lost and only nerve endings remain in the residual limb. Initially, body-powered prostheses allow certain sensory feedback due to the movement of cable-driven mechanisms and socket–residual limb interaction. Moreover, different techniques were developed regarding stimuli over human skin to induce artificial human sensory feedback to an amputee in the upper and/or lower residual limbs. This artificial human sensory feedback can be of two modes: (1) by using alternative feedback over human skin as vibrations, electrical stimulus, temperature, and so on (here the mode is sensory substitution feedback); and (2) by using a proportional stimulus according to the performed task for obtaining artificial sensory feedback (this mode is called modality matched feedback; see Fig. 3). The latter potentially requires a lower cognitive effort as the modality of the feedback signal does not require interpretation by the user (Schofield et al., 2014). In the last few years, both modes of sensory feedback prosthesis have been studied, however, nowadays the modality-matched feedback has garnered the attention of the scientific community due to its

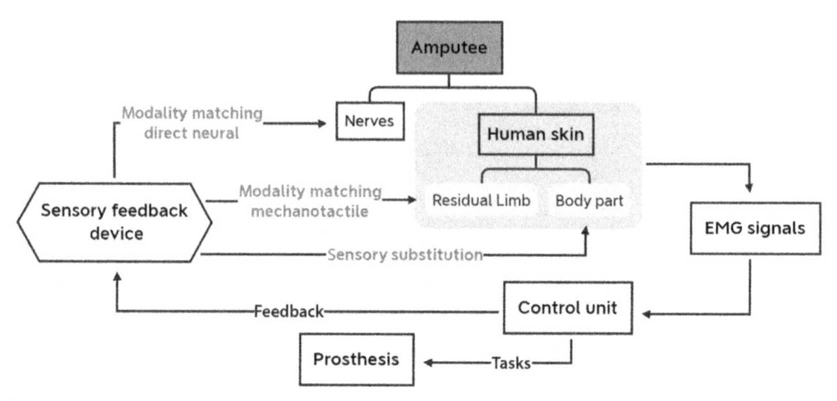

Fig. 3 Sensory substitution system scheme.

nearly natural feedback and the possibility to combine with techniques of targeted muscle reinnervation.

4.5.1 Sensory substitution feedback

Vibrotactile

Vibrotactile stimulation is evoked by mechanical vibration on the human skin at frequencies ranging between 10 and 500 Hz. The two main features of the stimulus are vibration amplitude and frequency, but other features like pulse duration, shape, and duty cycle can be modulated to convey different kinds of information (Thomas et al., 2019; Di Luzio et al., 2020). The amplitude discrimination threshold depends on several parameters, including the frequency and location on the body. Lower thresholds are found on glabrous skin as compared with hairy skin at frequencies in the range of 150–300 Hz (Antfolk et al., 2013). Antfolk et al. (2012) assessed the ability of trans-radial amputees to discriminate multi-site tactile stimuli in sensory discrimination tasks. The study compared different sensory feedback modalities, such as vibrotactile and mechanotactile stimuli (pressure or tangential forces on the skin, using an artificial hand prosthesis. The results demonstrated that pressure stimulation surpassed vibrotactile stimulation in multi-site sensory feedback discrimination.

Thomas et al. (2019) reported a comparison of the feedback intrinsically presented in body-powered prostheses (joint-torque feedback) to a commonly proposed modality feedback for myoelectric prostheses (vibrotactile feedback). His results suggested that even when haptic feedback was not a modality matched to the task, such as the case of vibrotactile feedback, performance with a myoelectric prosthesis can improve significantly.

This implied that it is possible to achieve the same results with vibrotactile feedback, which is cheaper and easier to implement than other forms of feedback. Markovic et al. (2018) evaluated the impact of these factors with a longitudinal assessment in six amputee subjects, using a clinical setup (socket, embedded control) and a range of tasks (box and blocks, block turn, clothespin, and cups relocation). To provide sensory feedback the study used a novel vibrotactile configuration. The study demonstrated, for the first time, the relevance of an advanced, multi-variable feedback interface for dexterous, multi-functional prosthesis control in a clinically relevant setting.

The aforementioned devices are made of rigid materials and components. Alternatively, a promising and interesting technology is soft robotics that exploits the use of soft and flexible materials for developing soft wearable devices with better adaptability and affordance with the human body. In this context, soft devices are created for generating vibrations and vibrotactile feedback for amputees. Sonar and Paik (2016) designed a soft pneumatic actuator capable of inducing vibrotactile feedback and sensing the force provided on the skin by using PZT sensors embedded in a matrix of silicone reaching a bandwidth of 56 Hz. Georgarakis et al. (2017) enhanced this actuator by using different 2D shapes in the inflation area, obtaining a bandwidth of 120 Hz, but with an amplitude of vibration in the order of 10 μm. More recently, Huaroto et al. (2018) explored the 3D shapes of the inflation area to reach much larger amplitude displacements in vibration mode with a significant bandwidth of about 70 Hz and displacements of 2 mm as required to generate kinesthetic illusions (Marasco et al., 2018).

Electrotactile

Electrotactile stimulation communicates nontactile information via electrical stimulation of the sense of touch. This information can be transmitted by using electrodes over/under the skin (noninvasive and invasive electro stimulation, respectively) (Szeto and Saunders, 1982). Electrotactile (or electrocutaneous) stimulation evokes sensations within the skin by stimulating afferent nerve endings through a local electrical current (Antfolk et al., 2013). Typical currents range within 1–20 mA with pulse frequencies ranging from 1 Hz to 5 kHz; biphasic pulses produce more comfortable sensations (Szeto and Saunders, 1982). Moreover, several studies compare this technique with vibrotactile stimulation, which is naturally noninvasive (Kaczmarek et al., 1991). For instance, Witteveen et al. (2012) investigated a longitudinal and transversal orientation of an array of four feedback conditions: no feedback, visual feedback, feedback through vibrotactile or electrotactile stimulation, and addition of an extra stimulator for touch feedback.

As a result, vibrotactile stimulation had much better performance compared to the nonfeedback conditions; and the addition of touch feedback further increased the performance but at the cost of increasing the duration. A benefit of electrocutaneous systems is that they often require less power than mechanical systems (e.g., vibrotactile devices). However, the contamination of EMG signals is probably due to the electrical stimulation of muscles.

Others

There are other substitution techniques such as vision substitution (Bach-y-Rita et al., 1969), auditory substitution (Lundborg et al., 1999), and others. These techniques currently open the paradigm of associating an external stimulus as sensory feedback called synesthesia (Cytowic, 2002). Two types of synesthesia can be distinguished: strong synesthesia is characterized by a vivid image in one sensory modality in response to stimulation in another one, and weak synesthesia is characterized by cross-sensory correspondences expressed through language, perceptual similarity, and perceptual interactions during information processing (Martino and Marks, 2001). Synesthetic learning can also be present in vibrotactile stimuli, for instance, Huaroto et al. (2018) reported, after 2 days of training, WD adherence and synesthetic learning due to a pressure stimulus on healthy individuals using a soft vibrotactile device.

4.5.2 Modality-matched feedback

In modality-matched methods, the information from the user is matched with sensations, for example, touch with a prosthesis is felt as touch on the user's skin although with mismatched locations. As a result, the user must still dedicate attention to interpret the feedback signal. In some applications, modality-matched feedback is preferred to sensory substitution because it does not require a significant interpretation by the user (Schofield et al., 2014). For instance, Meek et al. (1989) explored grip pressure feedback with a myoelectric-controlled prosthetic arm. They mounted a servo-controlled "pusher" on a socket. The pusher pressed into the skin an amount proportional to the force in the terminal device, a method that the authors termed "Extended Physiologic Taction" (EPT). Additionally, to achieve intuitive haptic feedback it is beneficial to satisfy two conditions: (1) somatotopic matching and (2) modality matching (Kim et al., 2009). Kuiken et al. developed Targeted Reinnervation (TR) surgery that reroutes severed nerves from the amputated limb to residual muscles and skin (Kuiken et al., 2004, 2009; Dumanian et al., 2019). The results of these studies provided a kind of somatotopic matching. So, in this context, some soft wearable devices were developed to reproduce the modality matching in order to complement the somatotopic matching (Huaroto et al., 2018, 2019). In effect, the next generation of prosthesis robotics should satisfy these two aforementioned conditions and enable direct connection to the nervous human system. Hence, the next sections introduce the most important technologies developed up to now as modality-matched feedback devices.

Mechanotactile

One of the first devices developed to enable mechanotactile feedback was proposed by Meek et al. (1989). Their device consisted of a "tactor" that penetrates into the skin as a noninvasive stimulus using a force of about 9 N, and a skin surface penetration of about 1 cm. Many years later, other devices were proposed, such as Kim's tactor (see Fig. 3) (Kim et al., 2009) and Casinis' device, which was capable of providing a cutaneous stimulation to a forearm in tangential direction by using two belts and two DC motors with encoders to record the torsional angle (Casini et al., 2015). Most rigid devices used for mechanotactile feedback are uncomfortable, have visual discomfort, avoid the inherent suction between residual limb and liner, and are likely to contaminate EMG signals (due to vibrations of DC motors) (Kim et al., 2009). Currently, a new generation of robotics, soft wearable robots, has opened the possibility of introducing soft and flexible, smart and responsive materials in prosthetic applications with the possibility of avoiding the loss of limb fixation suction due to holes in the liner to pass touch/vibration to the residual limb, and the contamination of EMG signal associated with rigid devices based on DC motors (Huaroto et al., 2018). Some soft wearable devices have been recently developed that are capable of exerting pressure and stretching on the skin (Agharese et al., 2018; Suarez et al., 2018, and Huaroto et al., 2019).

Direct-neural

In the previous sections, we discussed noninvasive human stimulation; however, nowadays several studies of direct-neural stimulation examined how to connect a wearable robotic device directly to the human nervous system (see Fig. 3).

Peripheral nerve stimulation is the most common direct-neural stimulation. This stimulation relies on the principle that, following upper limb amputation, the original afferent neural pathways are proximally preserved and can be exploited for interfacing with prostheses (Micera et al., 2010; Dhillon and Horch, 2005; Ortiz-Catalan et al., 2014; Raspopovic and Petrini, 2018). Direct-neural methods deliver feedback so that an amputee senses the stimulus as though it were applied to the same corresponding location of their missing limb (Schofield et al., 2014). This kind of modality-matched feedback is the most natural and currently is gaining more attention in the scientific community that is developing novel techniques to modulate intraneural signals such as biomimetic techniques of frequency modulation (Valle et al., 2018).

Another approach is by cortical control (Velliste et al., 2008), where cortical electrical signal activity is acquired and processed in order to distinguish desired human movements. These desires are then converted into control signals for prosthetic movements as presented in Johannes et al. (2020).

4.6 Summary of wearable devices

There are some reviews of mechatronic WDs (Maciejasz et al., 2014; Jarrassé et al., 2014; Islam et al., 2017; and Gopura et al., 2016), but an update is presented in Table 1, which shows an overview of WDs to describe their classification in the sense of human-machine bidirectional communication and somatosensory feedback.

5 Challenges

Regarding the aforementioned technologies, we address some important challenges to achieve more naturally integrated mechatronic WDs for amputees.

- Materials science: The need for including novel materials and intelligent mechanical structures for sockets, liners, and prostheses is under consideration because patient adherence with WDs depends on a natural integration of the machine and the human skin. This includes an anatomical fit over the residual limb and the physiological behavior of materials in contact with the skin (in a noninvasive approach). In addition, for an invasive approach, the materials must be compatible with the internal tissues, avoiding any rejection (biomaterials).

Table 1 Overview of WD.

Device, Researcher	Main type of WD	Description	Remarks
CADEN-7, Perry et al. (2007)	Assistive exoskeleton	DOFs: 7, Brushed motors with cable-driven reduction pulleys	*Strength:* Full range of motion, low inertias *Weakness:* Hysteresis of cable mechanisms, no clinical test, no sensory feedback
Open Fingerpad eXoskeleton (OFX), Heo and Kim (2014)	Assistive exoskeleton	Finger exoskeleton allowing free fingertip manipulation and grasping force estimation by means of load cells; it has a pneumatic cylinder for assistance	*Strength:* Dexterity due to direct human fingertip environment interaction *Weakness:* Noncompliant, no sensory feedback, no clinical study
Polygerinos et al. (2014)	Assistive exoskeleton	Soft robotic glove, fluidic actuation, compliant materials	*Strength:* Range of motion evaluation, lightweight, portable and low-cost components *Weakness:* 2.2 s to steady state, pressurized fluid source needed, no sensory feedback
Merchant et al. (2018)	Assistive exoskeleton	Five joints; weight (kg): 2.6; additive rapid prototyping manufacturing	*Strength:* Tested in 15 volunteers, 2% steady-state position error *Weakness:* No hand manipulation tasks, no sensory feedback, no clinical study
EXO-UL, Shen et al. (2020)	Assistive exoskeleton	DOFs: 8, Attachable to a frame and chair; Control method: from EMG control to force/torque control	*Strength:* VR module for training, interchangeable hand exoskeleton, sensory feedback, clinical study *Weakness:* Expensive

Continued

Table 1 Overview of WD—cont'd

Device, Researcher	Main type of WD	Description	Remarks
Armeo Power (2020)	Assistive exoskeleton	DOFs: 6; Control method: impedance control	*Strength:* Commercially available, robust control, clinical study *Weakness:* Limited shoulder range of motion, no sensory feedback
Massa et al. (2002)	Underactuated prosthetic hand	Three-fingers prosthetic robot for grasping tasks	*Strength:* Underactuated *Weakness:* Noncompliant, no clinical study, no sensory feedback
Laliberté et al. (2010)	Underactuated prosthetic hand	DOFs: 15, five fingers robotic hand, additive rapid prototyping manufacturing	*Strength:* Out-of-the-plane thumb movements *Weakness:* Elastic tendons present hysteresis, No sensory feedback, no clinical study
Victoria Hand, Victoria Hand (2019)	Body-powered prosthetic arm	For trans-radial, trans-radial suspension or trans-humeral level of amputation	*Strength:* Commercially available, personalized, inexpensive, dynamic wrist *Weakness:* Not autonomous, no sensory feedback
Schulz et al. (2001)	Electrically powered prosthetic hand	DOFs: 13, Five-fingered robotic hand, 18 small-size flexible fluidic actuators, maximum force of 6 N at 50 kPa	*Strength:* High power capacity, self-adaptability *Weakness:* No clinical study, no sensory feedback
Bahari et al. (2011)	Electrically powered prosthetic hand	DOFs: 14, additive rapid prototyping manufacturing; Control methods: joystick control, Graphical User Interface, and autonomous operation	*Strength:* Grasping function tested, multi-fingered dexterity *Weakness:* No clinical study, no sensory feedback

Andrianesis and Tzes (2015)	Electrically powered prosthetic hand	SMA actuators, five-fingered prosthetic hands, additive rapid prototyping manufacturing	*Strength:* Personalized design, fingertip force sensors for feedback, silent *Weakness:* Long time responses, bandwidth of 0.2 Hz, no clinical study, no sensory feedback
Gretsch et al. (2015)	Electrically powered prosthetic arm	For patients with transradial limb amputation, shoulder controller, open and close all five fingers, use of 3D printing	*Strength:* Independent thumb movements, estimated cost $300, tested by 13-year-old female patient *Weakness:* For limited tasks, no clinical study, no sensory feedback
Johannes et al. (2020)	Electrically powered prosthetic arm	17 actuators driving 26 articulated joints, 4.7 kg; Control method: cortical control	*Strength:* Vibration sensing, heat flux sensing, contact sensing, modularity, accommodates varying user amputation levels, clinical studies, sensory feedback *Weakness:* Heavy, nonaffordable
Popov (1965)	Myoelectric prosthetic hand	Weight (g): 800, for grasping tasks, objects up to 25 mm in diameter and 5 kg	*Strength:* First myoelectric prostheses *Weakness:* Limited number of tasks, no clinical study, no sensory feedback
Bebionic, Ottobockus (2019)	Myoelectric prosthetic hand	DOFs: 6; Weight (g): 390–598; Maximum grasp force (N): 140; Control method: myoelectric control with 14 grip patterns	*Strength:* Commercially available, and dexterous fingers movements *Weakness:* Expensive, limited to two hand sizes and three different types of wrist couplings, passive thumb movements, no sensory feedback
Hero Arm, OpenBionic (2019)	Myoelectric prosthetic arm	DOFs: 5–6; Weight (g): 280–346; Control method: myoelectric control with personalized grip patterns	*Strength:* Commercially available, sensory feedback, personalized 3D-printed design *Weakness:* Long time response due to proportional control

Continued

Table 1 Overview of WD—cont'd

Device, Researcher	Main type of WD	Description	Remarks
I-Limb Ultra (2019)	Myoelectric prosthetic arm	DOFs: 6; Weight (g): 507–515; Maximum grasp force (N): 136; Control method: myoelectric control with 24 grip patterns	*Strength:* Commercially available, dexterity
LUKE Arm, Mobius Bionics (2019)	Myoelectric prosthetic arm	DOFs: 6; Weight (g): 1400	*Strength:* Commercially available, sensory feedback, VR training *Weakness:* Controlled by foot movements
Michelangelo Hand, Ottobockus (2019)	Myoelectric prosthetic arm	DOFs: 4; Weight (g): 420; Maximum grasp force (N): 70	*Strength:* Commercially available, clinical study, abduction/adduction fingers movements *Weakness:* Very expensive, no sensory feedback
TASKA Hand, Taska Prosthetics (2019)	Myoelectric prosthetic arm	DOFs: 8; Control method: myoelectric control with 23 grip patterns	*Strength:* Commercially available, waterproof, flexible interface material *Weakness:* Passive wrist flexion and rotation, no sensory feedback
VINCENT evolution 3, VINCENT evolution 3 (2019)	Myoelectric prosthetic arm	DOFs: 8; Weight (g): 386; Control method: myoelectric control with 14 grip patterns	*Strength:* Commercially available, sensory feedback

HapWRAP, Agharese et al. (2018)	Sensory feedback	Soft pneumatic robot made of low-density polyethylene, pouches inflated at 10 kPa and 1.5 LPM	*Strength:* Accessible low-cost materials, human-user studied *Weakness:* 6 h per pouch manufacturing time, weakening after few cycles of operation, no clinical test, no sensory feedback
Sonar and Paik (2016)	Sensory feedback	Vibratory soft pneumatic actuator with piezoelectric sensors, at 70 kPa and less than 90 Hz it generates a perceivable amplitude	*Strength:* Allows bidirectional tactile information (sensory feedback) *Weakness:* Prototype, noise generation, no clinical study
Georgarakis et al. (2017)	Sensory feedback	Soft pneumatic actuator for wrist proprioception based on vibratory stimulation, 0.5 mm (amplitude) and 20–120 Hz at 35 kPa, 1 mm diameter each actuator	*Strength:* Compliant mechanisms, feasibility human-user studied *Weakness:* Noise generation, dependent on pressurized air source.
Huaroto et al. (2018)	Sensory feedback	Soft pneumatic actuator, maximum force of 12.5 N at 70 kPa, free displacement of 4.5 mm at 50 kPa and a bandwidth of 70 Hz	*Strength:* Can be incorporated in liner prostheses, adaptable, sensory feedback *Weakness:* Noise generation, no clinical study

- Control techniques: With AI approaches, new methodologies for controlling prostheses in a natural way and with fast response should be considered in novel designs.
- Design methodologies: Currently there is not a defined standard methodology for envisaging prosthetic robots. Hence, there is a need to acquire standard methods and tools for manufacturing prostheses.
- Digital manufacturing: With Industry 4.0, the conventional ways of designing prostheses and their sub-technologies should be changed, including additive and subtractive computerized manufacture and intelligent approaches for designing prosthetic devices with personalized requirements.
- Energy autonomy: The growing development of materials in the field of batteries should focus on decreasing the weight and dimensions of portable batteries, and increasing self-energy-generating sources for prosthetic mechatronic WDs.

6 Conclusion

Overviewing the history and state of the art of wearable devices, mechatronics and robotics are still in their way of maturity, although great efforts have been made by the scientific community and companies. We do not yet see every upper-limb amputee using a prosthetic WD, especially in developing countries. The high cost of most advanced prostheses is a barrier, as is the lack of user adherence. More emphasis is needed on improving the bilateral communication between amputees and their prostheses in order to naturally achieve their integration and use. To meet this challenge, more clinical trials are needed. Sensory feedback prosthetics is a fascinating field that is not yet well explored; it opens up opportunities to achieve more intelligent WD for the benefit of users.

References

Agharese, N., Cloyd, T., Blumenschein, L.H., Raitor, M., Hawkes, E.W., Culbertson, H., Okamura, A.M., 2018. HapWRAP: soft growing wearable haptic device. In: 2018 IEEE International Conference on Robotics and Automation (ICRA). IEEE, pp. 5466–5472.

Al-Ajam, Y., Lancashire, H., Pendegrass, C., Kang, N., Dowling, R.P., Taylor, S.J., Blunn, G., 2013. The use of a bone-anchored device as a hard-wired conduit for transmitting EMG signals from implanted muscle electrodes. IEEE Trans. Biomed. Eng. 60 (6), 1654–1659.

Andrianesis, K., Tzes, A., 2015. Development and control of a multifunctional prosthetic hand with shape memory alloy actuators. J. Intell. Robot. Syst. 78 (2), 257–289.

Antfolk, C., D'Alonzo, M., Controzzi, M., Lundborg, G., Rosén, B., Sebelius, F., Cipriani, C., 2012. Artificial redirection of sensation from prosthetic fingers to the phantom hand map on transradial amputees: vibrotactile versus mechanotactile sensory feedback. IEEE Trans. Neural Syst. Rehabil. Eng. 21 (1), 112–120.
Antfolk, C., D'alonzo, M., Rosen, B., Lundborg, G., Sebelius, F., Cipriani, C., 2013. Sensory feedback in upper limb prosthetics. Expert Rev. Med. Devices 10 (1), 45–54.
Armeo Power, Hocoma, 2020. Available from: https://www.hocoma.com/solutions/armeo-power/. (Accessed 12 January 2020).
Armstrong, T.W., Williamson, M.L., Elliott, T.R., Jackson, W.T., Kearns, N.T., Ryan, T., 2019. Psychological distress among persons with upper extremity limb loss. Br. J. Health Psychol. 24, 746–763.
Atkins, D.J., Heard, D.C.Y., Donovan, W.H., 1996. Epidemiologic overview of individuals with upper limb loss and their reported research priorities. J. Prosthet. Orthot. 8 (1), 2–11.
Bach-y-Rita, P., Collins, C.C., Saunders, F.A., White, B., Scadden, L., 1969. Vision substitution by tactile image projection. Nature 221 (5184), 963–964.
Bahari, M., Jaffar, A., Low, C.Y., Jaafar, R., Yussof, H., 2011. Design and development of a multifingered prosthetic hand. Int. J. Soc. Robot. 4, 59–66.
Bebionic Hand, 2019. Available from: https://www.ottobockus.com/prosthetics/upper-limb-prosthetics/solution-overview/bebionic-hand/. (Accessed 3 September 2019).
Biddiss, E., Chau, T., 2007. Upper-limb prosthetics: critical factors in device abandonment. Am. J. Phys. Med. Rehabil. 86 (12), 977–987.
Birglen, L., Laliberté, T., Gosselin, C., 2008. Underactuated robotic hands. In: Springer Tracts in Advanced Robotics. vol. 40. Spinger.
Bonfiglio, A., De Rossi, D. (Eds.), 2011. Wearable Monitoring Systems. Springer, New York, pp. 3–4.
Bottomley, A., Kinnier Wilson, A.B., Nightingale, A., 1963. Muscle substitutes and myo electric control. Radio Electron. Eng. 26 (6), 439–448.
Caldwell, D.G., Tsagarakis, N., Wardle, A., 1997. Mechano thermo and proprioceptor feedback for integrated haptic feedback. In: 1997 IEEE International Conference on Robotics and Automation (ICRA). vol. 3. IEEE, pp. 2491–2496.
Casini, S., Morvidoni, M., Bianchi, M., Catalano, M., Grioli, G., Bicchi, A., 2015. Design and realization of the cuff-clenching upper-limb force feedback wearable device for distributed mechano-tactile stimulation of normal and tangential skin forces. In: 2015 IEEE/RSJ International Conference on Intelligent Robots and Systems (IROS). IEEE, pp. 1186–1193.
Castellini, C., 2020. Upper limb active prosthetic systems—overview. In: Wearable Robotics. Academic Press, pp. 365–376.
Cheesborough, J., Smith, L., Kuiken, T., Dumanian, G., 2015. Targeted muscle Reinnervation and advanced prosthetic arms. Semin. Plast. Surg. 29 (1), 62–72.
Childress, D.S., 1985. Historical aspects of powered limb prostheses. Clin. Prosthet. Orthot. 9 (1), 2–13.
Cipriani, C., Zaccone, F., Stellin, G., Beccai, L., Cappiello, G., Carrozza, M.C., Dario, P., 2006. Closed-loop controller for a bio-inspired multi-fingered underactuated prosthesis. In: 2006 IEEE International Conference on Robotics and Automation (ICRA), IEEE. pp. 2111–2116.
Clark, D.C., DeLeys, N.J., Matheis, C.W., 1962. Exploratory Investigation of the Man-Amplifier Concept. U.S Air Force AMRL-TDR-62-89, AD-390070.
Cordella, F., Ciancio, A.L., Sacchetti, R., Davalli, A., Cutti, A.G., Guglielmelli, E., Zollo, L., 2016. Literature review on needs of upper limb prosthesis users. Front. Neurosci. 10, 209.
Culbertson, H., Schorr, S.B., Okamura, A.M., 2018. Haptics: the present and future of artificial touch sensation. Annu. Rev. Control Robot. Auton. Syst. 1, 385–409.

Cytowic, R.E., 2002. Synesthesia: A Union of the Senses. MIT press.
Dahlheim, W., 1915. Pressluft Hand für kreigsbeschädigte Industriearbeiter Z. komprimierte und flüssige Gase. German Patent.
Daniels, C.A., Olsen, C.H., Scher, A.I., McKay, P.L., Niebuhr, D.W., 2019. Severe upper limb injuries in US military personnel: incidence, risk factor and outcomes. Mil. Med. 185, e146–e153.
Dewald, H.A., Lukyanenko, P., Lambrecht, J.M., Anderson, J.R., Tyler, D.J., Kirsch, R.F., Williams, M.R., 2019. Stable, three degree-of-freedom myoelectric prosthetic control via chronic bipolar intramuscular electrodes: a case study. J. Neuroeng. Rehabil. 16, 147.
Dhillon, G.S., Horch, K.W., 2005. Direct neural sensory feedback and control of a prosthetic arm. IEEE Trans. Neural Syst. Rehabil. Eng. 13 (4), 468–472.
Di Luzio, F.S., Lauretti, C., Cordella, F., Draicchio, F., Zollo, L., 2020. Visual vs vibrotactile feedback for posture assessment during upper-limb robot-aided rehabilitation. Appl. Ergon. 82, 102950.
Dosen, S., Cipriani, C., Kostic, M., Controzzi, M., Carrozza, M., Popovic, D., 2010. Cognitive vision system for control of dexterous prosthetic hands: experimental evaluation. J. Neuroeng. Rehabil. 7, 42.
Dumanian, G., Potter, B., Mioton, L., Ko, J., Cheesborough, J., Souza, J., Ertl, W., Tintle, S., Nanos, G., Valerio, I., Kuiken, T., Apkaian, V., Porter, K., Sumanas, J., 2019. Targeted muscle reinnervation treats neuroma and phantom pain in major limb amputees: a randomized clinical trial. Ann. Surg. 270 (2), 238–246.
Eklund, G., 1972. Position sense and state of contraction; the effects of vibration. J. Neurol. Neurosurg. Psychiatry 35 (5), 606–611.
Fougner, A., Stavdahl, O., Kyberd, P.J., Losier, Y.G., Parker, P.A., 2012. Control of upper limb prostheses: terminology and proportional myoelectric control-a review. IEEE Trans. Neural Syst. Rehabil. Eng. 20 (5), 663–677.
Geethanjali, P., 2016. Myoelectric control of prosthetic hands: state-of-art review. Med. Dev. 27 (9), 247–255.
Geiss, L.S., Li, Y., Hora, I., Albright, A., Rolka, D., Gregg, E.W., 2019. Resurgence of diabetes-related nontraumatic lower-extremity amputation in the young and middle-aged adult US population. Diabetes Care 42 (1), 50–54.
Georgarakis, A.M., Sonar, H.A., Rinderknecht, M.D., Lambercy, O., Martin, B.J., Klamroth-Marganska, V., Paik, J., Riener, R., Duarte, J.E., 2017. A novel pneumatic stimulator for the investigation of noise-enhanced proprioception. In: 2017 International Conference on Rehabilitation Robotics (ICORR). IEEE, pp. 25–30.
Goodwin, G.M., McCloskey, D.I., Matthews, P.B.C., 1972. The contribution of muscle afferents to keslesthesia shown by vibration induced illusions of movement and by the effects of paralysing joint afferents. Brain 95 (4), 705–748.
Gopura, R.A.R.C., Bandara, D.S.V., Kiguchi, K., Mann, G.K., 2016. Developments in hardware systems of active upper-limb exoskeleton robots: a review. Robot. Auton. Syst. 75, 203–220.
Gretsch, K., Lather, H., Peddada, K., Deeken, C., Wall, L., Goldfarb, C., 2015. Development of novel 3D-printed robotic prosthetic for transradial amputees. Prosthetics Orthot. Int. 40 (3), 400–403.
Hahne, J.M., Biessmann, F., Jiang, N., Rehbaum, H., Farina, D., Meinecke, F.C., Muller, K.R., Parra, L.C., 2014. Linear and nonlinear regression techniques for simultaneous and proportional control. IEEE Trans. Neural Syst. Rehabil. Eng. 22 (2), 269–279.
Heger, H., Millstein, S., Hunter, G.A., 1985. Electrically powered prostheses for the adult with an upper limb amputation. J. Bone Joint Surg. 67 (2), 278–281 British volume.
Heo, P., Kim, J., 2014. Power-assistive finger exoskeleton with a palmar opening at the Fingerpad. IEEE Trans. Biomed. Eng. 61 (11), 2688–2697.

Hero Arm, 2019. Open Bionics. Available from: https://openbionics.com/d100161_03_hero-arm-user-manual/. (Accessed 3 September 2019).
Hollies Hand Version 4, Anthromod, 2015. Available from: http://www.thingiverse.com/thing:696343. (Accessed 3 September 2019).
Huaroto, J.J., Suarez, E., Krebs, H.I., Marasco, P.D., Vela, E.A., 2018. A soft pneumatic actuator as a haptic wearable device for upper limb amputees: toward a soft robotic liner. IEEE Robot. Autom. Lett. 4 (1), 17–24.
Huaroto, J.J., Ticllacuri, V., Suarez, E., Ccorahua, R., Vela, E.A., 2019. A soft pneumatic haptic actuator mechanically programmed for providing Mechanotactile feedback. MRS Adv. 4 (19), 1131–1136.
Hung, K., Zhang, Y.T., Tai, B., 2004. Wearable medical devices for tele-home healthcare. In: Proceedings of the 26h Annual International Conference of the IEEE EMBS, vol. 7, pp. 5384–5387.
I-Limb Ultra, 2019. Available from: https://www.ossur.com/en-us/prosthetics/arms/i-limb-ultra. (Accessed 9 March 2019).
Islam, M.R., Spiewak, C., Rahman, M.H., Fareh, R., 2017. A brief review on robotic exoskeletons for upper extremity rehabilitation to find the gap between research porotype and commercial type. Adv. Robot. Autom. 6 (3), 2.
Jacobsen, S.C., Knutti, D.F., Sarcos, L.C., 1999. Body-Powered Prosthetic Arm. U.S. Patent 5,888,235.
Jarrassé, N., Proietti, T., Crocher, V., Robertson, J., Sahbani, A., Morel, G., Roby-Brami, A., 2014. Robotic exoskeletons: a perspective for the rehabilitation of arm coordination in stroke patients. Front. Hum. Neurosci. 8, 947.
Johannes, M., Faulring, E., Katyal, K., Helder, J., Makhlin, A., Moyer, T., Wahl, D., Solberg, J., Clark, S., Armiger, R., Lontz, T., Geberth, K., Moran, C., Wester, B., Van Doren, T., Santos-Munne, J., 2020. The modular prosthetic limb. In: Wearable Robotics. Academic Press, pp. 393–444.
Johnson, K.O., 2001. The roles and functions of cutaneous mechanoreceptors. Curr. Opin. Neurobiol. 11 (4), 455–461.
Kaczmarek, K.A., Webster, J.G., Bach-y-Rita, P., Tompkins, W.J., 1991. Electrotactile and vibrotactile displays for sensory substitution systems. IEEE Trans. Biomed. Eng. 38 (1), 1–16.
Kazerooni, H., 1989. Human-robot interaction via the transfer of power and information signals. IEEE Trans. Syst. Man Cybern. 20 (2), 450–463.
Kim, K., Colgate, J.E., Santos-Munné, J.J., Makhlin, A., Peshkin, M.A., 2009. On the design of miniature haptic devices for upper extremity prosthetics. IEEE/ASME Trans. Mechatron. 15 (1), 27–39.
Kristjansdottir, F., Dahlin, L.B., Rosberg, H.E., Carlsson, I.K., 2019. Social participation in persons with upper limb amputation receiving an esthetic prosthesis. J. Hand Ther. https://doi.org/10.1016/j.jht.2019.03.010 in press.
Kuiken, T., 2006. Targeted reinnervation for improved prosthetic function. Phys. Med. Rehabil. Clin. N. Am. 17 (1), 1–13.
Kuiken, T.A., Dumanian, G.A., Lipschutz, R.D., Miller, L.A., Stubblefield, K.A., 2004. The use of targeted muscle reinnervation for improved myoelectric prosthesis control in a bilateral shoulder disarticulation amputee. Prosthetics Orthot. Int. 28 (3), 245–253.
Kuiken, T., Miller, L.A., Lipschutz, R.D., Lock, B.A., Stubblefield, K., Marasco, P.D., Zhou, P., Dumanian, G., 2007. Targeted reinnervation for enhanced prosthetic arm function in a woman with a proximal amputation: a case study. Lancet 369 (9559), 371–380.
Kuiken, T.A., Li, G., Lock, B.A., Lipschutz, R.D., Miller, L.A., Stubblefield, K.A., Englehart, K.B., 2009. Targeted muscle reinnervation for real-time myoelectric control of multifunction artificial arms. JAMA 301 (6), 619–628.

Kuiken, T.A., Barlow, A.K., Hargrove, L., Dumanian, G.A., 2017. Targeted muscle reinnervation for the upper and lower extremity. Tech. Orthop. 32 (2), 109–116.

Laliberté, T., Baril, M., Guay, F., Gosselin, C., 2010. Towards the design of a prosthetic underactuated hand. Mech. Sci. 1, 19–26.

LUKE Arm, 2019. Available from: http://www.mobiusbionics.com/wp-content/uploads/2017/08/Mobius-Bionics-LUKE-Product-Spec-Sheet.pdf. (Accessed 3 September 2019).

Lundborg, G., Rosén, B., Lindberg, S., 1999. Hearing as substitution for sensation: a new principle for artificial sensibility. J. Hand. Surg. 24 (2), 219–224.

Maciejasz, P., Eschweiler, J., Gerlach-Hahn, K., Jansen-Troy, A., Leonhardt, S., 2014. A survey on robotic devices for upper limb rehabilitation. J. Neuroeng. Rehabil. 11 (1), 3.

Marasco, P.D., Hebert, J.S., Sensinger, J.W., Shell, C.E., Schofield, J.S., Thumser, Z.C., Nataraj, R., Beckler, D.T., Dawson, M.R., Blustein, D.H., Gill, S., 2018. Illusory movement perception improves motor control for prosthetic hands. Sci. Transl. Med. 10(432), eaao6990.

Markovic, M., Schweisfurth, M.A., Engels, L.F., Bentz, T., Wüstefeld, D., Farina, D., Dosen, S., 2018. The clinical relevance of advanced artificial feedback in the control of a multi-functional myoelectric prosthesis. J. Neuroeng. Rehabil. 15 (1), 28.

Marquardt, E., 1965. The Heidelberg pneumatic arm prosthesis. J. Neuroeng. Rehabil. British volume. 47 (3), 425–434.

Martino, G., Marks, L.E., 2001. Synesthesia: strong and weak. Curr. Dir. Psychol. Sci. 10 (2), 61–65.

Massa, B., Roccella, S., Carrozza, M.C., Dario, P., 2002. Design and development of an underactuated prosthetic hand. In: 2002 IEEE International Conference on Robotics and Automation (ICRA). IEEE, vol. 4. pp. 3374–3379.

Mathiowetz, V., Volland, G., Kashman, N., Weber, K., 1985. Adult norms for the box and block test of manual dexterity. Am. J. Occup. Ther. 39 (6), 386–391.

Mazumdar, A., 2004. Powered Upper Limb Prostheses Control, Implementation and Clinical Application. Springer.

Meek, S.G., Jacobsen, S.C., Goulding, P.P., 1989. Extended physiologic taction: design and evaluation of a proportional force feedback system. J. Rehabil. Res. Dev. 26 (3), 53–62.

Meier, R.H., 2004. History of arm amputation, prosthetic restoration and arm amputation rehabilitation. In: Functional Restoration of Adults and Children with Upper Limb Amputation. Springer, pp. 1–8.

Merchant, R., Cruz-Ortiz, D., Ballesteros-Escamilla, M., Chairez, I., 2018. Integrated wearable and self-carrying active upper limb orthosis. Proc. Inst. Mech. Eng. H J. Eng. Med. 232 (2), 172–184.

Micera, S., Carpaneto, J., Raspopovic, S., 2010. Control of hand prostheses using peripheral information. IEEE Rev. Biomed. Eng. 3, 48–68.

Michelangelo Hand, 2019. Available from: https://accessprosthetics.com/wp-content/uploads/2017/06/michelangelo-technology.pdf. (Accessed 3 September 2019).

Miller, L.A., Stubblefield, K.A., Lipschutz, R.D., Lock, B.A., Kuiken, T.A., 2008. Improved myoelectric prosthesis control using targeted reinnervation surgery: a case series. IEEE Trans. Neural Syst. Rehabil. Eng. 16 (1), 46–50.

Motti, V.G., 2020. Introduction to wearable computers. In: Wearable Interaction. Springer, Cham, pp. 1–39.

Muceli, S., Farina, D., 2011. Simultaneous and proportional estimation of hand kinematics from EMG during mirrored movements at multiple degrees-of-freedom. IEEE Trans. Neural Syst. Rehabil. Eng. 20 (3), 371–378.

Muilenburg, A.L., LeBlanc, M.A., 1989. Body-powered upper-limb components. In: Comprehensive Management of the Upper-Limb Amputee. Springer, New York, NY, pp. 28–38.

Nemah, M.N., Low, C.Y., Aldulaymi, O.H., Ong, P., Qasim, A.A., 2019. A review of non-invasive haptic feedback stimulation techniques for upper extremity prostheses. Int. J. Integr. Eng. 11(1).

Ortiz-Catalan, M., Håkansson, B., Brånemark, R., 2014. An osseointegrated human-machine gateway for long-term sensory feedback and motor control of artificial limbs. Sci. Transl. Med. 6 (257), 257re6.

Ostlie, K., Lesjo, I.M.M., Franklin, R.J., Garfelt, B., Skieldal, O.H., Magnus, P., 2012. Prosthesis use in adult acquired major upper-limb amputees: patterns of wear, prosthetic skills and the actual use of prostheses in activities of daily life. Disabil. Rehabil. Assist. Technol. 7, 479–493.

Patel, S., Park, H., Bonato, P., Chan, L., Rodgers, M., 2012. A review of wearable sensors and systems with application in rehabilitation. J. Neuroeng. Rehabil. 9, 21.

Perry, J., Rosen, J., Burns, S., 2007. Upper-limb powered exoskeleton design. IEEE/ASME Trans. Mechatron. 12 (4), 408–417.

Polygerinos, P., Wang, Z., Galloway, K., Wood, R., Walsh, C., 2014. Soft robotic glove for combined assistance and at-home rehabilitation. Robot. Auton. Syst. 73, 135–143.

Pons, J.L., 2008. Wearable Robots: Biomechatronic Exoskeletons. John Wiley & Sons.

Popov, B., 1965. The bio-electrically controlled prosthesis. J. Bone Joint Surg. 14, 421–424.

Proske, U., Gandevia, S.C., 2012. The proprioceptive senses: their roles in signaling body shape, body position and movement, and muscle force. Physiol. Rev. 92 (4), 1651–1697.

Radder, B., Prange-Lasonder, G., Kottink, A., Holmberg, J., Sletta, K., van Dijk, M., Meyer, T., Buurke, J.H., Rietman, J.S., 2018. The effect of a wearable soft-robotic glove on motor function and functional performance of older adults. Assist. Technol. 30, 1–7.

Rahman, M.H., Ochoa-Luna, C., Rahman, J., Saad, M., Archambault, P., 2014. Force-position control of a robotic exoskeleton to provide upper extremity movement assistance. Int. J. Model. Identif. Control. 21 (4), 390–400.

Raskovic, D., Martin, T., Jovanov, E., 2004. Medical monitoring applications for wearable computing. Comput. J. 47(4).

Raspopovic, S., Petrini, F.M., 2018. A computational model for the design of lower-limb sensorimotor neuroprostheses. In: International Conference on NeuroRehabilitation. Springer, Cham, pp. 49–53.

Resnik, L., Meucci, M.R., Lieberman-Klinger, S., Fantini, C., Kelty, D.L., Disla, R., Sasson, N., 2012. Advanced upper limb prosthetic devices: implications for upper limb prosthetic rehabilitation. Arch. Phys. Med. Rehabil. 93 (4), 710–717.

Resnik, L., Ekerholm, S., Borgia, M., Clark, M.A., 2019. A national study of veterans with major upper limb amputation: survey methods, participants, and summary findings. PLoS One 14(3), e0213578.

Roche, A.D., Lakey, B., Mendez, I., Vujaklija, I., Farina, D., Aszmann, O.C., 2019. Clinical perspectives in upper limb prostheses: an update. Curr. Surg. Rep. 7 (3), 5.

Schlesinger, G., 1919. Der mechanische aufbau der künstlichen glieder. In: Ersatzglieder und Arbeitshilfen. Springer, Berlin, Heidelberg, pp. 321–661.

Schofield, J.S., Evans, K.R., Carey, J.P., Hebert, J.S., 2014. Applications of sensory feedback in motorized upper extremity prosthesis: a review. Expert Rev. Med. Devices 11 (5), 499–511.

Schulz, S., F'ylatiuk, C., Bretthauer, G., 2001. A new ultralight anthropomorphic hand. In: 2001 IEEE Conference on Robotics and Automation (ICRA). IEEE, pp. 2437–2441.

Scott, R.N., 1992. Myoelectric control of prostheses, a brief history. In: 1992 Myoelectric Controls/Powered Prosthetics Symposium, Fredericton, Canada.

Shen, Y., Rosen, J., 2020. EXO-UL upper limb robotic exoskeleton system series: from 1 DOF single-arm to (7+1) DOFs dual-arm. In: Wearable Robotics. Academic Press, pp. 91–103.

Shen, Y., Ferguson, P.W., Rosen, J., 2020. Upper limb exoskeleton systems—overview. In: Wearable Robotics. Academic Press, pp. 1–22.
Sonar, H.A., Paik, J., 2016. Soft pneumatic actuator skin with piezoelectric sensors for vibrotactile feedback. Front. Robot. AI 2, 38.
Suarez, E., Huaroto, J.J., Reymundo, A.A., Holland, D., Walsh, C., Vela, E., 2018. A soft pneumatic fabric-polymer actuator for wearable biomedical devices: proof of concept for lymphedema treatment. In: 2018 IEEE International Conference on Robotics and Automation (ICRA). IEEE, pp. 5452–5458.
Szeto, A.Y., Saunders, F.A., 1982. Electrocutaneous stimulation for sensory communication in rehabilitation engineering. IEEE Trans. Biomed. Eng. 4, 300–308.
Talon Flextensor 1.0, 2014. Available from: https://www.thingiverse.com/thing:413983. (Accessed 3 September 2019).
TASKA Hand, 2019. TASKA Prosthetics. Available from: http://www.taskaprosthetics.com/. (Accessed 3 September 2019).
Ten Kate, J., Smit, G., Breedveld, P., 2017. 3D-printed upper limb prostheses: a review. Disabil. Rehabil. Assist. Technol. 12 (3), 300–314.
Thomas, N., Ung, G., McGarvey, C., Brown, J.D., 2019. Comparison of vibrotactile and joint-torque feedback in a myoelectric upper-limb prosthesis. J. Neuroeng. Rehabil. 16 (1), 70.
Upshaw, B., Sinkjaer, T., 1998. Digital signal processing algorithms for the detection of afferent nerve activity from cuff electrodes. IEEE Trans. Rehabil. Eng. 6 (2), 172–181.
Valle, G., Mazzoni, A., Iberite, F., D'Anna, E., Strauss, I., Granata, G., Controzzi, M., Clemente, F., Rognini, G., Cipriani, C., Stieglitz, T., 2018. Biomimetic intraneural sensory feedback enhances sensation naturalness, tactile sensitivity, and manual dexterity in a bidirectional prosthesis. Neuron 100 (1), 37–45.
Veale, A., Staman, K., Van der Kooij, H., 2018. Realizing soft high torque actuators for complete assistance wearable robots. In: International Symposium on Wearable Robotics, pp. 39–43.
Velliste, M., Perel, S., Chance, M., Whitford, A., Schwartz, A., 2008. Cortical control of a prosthetic arm for self-feeding. Nature 453, 1098–1101.
Victoria Hand, 2019. Available from: https://www.victoriahandproject.com/vhp-hands. (Accessed 3 September 2019).
VINCENT evolution 3, 2019. Available from: https://vincentsystems.de/en/. (Accessed 3 September 2019).
Voloshina, A.S., Collins, S.H., 2020. Lower limb active prosthetic systems—overview. In: Wearable Robotics. Academic Press, pp. 469–486.
Weihe, W., 1998. In memorian reinhold reiter. Int. J. Biometeorol. 43, 96–98.
Weir, R.F., Troyk, P.R., DeMichele, G.A., Kerns, D.A., Schorsch, J.F., Mass, H., 2009. Implantable myoelectric sensors (IMESs) for intramuscular electromyogram recording. IEEE Trans. Biomed. Eng. 56 (1), 159–171.
Witteveen, H.J., Droog, E.A., Rietman, J.S., Veltink, P.H., 2012. Vibro-and electrotactile user feedback on hand opening for myoelectric forearm prostheses. IEEE Trans. Biomed. Eng. 59 (8), 2219–2226.
Xiloyannis, M., Chiaradia, D., Frisoli, A., Masia, L., 2019. Physiological and kinematic effects of soft exosuit on arm movements. J. Neuroeng. Rehabil. 16(29).
Yang, Y., Sahabi, A., 2016. Modular Wearable Sensor Device. US9277864B2.
Zero Point Frontiers, 2013. Available from: https://www.zeropointfrontiers.com/2013/08/zero-point-frontiers-prints-low-cost-prosthetic-hand/. (Accessed 3 September 2019).
Zuo, K.J., Olson, J.L., 2014. The evolution of functional hand replacement: from iron prostheses to hand transplantation. Plast. Surg. (Oakv) 22 (1), 44–51.

CHAPTER 9

Exoskeletons in upper limb rehabilitation: A review to find key challenges to improve functionality*

Md Rasedul Islam[a], Brahim Brahmi[b], Tanvir Ahmed[c], Md. Assad-Uz-Zaman[c], Mohammad Habibur Rahman[c]
[a]Richard J. Resch School of Engineering, University of Wisconsin-Green Bay, Green Bay, WI, United States
[b]Musculoskeletal Biomechanics Research Lab, McGill University, Montreal, QC, Canada
[c]Bio-Robotics Lab, University of Wisconsin-Milwaukee, Milwaukee, WI, United States

1 Introduction

Approximately 795,000 strokes occur annually in the United States, according to a report by the American Heart Association (Benjamin et al., 2017). The global scenario of stroke is quite similar to that in the United States; according to a report by the World Health Organization, every year 15 million people suffer a stroke worldwide, among which two-thirds survive (Stroke Statistics, 2019). The recovery in stroke occurrences shows that a large number of patients end up with impairments; 25% of survivors are left with minor impairments, while 40% of survivors appear to experience moderate-to-severe impairments requiring special care (American Stroke Association, 2019). Stroke is one of the main causes of serious long-term disability due to it often impairing the upper limbs of the body. In addition, upper limb impairments can occur due to sports injuries, trauma, occupational injuries, and spinal cord injuries. Since upper limbs are involved in performing a wide variety of daily activities, it is imperative to return independence back to upper limb-impaired patients as quickly as possible. Rehabilitation programs are the main method of promoting functional recovery in individuals with upper limb dysfunction (ULD), requiring a long commitment by both the clinician and patient.

To rehabilitate upper limb impairments, extensive task-oriented repetitive movement has proved to be a safe and effective method that often

* Fully documented templates are available in the elsarticle package on CTAN.

Control Theory in Biomedical Engineering
https://doi.org/10.1016/B978-0-12-821350-6.00009-3

depends on one-on-one physical interaction with the therapist (Poli et al., 2013; Demircan et al., 2020). As cases of ULD are increasing, robot-aided therapeutic intervention has the potential to be an effective solution in this regard. Moreover, there are recent studies corroborating that repetitive robot-assisted rehabilitation programs decrease upper limb motor impairment significantly (Amirabdollahian et al., 2007; Gandolfi et al., 2018; Veerbeek et al., 2016; Kim et al., 2017; Lee et al., 2017; Sale et al., 2014; Yoo and Kim, 2015). Robot-aided therapy has advantages over conventional manual therapy, as the former is capable of providing therapy to patients for a longer period of time, is a more precise training method, and results in better quantitative feedback (Teasell and Kalra, 2004). There are two kinds of robot-aided rehabilitative devices based on the mapping of a device's joint onto human anatomical joints (e.g., end-effector type and exoskeleton type). End-effector-type devices (e.g., MIT-MANUS-commercialized as Inmotion as shown in Fig. 1A (Hogan et al., 1992; Krebs et al., 2007, 2016), Gentle/S (Coote et al., 2008), ARM Guide (Reinkensmeyer et al., 2000)) are suitable for end-point exercises as they are unable to provide individual joint movement, meaning they cannot map onto human anatomical joints.

Exoskeleton type devices have advantages over end-effector-type devices, as they have complete control over a patient's individual joint movement and applied torque, better guidance of motion, relatively larger range of motion (ROM), and better quantitative feedback. To date numerous research prototypes of exoskeletons have been developed for human

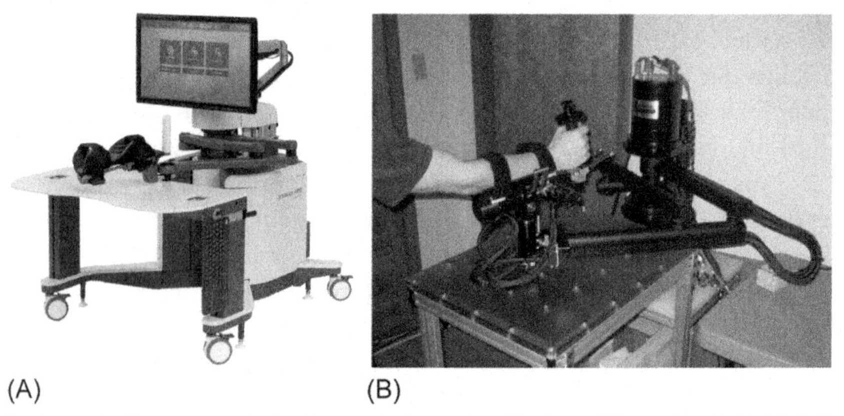

Fig. 1 End-effector-type device for upper limb rehabilitation. (A) Inmotion ARM; (B) MIT-MANUS (Krebs et al., 2007). *((A) Courtesy to Bionik Lab.)*

upper limb rehabilitation (e.g., CADEN-7 (Perry et al., 2007), ARMIN (Nef et al., 2009a), SUFUL-7 (Gopura et al., 2009), MARSE-7 (Rahman et al., 2014), 6-REXOS (Gunasekara et al., 2015), RehabArm (Liu et al., 2016), CAREX-7 (Cui et al., 2017), CABexo (Xiao et al., 2017), BLUE SABINO (Perry et al., 2019), u-Rob (Islam et al., 2020), PWRR (Zhang et al., 2020), and so on; see Figs. 2–4). However, their usage in clinical settings at hospitals and outpatient centers is still limited; only one exoskeleton is commercially available (see Fig. 5A). To make an exoskeleton suitable for clinical usage, researchers have been looking for better solutions in terms of light weight, compactness, low power-to-weight ratio, lightweight reducers for power transmission, easy don/doff, quick don/doff, alignment with human joints, modularity of kinematic structure, fast computation, sampling, control algorithm, and modularity of control.

In addition to rehabilitation, researchers have designed and developed robotic exoskeletons for other applications as well. For instance, this exoskeleton was designed and developed to augment human strength while handling heavy loads in the unstructured environment. Marcheschi et al. (2011) developed an exoskeleton to use as a body extender. In BLEEX, Kazerooni (2005) successfully improved maneuverability, mechanical robustness and durable outfit to surpass typical human limitations while carrying heavy load. Zhu et al. (2014) used electromyography signals of agonist muscle to control their exoskeleton for power augmentation. Walsh et al. (2006) developed an under-actuated exoskeleton that augmented the human power for lifting heavy loads. The purpose of this chapter is to review the hardware design and control aspects of existing exoskeletons for upper limb rehabilitation to find challenges that need to be overcome for improved functionality. To limit the scope of this review, end-effector-type robotic devices are excluded. Moreover, upper limb exoskeletons that support either shoulder, elbow, and/or wrist rehabilitation are reviewed, while exoskeletons for finger rehabilitation are not considered. Furthermore, the authors do not guarantee to include all the upper limb exoskeletons. This chapter is organized as follows: Section 2 illustrates existing (either research prototype or commercial version) exoskeletons for upper limb rehabilitation; Section 3 describes the design requirements and challenges for such exoskeletons; Section 4 reviews control approaches used in upper limb exoskeletons; Section 5 presents an overall discussion, and the chapter concludes with Section 6. Throughout this chapter, "exoskeletons" is used interchangeably with "upper limb exoskeleton for rehabilitation."

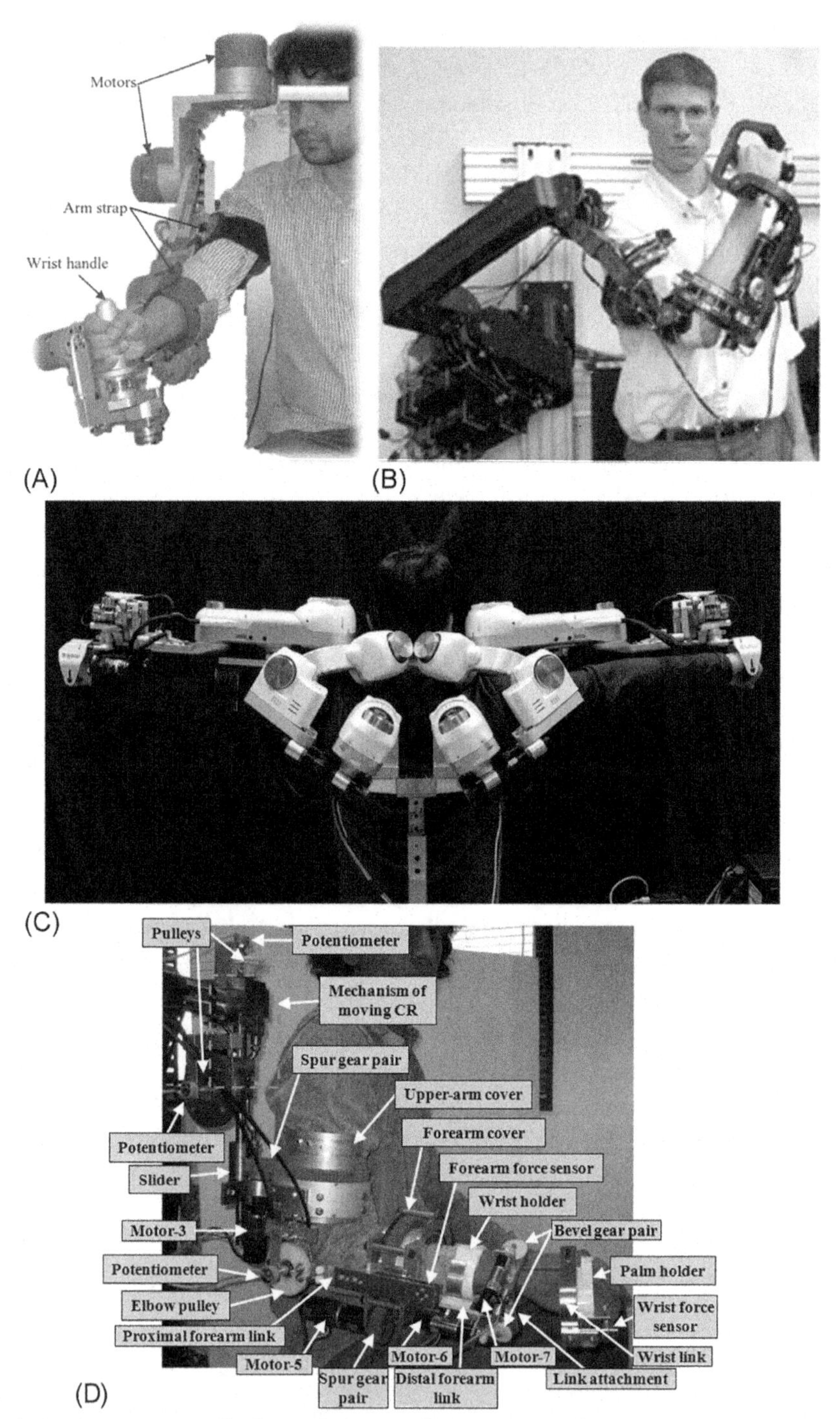

Fig. 2 Existing upper limb exoskeletons. (A) MARSE-7 (Rahman et al., 2013a, 2014); (B) CADEN-7 (Perry et al., 2007); (C) Harmony (Kim and Deshpande, 2017); (D) SUFUL-7 (Gopura et al., 2009).

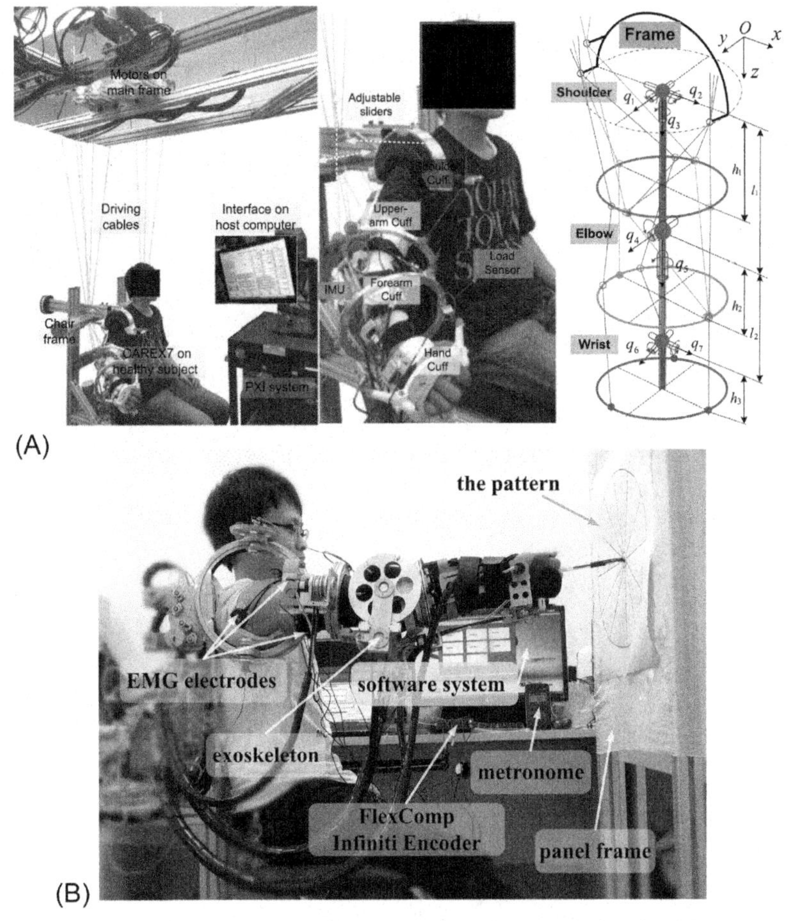

Fig. 3 Cable-driven-type exoskeletons. (A) CAREX-7 (Cui et al., 2017); (B) CABXLeo-7 (Xiaoet al., 2018).

2 Existing upper limb exoskeletons

Though development of end-effector-type devices for upper limb rehabilitation began in early 1990, exoskeleton-type devices started to be developed in early 2000. In this section, we present case studies of selected exoskeletons. We also include Table 1 that summarizes several exoskeletons.

SUEFUL-7 (Gopura et al., 2009) as shown in Fig. 2D, a 7-DOF upper limb motion-assist exoskeleton robot was developed by Gopura et al. (2009) in Dr. Kazuo Kiguchi's Laboratory at Saga University, Japan, in 2009. This robot was designed for providing motion assist to physically weak

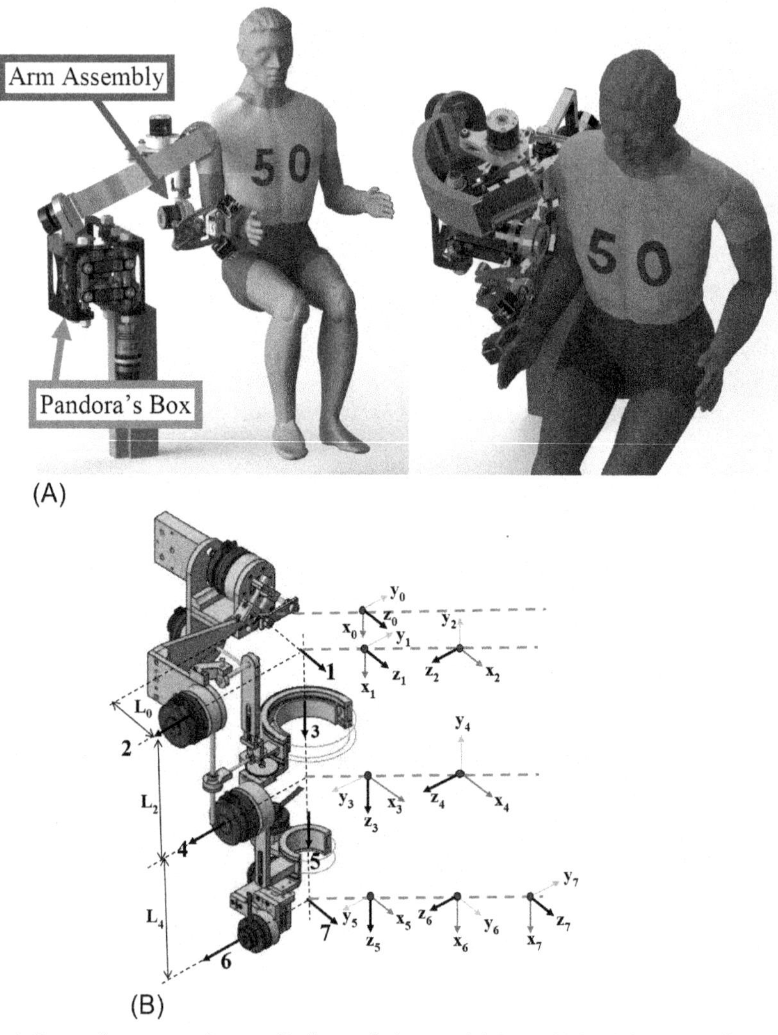

Fig. 4 Recently proposed upper limb exoskeletons. (A) BLUE SABINO (Perry et al., 2019); (B) u-Rob (Islam et al., 2019, 2020).

individuals in their activities of daily living. Design considerations for SUEFUL-7 include moving center of rotation (CR) of shoulder joint and axes deviation of wrist. In this robotic exoskeleton, the shoulder vertical and horizontal flexion/extension, elbow flexion/extension movements are driven by pulleys and cable drive while the actuators are placed on the stationary frames. Whereas, for delivering shoulder internal/external rotation, forearm supination/pronation, wrist flexion/extension, and wrist radial/

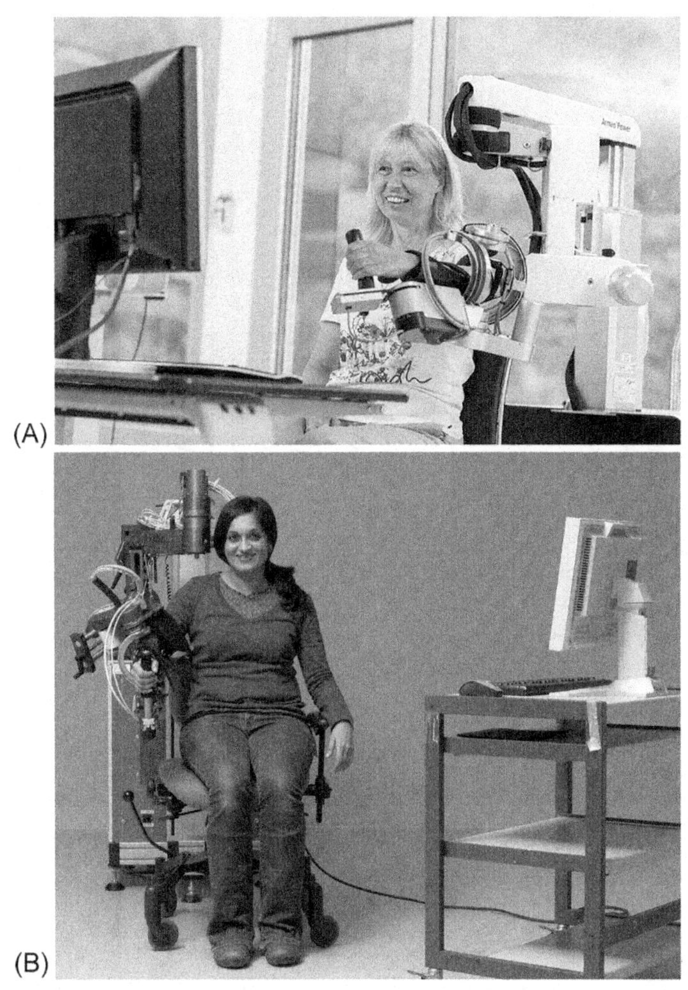

Fig. 5 Commercially available exoskeleton for upper limb rehabilitation. (A) Armeopower exoskeleton; (B) ARMIN (Nef et al., 2009a, b), research prototype of Armeopower. *((A) Courtesy to Hocoma.)*

ulnar deviation movements to the robot, the actuators are mounted on the robot itself and either directly connected or via gear drive. A slider crank mechanism was used for compensating CR of the shoulder joint. The robot weighs about 5 kg and is intended to be wheelchair mounted under the impression that physically weak individuals use wheelchairs. In order to control SUEFUL-7, an EMG-based neuro-fuzzy control was applied. This exoskeleton was tested with healthy subjects only.

Table 1 Summary table of existing upper limb rehabilitative exoskeletons.

Device/researcher	Specifications	Remarks
Commercially available		
ARMIN (research version of commercially available Armeopower) (Nef et al., 2009b; Guidali et al., 2011)	**DOFs:** 6 **Motion:** Shoulder (Ab/Ad, F/E, IR/ER), elbow (F/E), forearm (P/S), wrist (F/E) **Actuation:** DC motors with strain wave gear reduction **Control:** (Method: PD, CTC, impedance control; input: position, wrist force; output: torque) **Clinical test:** Yes	**Strength:** CR mechanism for shoulder **Weakness:** Lower ROM than natural ROM in shoulder
Research prototypes		
Kiguchi et al. (2003)	**DOF:** 1 **Motion:** Elbow (F/E) **Actuation:** DC motor coupled with ballscrew drive shaft **Control:** (Method: neuro-fuzzy; input: biceps and triceps EMG, wrist force; output: torque) **Clinical test:** No	**Strength:** Adjustability of the assist level; consideration of patient's physiological condition **Weakness:** Controller adaption is required before each experiment; high computation required for neuro-fuzzy
SRE (Tsagarakis and Caldwell, 2003)	**DOFs:** 7 **Motion:** Shoulder (Ab/Ad, F/E, IR/ER), elbow (F/E), forearm (P/S), wrist (R/U, F/E) **Actuation:** Pneumatic muscle actuators with cable-driven pulley reduction **Control:** (Method: impedance control; input: position and torque; output: torque) **Clinical test:** No	**Strength:** Lightweight; high power-to-weight ratio; inherent compliance **Weakness:** Cable could stretch and slip
Cheng et al. (2004)	**DOF:** 1 **Motion:** Elbow (F/E) **Actuation:** Servo motor coupled with a regular shaft **Control:** (Method: proportional EMG control; input: biceps and triceps EMG; output: torque) **Clinical test:** Yes (5)	**Strength:** Nonlinear damping was added; adaptive filter for torque signal **Weakness:** Oscillation occurs if subject exerts an excessive force, making system unable to deal with uncertainties and disturbances

Song et al. (2007)	**DOF:** 1 **Motion:** Wrist (F/E) **Actuation:** Rotary-type soft actuators using pneumatic pressure **Control:** (Method: proportional EMG control; input: flexor carpi radialis and extensor carpi radialis EMG, torque, and servo angle; output: motor current (torque control), position (position control)) **Clinical test:** No	**Strength:** Wrist has a support during experiment as it rests on a planer surface; two control modes (Torque and position) **Weakness:** Before every trial gain for EMG signals need to be adjusted
CADEN-7 (Perry et al., 2007)	**DOFs:** 7 **Motion:** Shoulder (Ab/Ad, F/E, IR/ER), elbow (F/E), forearm (P/S), wrist (R/U, F/E) **Actuation:** Brushed motors with cable-driven reduction pulleys **Clinical test:** No	**Strength:** Hard to reach singular position; full ROM **Weakness:** Cable could stretch or slip, which might affect bearings, and cause transmission backlash and cable derailment
Kiguchi et al. (2008)	**DOFs:** 3 **Motion:** Shoulder (F/E, Ab/Ad), elbow (F/E) **Actuation:** DC servo with harmonic gearing **Control:** (Method: neuro-fuzzy; input: EMG signals, wrist force; output: torque) **Clinical test:** No	**Strength:** CR mechanism for shoulder **Weakness:** Mechanism used did not realize shoulder elevation/depression; controller adaption is required before each experiment
SUEFUL-7 (Gopura et al., 2009)	**DOFs:** 7 **Motion:** Shoulder (vertical and horizontal F/E, IR/ER), elbow (F/E), forearm (P/S), wrist (R/U, F/E) **Actuation:** DC motors with spur gear, bevel gears, and cable-driven pulley **Control:** (Method: impedance control based on EMG; input: EMG signals, wrist force; output: torque) **Clinical test:** No	**Strength:** CR mechanism for shoulder **Weakness:** No means for maintaining cable tension; gear could have backlash; controller adaption is required before each experiment

Continued

Table 1 —cont'd

Device/researcher	Specifications	Remarks
MARSE-7 (Rahman et al., 2013a, 2014)	**DOFs:** 7 **Motion:** Shoulder (vertical and horizontal F/E, IR/ER), elbow (F/E), forearm (P/S), wrist (R/U, F/E) **Actuation:** Brushless motors with strain wave gear reduction **Control:** (Method: CTC, SMC, PID; input: EMG, wrist force; output: torque) **Clinical test:** No	**Strength:** Full ROM, robust control **Weakness:** Absence of CR mechanism for shoulder
6-REXOS (Gunasekara et al., 2015)	**DOFs:** 4 **Motion:** Elbow (F/E), forearm (P/S), wrist (R/U, F/E) **Actuation:** DC motors with spur and bevel gears reduction **Clinical test:** No	**Strength:** Passive DOF to allow motion of CR of elbow and wrist joint **Weakness:** No bearing at semicircular surfaces, which might produce backlash
Triwiyanto et al. (2016)	**DOF:** 1 **Motion:** Elbow (F/E) **Actuation:** Cable-driven DC servo pulley **Control:** (Method: proportional EMG; input: biceps EMG; output: servo position) **Clinical test:** No	**Strength:** Zero-crossing shows promising prediction of elbow motion **Weakness:** No measure taken to ensure proper cable tension; no use of antagonist muscles
Rehab-Arm (Liu et al., 2016)	**DOFs:** 7 **Motion:** Shoulder (F/E, Ab/Ad, IR/ER), elbow (F/E), forearm (P/S), wrist (R/U, F/E) **Actuation:** Cylindrical-shaped micromotor with reduction bevel gears, belt, and pulleys **Control:** (Method: fuzzy-PID; input: joint position; output: torque and velocity) **Clinical test:** No	**Strength:** Lightweight; six motors to actuate seven DOFs **Weakness:** Constant velocity; unable to provide simultaneous movement of forearm and wrist

CAREX-7 (Cui et al., 2017)	**DOFs:** 7 **Motion:** Shoulder (Ab/Ad, F/E, IR/ER), elbow (F/E), forearm (P/S), wrist (R/U, F/E) **Actuation:** DC motors with spur gear, bevel gears, and cable-driven pulley **Control:** (Method: CTC; input: position, velocity, and wrist force; output: torque) **Clinical test:** No	**Strength:** Cable-driven parallel mechanism allows little to no kinematic mismatch between exoskeleton and its wearer; robust control **Weakness:** Cable could stretch and slip
CABXLexo-7 (Xiao et al., 2018)	**DOFs:** 7 **Motion:** Shoulder (Ab/Ad, F/E, IR/ER), elbow (F/E), forearm (P/S), wrist (R/U, F/E) **Actuation:** DC motors with cable-driven epicyclic gear train reduction **Clinical test:** No	**Strength:** Lightweight; stiff kinematic structure **Weakness:** Gear transmission could have backlash
Islam et al. (2019)	**DOF:** 3 **Motion:** Shoulder (Ab/Ad, F/E, IR/ER) **Actuation:** Brushless motors with strain wave gear reduction **Control:** (Method: CTC; input: position, velocity; output: torque) **Clinical test:** No	**Strength:** Ergonomic CR mechanism **Weakness:** Only three DOFs
u-Rob(Islam et al., 2020)	**DOF:** 7 **Motion:** Shoulder (Ab/Ad, F/E, IR/ER), elbow (F/E), forearm (P/S), wrist (R/U, F/E) **Actuation:** Brushless motors with strain wave gear reduction **Control:** (Method: FSMC input: position, velocity; output: torque) **Clinical test:** No	**Strength:** Ergonomic CR mechanism

Notes: Clinical test: The number in the bracket in clinical test shows number of patients that participated in the test.
DOFs, degrees of freedom; *ROM*, range of motion; *EMG*, electromyography signals; *Ab*, abduction; *Ad*, adduction; *F*, flexion; *E*, extension; *P*, pronation; *S*, supination; *R*, radial deviation; *U*, ulnar deviation; *PD*, proportional derivative; *PID*, proportional integral derivative; *CTC*, computed torque control; *SMC*, sliding mode control; *FSMC*, fractional sliding mode control.

CABXLexo-7 (Xiao et al., 2018) as shown in Fig. 3B, a 7-DOF cable-driven upper limb exoskeleton (CABXLexo-7) successor to 6-DOF CABexo (Xiao et al., 2017), was developed by Feiyun et al. from Harbin Institute of Technology (HIT), China and Hefei University of Technology (HFUT), China, as a joint effort in 2017 (Xiao et al., 2018). In order to create a lightweight exoskeleton robot capable of providing all seven DOFs, the development team put all the actuators on the stationary board and transmitted power through a cable-conduit system using two types of cable-driven differential mechanisms and using a tension device to work with the cable slag problem. The resulting weight of the moving robot is 3.5 kg. However, this robot does not consider the movement of CR of the human shoulder. Experimentation was carried out using surface electromyography on with five healthy individuals. This robot was designed for providing motion assist to poststroke patients (Xiao et al., 2018). CABXLexo-7 is yet to go through clinical trials.

ARMin-III (Nef et al., 2009a, b) as shown in Fig. 5B (successor of ARMIn and ARMin-II, and predecessor of commercial exoskeleton Armeopower as shown in Fig. 5A), developed at ETH Zurich, Switzerland, is one of the early and well-known robotic exoskeletons with high degrees of freedom for upper extremity rehabilitation. The very first version ARMin (Nef et al., 2007) was designed with four DOFs intended to provide rehabilitation in the human shoulder (giving mobility for shoulder abduction-adduction, flexion-extension, and internal-external rotation) and elbow (flexion-extension). Then, the 7-DOF ARMin-II was developed with five adjustable lengths segments to provide better patient cooperative rehabilitation. Unlike ARMin, the shoulder axis of rotation is not fixed in ARMin-II, allowing passive elevation/depression and protraction/retraction of the glenohumeral (GH) joint during shoulder vertical flexion-extension. ARMin-II also includes ergonomic shoulder actuation to provide as much natural movement as possible for shoulder rehabilitation. The advancement of the ARMin rehabilitative exoskeleton went through several stages of development and is now commercially available (known as ArmeoPower developed by Hocoma AG, Volketswil, Switzerland) for use in human upper extremity rehabilitation at clinical settings in hospitals.

ETS-MARSE (Rahman et al., 2013a) as shown in Fig. 2A, a 7-DOF upper limb exoskeleton for whole arm, used a novel power transmission mechanism to assist shoulder internal/external rotation and forearm pronation supination (Rahman et al., 2012, 2014). Since it is somewhat difficult to fit a shaft along the axis of rotation of above cases (axis of humerus and radius), the developer of ETS-MARSE used an antibacklash spur gear meshed with open-type semicircular gear and bearing assembly.

Harmony (Kim and Deshpande, 2015, 2017) as shown in Fig. 2C, a recent robotic exoskeleton in the field of upper limb rehabilitation, has been developed intending to enable the patient to do bilateral arm training. This system is comprised of a dual arm with four-bar linkage, which makes it capable of providing naturalistic shoulder movement. Unlike, ARMin-III (where shifting of shoulder CR was considered only for vertical flexion-extension), here the four-bar linkage mechanism moved shoulder CR during either shoulder abduction-adduction or vertical flexion-extension, which made it more anatomical-like (Kim and Deshpande, 2017). The range of motion of the robot differs based on the way its other joints are configured. For instance, ROM of shoulder abduction increased when it was performed simultaneously with shoulder external rotation.

3 Design requirements and challenges

Unlike industrial and other genres of robots, upper limb exoskeletons by their nature are complex in structure, requiring more sophistication in design. Researchers have used different measures and features in upper limb exoskeleton design to enhance functionality. However, there are still limitations that need to be addressed.

Safety

Since upper limb exoskeletons have close interactions with wearers, safety is paramount. Human-exoskeleton interaction (HEI) must be designed so as to ensure safe operation. For safe running of exoskeletons, a HEI should include safety measures in mechanical, electronic, and control design. Mechanically, safety is ensured by placing physical stoppers in the exoskeleton's structure to prevent it going beyond natural ROM; safety can also be ensured by designing links and exoskeleton robot parts in a way where adjacent links act as physical stoppers in extremes. Electronically, by setting current and voltage limits in motors, exoskeleton joints can be refrained from going beyond permissible ROM. In control design, saturation can be set for torque, force, velocity, and position to ensure the wearer's safety during exoskeleton failure.

Comfort of wearing

As a patient wears the exoskeleton the entire time during the rehab session, which ranges from a half-an hour to 2 hours, any kind of discomfort is unexpected. A HEI needs to ensure the patient's comfort. Exoskeleton joints correspond to human joints and as such there are reactive forces and torques between exoskeleton joints and human joints. The weight of the exoskeleton contributes to producing reactive forces and torques in the human joint. Therefore, the lesser the weight, the more comfortable for the wearer. When it comes to wearing, an open-type structure (e.g., CADEN-7 (Perry et al., 2007), ARMIN (Nef et al., 2009a), SUFUL-7 (Gopura et al., 2009), MARSE-7 (Rahman et al., 2014)) is always preferred in exoskeleton design. Open-type structure is advantageous because of easy don/doff, comfortable fitting, and better compliance. In addition, it is expected that the exoskeleton is connected to its wearer with flexible straps/links in between.

Alignment of exoskeleton joints with human joints

To perform exercises during rehabilitation, forces and torques generated in exoskeleton joints must successfully be transferred to human joints. This transfer does not happen properly if the exoskeleton joints are not aligned with the human joints. In addition, misalignment might cause the exoskeleton wearer pain and/or discomfort during rehabilitation (Schiele and van der Helm, 2006; Stienen et al., 2009). Therefore, to provide better compliance and for successful transfer of forces and torques, exoskeleton joints need to be aligned with corresponding human joints. The human shoulder joint, also known as the GH joint, is the most biomechanically complex joint and has many articulations that eventually leads to three general motions (i.e., vertical flexion/extension, abduction/adduction, and upper-arm internal/external rotation) (Schenkman and de Cartaya, 1987). The CR of the shoulder joint does not remain fixed during shoulder movements and has two additional movements: elevation/depression in the frontal plane and protraction and retraction in the sagittal plane (Halder et al., 2000). In the literature, we found many exoskeletons where the shoulder joint is simplified and modeled as a three-DOF ball-and-socket joint by ignoring motion of CR joint (Rahman et al., 2014; Tang et al., 2014; Mahdavian et al., 2015; Liu et al., 2016; Stroppa et al., 2017; Madani et al., 2017). However, a few exoskeletons have added extra DOFs in modeling the shoulder joint to realize motion of CR (Kiguchi et al., 2003, 2008; Nef et al., 2009b; Gopura et al., 2009; Kim and Deshpande, 2017). These adjustments come with the tradeoff of limited ROM and complex design. Apart from the shoulder joint, misalignment may occur at the elbow and wrist joints (Rocon et al., 2008). Gunasekara et al. (2015) in their 6-REXOS exoskeleton has proposed two additional DOFs to compensate for elbow and wrist joint misalignment. An exoskeleton that has a cable-driven parallel mechanism, such as CAREX-7 (Cui et al., 2017), is inherently aligned and has great kinematic matching with humans. A review on cable-driven exoskeletons was recently published (Sanjuan et al., 2020). Li et al. (2019) made an exoskeleton comprised of a novel mechanism that makes the exoskeleton self-aligning during its maneuver (Li et al., 2019). Furthermore, to align axes of an exoskeleton's joint for upper arm internal/external rotation and forearm pronation/supination is a challenging task as those axes are longitudinal axes of corresponding limb segments; therefore, placing motors along those axes is impossible. Perry et al. (2007) used curve rail bearing to provide rotational motion along the upper arm and forearm. Rahman et al. (2014) developed an innovative gear mechanism where motion is transmitted from an antibacklash gear (mounted on a motor shaft) to an open-type, custom-made meshing ring gear attached rigidly to the upper arm cup. In harmony robot, a parallelogram mechanism along with belt and pulley drive has been used to produce forearm pronation/supination (Kim and Deshpande, 2017). Though progress has been made, designing an exoskeleton that is ergonomic with full natural ROM and capable of evading misalignment in joints is still a great challenge for researchers. To provide better HEI, kinematic matching between the exoskeleton and the human is a must in upper limb rehabilitative exoskeletons.

Actuation

Actuators are the main elements that contribute more to exoskeleton weight than any other elements. Various kind of actuators have been used in upper limb exoskeletons, such as electric actuators (Kiguchi et al., 2008; Nef et al., 2009a; Gopura et al., 2009; Rahman et al., 2014; Cui et al., 2017; Xiao et al., 2018; Islam et al., 2019), pneumatic actuators (Song et al., 2007; Sutapun and Sangveraphunsiri, 2015; Balasubramanian et al., 2008), hydraulic actuators (Stienen et al., 2009; Otten et al., 2015; Liu et al., 2016), pneumatic muscle actuators (Tsagarakis and Caldwell, 2003; Tu et al., 2017; Irshaidat et al., 2019; Liu et al., 2020), and series elastic actuators (Kim and Deshpande, 2017). In electrical actuation, DC brushed, DC brushless, and servomotors are frequently used to generate required joint torques in exoskeleton.

Most of the existing prototypes of rehabilitation exoskeletons have used electric actuators as they are advantageous in terms of high power, commercial availability, reliability, easy mounting, and little to no maintenance. In pneumatic actuation, compressed air is used to produce mechanical motion. Pneumatic actuators have inherently low impedance and considerably reduce weight of the exoskeleton. However, they require a compressor to be installed on, and an air tube and valve, which limits their portability. On top of that, they require regular maintenance to keep the right air pressure. The major limitation of pneumatic actuators is the bandwidth they are operating on, which is relatively low (5 Hz). This low bandwidth limits the rate at which they can respond to command signals (Lo and Xie, 2012). In hydraulic actuation, pressure of liquid is used to produce required energy for actuation. The commercially available hydraulic actuators are large, bulky, and noisy, and are not suitable to use in clinical settings. Moreover, they have high impedance and liquid leakage. Pneumatic muscle actuators are composed of stretchable bladders and flexible braided mesh. These muscle-like actuators vary their diameter during actuation and this variation produces tension at their ends, ultimately leading to slow and nonlinear responses (Chou and Hannaford, 1996). In series elastic actuators (e.g., Harmony robot; Kim and Deshpande, 2017), a passive mechanical spring is placed between the motor and load to reduce interface stiffness to provide greater shock tolerance with the tradeoff of operating bandwidth (Pratt and Williamson, 1995). The desired characteristics for choosing actuators for upper limb exoskeleton are: (a) light weight, (b) high power-to-weight ratio, (c) high operating bandwidth, and (d) considerable impedance to deal with unwanted movements such as shock, (e) safe and easy operation, (f) reliability, (g) durability, and (h) low maintenance. For further study, the interested reader may want to see the review by Manna and Dubey (2018), which compares the different actuation systems used in upper limb exoskeletons.

Power transmission mechanism

Transmitting power from motors to exoskeleton joints is one of the key challenges in upper limb exoskeleton research. Continuous variable power transmission is required to provide therapy uninterruptedly to patients. The power transmission can be done by cable drive, wire rope drive, gear train transmission, harmonic drive, belt drive, and so on. Rotation produced at motors is considerably high and needs to be reduced before it transmits to exoskeleton joints. Many researchers have used cable-driven pulley reduction in power transmission mechanisms of their exoskeletons. Tsagarakis and Caldwell (2003) have used cable-driven pulley reduction for pneumatic muscle actuators. In the CADEN-7 exoskeleton, Perry et al. (2007) have used a cable-driven pulley reduction for brushed motors. Some research groups have used a hybrid approach such as CAREX-7 (Cui et al., 2017) and SUFUL-7 (Gopura et al., 2009), both of which a cable-driven pulley, spur gear, and bevel gear, and the CABXLexo, which uses cable drive with epicyclic gear train for power reduction (Xiao, 2019; Xiao et al., 2017, 2018). The main advantage of cable drive is that it can be fitted easily at spots within the exoskeleton, whereas it would be hard to place other drives. In addition, cable drives are beneficial in terms of low inertia, simplicity, and long range transmission. However, the main disadvantage of cable drive is that it can easily stretch and slip, leading to improper tension in the cable, and cause different joint movement than what is desired (Laschi and Cianchetti, 2014). Researchers also used gear drive in exoskeleton systems, such as 6-REXOS (Gunasekara et al., 2015), which embodied spur and bevel gears in its design for motor speed reduction. Such gear drive inherently contains some clearance between meshed teeth, causing friction and backlash (Walha et al., 2006; Zhou et al., 2019). Backlash is undesirable in exoskeleton design. Furthermore, gear drive requires lubrication, significant space for the gear box, and regular maintenance. In contrast, because of zero backlash, low maintenance, and smooth transmission, harmonic drive/strain wave gear reduction has gained popularity among exoskeleton researchers (Kiguchi et al., 2008; Nef et al., 2009a; Rahman et al., 2013a; Kim and Deshpande, 2017; Islam et al., 2019). The only limitation of such drives is their weight, as they need adapter parts to connect with both the motor output shaft and joint input shaft. An exoskeleton's power transmission mechanism should be (a) lightweight, (b) transmit quick, smooth, and backlash-free power, and (c) compact.

Singularity

Mechanical singularity occurs in an exoskeleton when any two joint axes are aligned with each other. As a result, in this scenario, a DOF is lost and ideally requires infinite torque to move further from this configuration. In upper limb, there are three joints (i.e., shoulder vertical flexion/extension, upper arm internal/external rotation, and forearm pronation/supination) whose axes could be aligned. For example, a singular position is seen when the upper arm internal/external rotation joint is at 0 degrees and forearm pronation/supination is at 0 degrees. The human upper limb has natural singularity, but it does not create difficulty in moving the limb away from a singular position. However, unlike the human upper limb, actuators in an exoskeleton require infinite torque to move itself from a singular position. Some researchers did not consider this issue because it is rare to encounter a singular position in a rehabilitation protocol (Perry et al., 2007; Carignan et al., 2007). Nonetheless, the exoskeleton should be rid of singularity, otherwise it can be stuck in a singular position. There are two areas where effort can be given to address singularity. Researchers might design an exoskeleton's structure so that it evades singular position, or a control algorithm might be developed as to avoid singular position in exoskeletons.

Backdrivability

For therapy where patients remain passive, actuators in an exoskeleton does not need to be backdrivable. In contrast, therapy where patients need to participate actively (i.e., patient contributes to the movement of the exoskeleton) in control sharing, actuators must be backdrivable. With intensive rehabilitation therapy, patients start regaining lost mobility in upper limbs. Hence, it is possible to engage them actively in the rehabilitation process. Therefore, upper limb exoskeletons must allow patients, when they are able, to move their limb on their own during therapy sessions (Garrec et al., 2008; Sutapun and Sangveraphunsiri, 2015).

Sensors

Sensors plays a big role in controlling exoskeletons. Incorrect sensor values may hurt the patient or even lead to injuries. Most of the existing exoskeletons use joint position and joint torque in their control approach. To get the position, IMU, potentiometer, and hall sensor were used in the exoskeleton system. Six-axis force sensor was used at the wrist to asses both the force and torque exerted by the user onto the exoskeleton. The 3-axis force sensor also can be used to get the Cartesian force at wrist; in this case Jacobian of the exoskeleton can be used to find the joint torques. This control technique requires more computation as it has to compute the system's Jacobian. In addition, EMG-based control uses muscle signals to detect the user's movement intention (Liu et al., 2020; Xiao, 2019; Priyadarshini et al., 2018; Accogli et al., 2017; Li et al., 2017; Peternel et al., 2016; Gopura et al., 2009). A noninvasive surface EMG electrode was used to obtain muscle activity. Researchers have also been working on using brain signals (EEG) to detect user intention (Bhagat et al., 2016). In a rehabilitative exoskeleton, unidirectional sensors are generally used. However, bidirectional sensors can be used to optimize the exoskeleton's maneuvering. The sensors of exoskeletons should have appropriate bandwidth, enough resolution, and high accuracy to be operated on a real-time controller.

4 Control approaches

The use of rehabilitation robots in the medical rehabilitation field has proved to be of great ability to improve patient quality of life, enhancing practical motions and assisting the patient in daily exercises. The exoskeleton robot is an articulated mechanical structure with several DOFs having the same anatomy of the human arm or leg. Unlike prostheses that replace a limb of the body, the exoskeleton robot clings to it externally and acts in parallel. This fixation allows the robot's wearer to move his/her arm in the workspace. The reachable workspace envelope depends on the number of DOFs available in the exoskeleton robot. It can be dedicated to a specific part of the body, such as the hand, arm, leg, or several limbs at the same time. Equipped with sensors and actuators, it measures the movements and forces of the user that allow the physiotherapist to accurately evaluate the patient's performance.

Usually, the design of these kind of robots is based on the anatomy of the human upper limb and is developed to faithfully represent the joints and movements of the upper limb movements. This robot system is able to provide the different levels of robotic assistance strategies used after neurological accidents. The most urgent, usually the first 6 weeks after the accident, is *passive physical therapy* (Sidney et al., 2013; Xie et al., 2016). In this type of therapy, the exoskeleton brings the patient's limb, which is completely passive, to realize a therapy task. Its advantage lies in the robot's ability to provide intensive therapy over a long period of time (Brahim et al., 2016a, b). The next types of therapy, *active-assisted and active modes*, allow the patient to voluntarily initiate movement. Then, the exoskeleton's wearer can perform a free motion (active mode) or an active-assisted movement where the robot corrects or guides this movement. In the latter case, the robot limits the tremors or corrects the trajectory. After detecting the initiation of a motion, usually voluntary, the exoskeleton will guide the achievement of the activity, often using an impedance and/or admittance control (Li et al., 2017; Ochoa Luna et al., 2015; Liu et al., 2020).

Additionally, these strategies can be used for evaluating or studying subject movements and performance improvement. In these modes, theoretically, the patient should not feel the presence of the exoskeleton robot. Therefore, the subject is completely active and the exoskeleton robot should not affect the movement.

However, these devices are still part of an emerging area and present many challenges. In fact, these robots have an additional complexity compared to conventional robotic manipulators due to their complex mechanical structure designed for human use, types of assistive motion, and the sensitivity of the interaction with a large diversity of human wearers. As a result, these conditions make the robot system vulnerable to dynamic uncertainties and external disturbances such as saturation, friction forces, backlash, and payload. Likewise, the interaction between the human and the exoskeleton subject the system to external disturbances due to different physiological conditions of the subjects, such as different weights of upper limbs for each patient (Brahmi et al., 2018a, b). During a rehabilitation movement, the nonlinear uncertain dynamic model and external forces can turn into an unknown function that can affect the performance of the exoskeleton robot.

The control of uncertain nonlinear dynamics is one of the challenges of nonlinear control engineering problems. In particular, a control system should be developed to ensure the stability of the system. In addition, its performance should not be affected by the disturbances generated from the variation of internal parameters of the system, unmodeled dynamics stimulation, and external disturbances (Slotine et al., 1991; Khalil and Grizzle, 1996). Many studies discussing the problem of modeling and control of exoskeleton robot manipulator based on centralized approaches have been given in Rahman et al. (2013b), Ueda et al. (2010), and Lee et al. (2012). Nevertheless, in the previously cited studies, the control design is model-based, in which the control law requires the full dynamic model of the exoskeleton robot. The estimation of the dynamic parameters is one of the open problems in exoskeleton manipulators, notably, with high DOFs (Krstic et al., 1995), and in the presence of human-robot interaction. Conventional control approaches consider that the dynamic model of the upper arm manipulator is known. However, in practice, it becomes very difficult to get the exact model and uncertainty may still exist. To overcome this problem, robust control approaches based on the Lyapunov theory are developed to ensure the stability of the full system (Khan et al., 2016a, b, 2017; Huang and Chien, 2010; Luna et al., 2016). However, these controllers are very complicated due to the complexity of the regression matrix (Huang and

Chien, 2010). As solution, a time delay estimation (TDE) is proposed (Youcef-Toumi and Shortlidge, 1991; Youcef-Toumi and Ito, 1990; Brahmi et al., 2017). With this method, it is sufficient to delay the output-input of the system only one step to provide a good approximation of the unknown uncertainties dynamic model of the exoskeleton robot. Nevertheless, the TDE approach suffers from time delay error (TDR) caused by the noisy measurements and hard nonlinear function of the robot model during delay constant, which degrades the approximation performance.

On the other hand, many other works have used decentralized control for these types of robotic systems as in Luna et al. (2016) and Ochoa Luna et al. (2015). A decentralized adaptive control, based on the virtual decomposition approach, was proposed, where the whole system was decomposed virtually into several individual subsystems. This decomposition makes the parameters, adaptation, and control law very easy. As an example of these works that applied on other type of robots, an adaptive tracking control design for an uncertain mobile manipulator dynamics based on appropriate reduced dynamic model was suggested in Aviles et al. (2012). An adaptive controller based on the backstepping technique (Brahmi et al., 2016) was implemented to the trajectory tracking of the wheeled mobile manipulator. Recently, approximation-based control strategies like fuzzy logic and neural networks have been used to learn the exoskeleton dynamic model (Chen et al., 2015; Li et al., 2015). However, through these approaches only uniformly ultimate boundedness of the tracking errors was achieved. Meanwhile, the estimated weights were not reached to their actual values. This might reduce convergence speed during weights training operation, which stops the approximation-based control for real-time implementation.

It is remarkable from a natural human movement (since the human upper limb is attached with the exoskeleton robot) that the human does not need accurate information about kinematics and dynamics of the arm (or any object carried by upper extremity) to reach an object in space. Due to that, many control strategies have been designed to solve the problem of kinematic and dynamic uncertainties (Arimoto, 1999; Cheah, 2006; Yazarel and Cheah, 2002; Huang and Chien, 2010; Cheah et al., 2005; Hutchinson et al., 1996). The main innovative point of these controllers is that the adaptation of the both kinematic/dynamic uncertainties has been provided, which allows the exoskeleton robot to perform the human-like motion and supplies to the control system more flexibility to handle the uncertainties and parameters variation. However, the aforementioned controllers are based on the classical regressor matrix. These types of controllers assume that the robot is linear in a set of physical parameters and find a control law able to ensure the stability of this linear system only around its operating points (Yao, 1996). In fact, the manipulator is highly nonlinear. So, the integration of this adaptation law may affect the stability of the system in the presence of even small disturbances (Yao, 1996). Adaptive visual or image-based tracking control (Hutchinson et al., 1996; Deng et al., 2002; Espiau et al., 1992; Gans et al., 2003; Malis and Chaumette, 2002; Liu et al., 2006) is one of the powerful approaches that has been developed to transact with the kinematic/dynamic uncertainties. This is due to their robustness practically to modeling and calibration errors (Deng et al., 2002). However, these controllers are concentrated on uncertainties in nonlinear transformation functions or image Jacobian matrix, but they ignored the uncertain kinematic/dynamic effects. Additionally, few stability analyses are provided in the literature for visual tracking control with the uncertainties of kinematics/dynamics and in the presence of uncertainties in visual system (camera) parameters (Cheah et al., 2006).

5 Discussion

Over the last two decades, numerous exoskeletons have been developed to rehabilitate people with upper limb disability and researchers have been extensively and constantly working to advance the hardware design and control approaches for such robot-aided therapeutic devices.

This chapter reviewed developments in hardware design to deliver better HEI in upper limb exoskeletons and advancements in control approaches to ensure safe and desired running of the exoskeleton. The developments of exoskeletons were compared based on safety, comfort of wearing, alignment, actuation, power transmission mechanism, singularity, and backdrivability. The main challenges while building an exoskeleton are that its joints should be aligned with the corresponding human anatomical joints properly, accommodating different size wearers/patients, providing safe and naturalistic movements. Most existing exoskeletons did not include mobility of shoulder, elbows and wrist joints during upper limb movements. To allow movements of shoulder, elbows and wrist joints, a more novel mechanism should be developed without compromising ROM.

Furthermore, exoskeletons need more sophisticated actuators that are light, compact, easily mountable, have high operating bandwidth, and are durable, reliable, and require low maintenance. The power transmission mechanism should also be novel with light weight, compactness, high gear-ratio, friction free, zero backlash, and quick responsiveness. One of the key novelties that researchers should include in exoskeleton design is modularity (also known as reconfigurability). A modular exoskeleton can be reconfigured by adding or removing parts/modules to or from it. For instance, a 7-DOF exoskeleton is redundant for an individual who has disability only in the shoulder. In this case, the elbow and wrist motion support parts could be removed from it, reducing the 7-DOF exoskeleton to a 3-DOF exoskeleton. Having quick attachable-detachable links in exoskeletons would allow it to be modular.

Despite the fact that an enormous amount of research has been done, development of a control strategy to provide effective rehabilitation is still evolving. To provide efficacious rehabilitation to upper limb-impaired patients, first, we must determine the patient's safe extreme ROM before starting therapy. This may also help to select appropriate and safe rehabilitation protocols for patients. For example, it is essential to know the safe ROM of a patient having muscle tone. In conventional therapy, this is done manually by a therapist who observes the patient. Second, the therapist can adjust his/her approach if the patient feels any pain or something unwanted happens. This is something exoskeletons should include in their control design. Third, to the best of the authors' knowledge, there is no such a controller that automatically selects a rehabilitation protocol for the patient and changes it depending on how the patient is doing exercise. Fourthly, when the patient starts sharing control, he/she has to move his/her impaired limb as well as exoskeleton before getting assistance. This seems to be a burden because the patient has already lost mobility. In such a case, it is worth compensating a portion of torque required to move the exoskeleton only.

6 Conclusion

In this chapter, the HEI, hardware design, safety, compactness, control technique, actuation, and power transmission mechanism of existing upper limb rehabilitative exoskeletons were reviewed. We found that most research prototypes of existing exoskeletons were not transferred into a commercial product, and perhaps the reason behind it is the lack of clinical testing. Therefore, in this review, the challenges that need to be addressed for improved functionality have been identified and discussed. Addressing these challenges in the development of exoskeletons will make them more functional for rehabilitation. In addition, recommendations were given as appropriate to improve the functionality of exoskeletons.

References

Accogli, A., Grazi, L., Crea, S., Panarese, A., Carpaneto, J., Vitiello, N., Micera, S., 2017. EMG-based detection of user's intentions for human-machine shared control of an assistive upper-limb exoskeleton. In: González-Vargas, J., Ibáñez, J., Contreras-Vidal, J.L., van der Kooij, H., Pons, J. (Eds.), Wearable Robotics: Challenges and Trends. Springer International Publishing, Cham, pp. 181–185.

American Stroke Association, 2019. https://www.stroke.org/we-can-help/survivors/stroke-recovery/first-steps-to-recovery/rehabilitation-therapy-after-a-stroke/.

Amirabdollahian, F., Loureiro, R., Gradwell, E., Collin, C., Harwin, W., Johnson, G., 2007. Multivariate analysis of the Fugl-Meyer outcome measures assessing the effectiveness of GENTLE/S robot-mediated stroke therapy. J. Neuroeng. Rehabil. 4 (1), 4. https://doi.org/10.1186/1743-0003-4-4.

Arimoto, S., 1999. Robotics research toward explication of everyday physics. Int. J. Robot. Res. 18 (11), 1056–1063.

Aviles, L.A.Z., Ortega, J.C.P., Hurtado, E.G., 2012. Experimental study of the methodology for the modelling and simulation of mobile manipulators. Int. J. Adv. Robot. Syst. 9 (5), 192.

Balasubramanian, S., Wei, R., Perez, M., Shepard, B., Koeneman, E., Koeneman, J., He, J., 2008. Rupert: an exoskeleton robot for assisting rehabilitation of arm functions. In: Virtual Rehabilitation, 2008, IEEE, pp. 163–167.

Benjamin, E.J., Blaha, M.J., Chiuve, S.E., Cushman, M., Das, S.R., Deo, R., de Ferranti, S.D., Floyd, J., Fornage, M., Gillespie, C., Isasi, C.R., Jimenez, M.C., Jordan, L.C., Judd, S.E., Lackland, D., Lichtman, J.H., Lisabeth, L., Liu, S., Longenecker, C.T., Mackey, R.H., Matsushita, K., Mozaffarian, D., Mussolino, M.E., Nasir, K., Neumar, R.W., Palaniappan, L., Pandey, D.K., Thiagarajan, R.R., Reeves, M.J., Ritchey, M., Rodriguez, C.J., Roth, G.A., Rosamond, W.D., Sasson, C., Towfighi, A., Tsao, C.W., Turner, M.B., Virani, S.S., Voeks, J.H., Willey, J.Z., Wilkins, J.T., Wu, J.H., Alger, H.M., Wong, S.S., Muntner, P., American Heart Association Council on Epidemiology and Prevention Statistics Committee, Stroke Statistics Subcommittee, 2017. Heart disease and stroke statistics-2017 update: a report from the American Heart Association. Circulation 135 (10), e146–e603. https://doi.org/10.1161/CIR.0000000000000485.

Bhagat, N.A., Venkatakrishnan, A., Abibullaev, B., Artz, E.J., Yozbatiran, N., Blank, A.A., French, J., Karmonik, C., Grossman, R.G., O'Malley, M.K., Francisco, G.E., Contreras-Vidal, J.L., 2016. Design and optimization of an EEG-Based brain machine interface (BMI) to an upper-limb exoskeleton for stroke survivors. Front. Neurosci. 10, 122. https://doi.org/10.3389/fnins.2016.00122.

Brahim, B., Maarouf, S., Luna, C.O., Abdelkrim, B., Rahman, M.H., 2016a. Adaptive iterative observer based on integral backstepping control for upper extremity exoskeleton robot. In: 2016 8th International Conference on Modelling, Identification and Control (ICMIC), IEEE, pp. 886–891.

Brahim, B., Rahman, M.H., Saad, M., Luna, C.O., 2016b. Iterative estimator-based nonlinear backstepping control of a robotic exoskeleton. Int. J. Mech. Aerosp. Ind. Mechatron. Manuf. Eng. 10 (8), 1313–1319.

Brahmi, A., Saad, M., Gauthier, G., Brahmi, B., Zhu, W.H., Ghommam, J., 2016. Adaptive backstepping control of mobile manipulator robot based on virtual decomposition

approach. In: 2016 8th International Conference on Modelling, Identification and Control (ICMIC), IEEE, pp. 707–712.
Brahmi, B., Saad, M., Luna, C.O., Archambault, P.S., Rahman, M.H., 2017. Sliding mode control of an exoskeleton robot based on time delay estimation. In: 2017 International Conference on Virtual Rehabilitation (ICVR), IEEE, pp. 1–2.
Brahmi, B., Saad, M., Luna, C.O., Archambault, P.S., Rahman, M.H., 2018a. Passive and active rehabilitation control of human upper-limb exoskeleton robot with dynamic uncertainties. Robotica 36 (11), 1757–1779.
Brahmi, B., Saad, M., Ochoa-Luna, C., Rahman, M.H., Brahmi, A., 2018b. Adaptive tracking control of an exoskeleton robot with uncertain dynamics based on estimated time-delay control. IEEE/ASME Trans. Mechatron. 23 (2), 575–585.
Carignan, C., Tang, J., Roderick, S., Naylor, M., 2007. A configuration-space approach to controlling a rehabilitation arm exoskeleton. In: 2007 IEEE 10th International Conference on Rehabilitation Robotics, pp. 179–187.
Cheah, C.C., 2006. Approximate Jacobian control for robot manipulators. In: Advances in Robot Control, Springer, pp. 35–53.
Cheah, C.-C., Liu, C., Slotine, J.-J.E., 2005. Adaptive Jacobian tracking control of robots based on visual task-space information. In: Proceedings of the 2005 IEEE International Conference on Robotics and Automation, 2005. ICRA 2005, IEEE, pp. 3498–3503.
Cheah, C.-C., Liu, C., Slotine, J.-J.E., 2006. Adaptive tracking control for robots with unknown kinematic and dynamic properties. Int. J. Robot. Res. 25 (3), 283–296.
Chen, W., Ge, S.S., Wu, J., Gong, M., 2015. Globally stable adaptive backstepping neural network control for uncertain strict-feedback systems with tracking accuracy known a priori. IEEE Trans. Neural Netw. Learn. Syst. 26 (9), 1842–1854.
Cheng, H.-S., Ju, M.-S., Lin, C.-C.K., 2004. Improving elbow torque output of stroke patients with assistive torque controlled by EMG signals. J. Biomech. Eng. 125 (6), 881–886. https://doi.org/10.1115/1.1634284.
Chou, C.-P., Hannaford, B., 1996. Measurement and modeling of McKibben pneumatic artificial muscles. IEEE Trans. Robot. Autom. 12 (1), 90–102. https://doi.org/10.1109/70.481753.
Coote, S., Murphy, B., Harwin, W., Stokes, E., 2008. The effect of the GENTLE/s robot-mediated therapy system on arm function after stroke. Clin. Rehabil. 22 (5), 395–405. https://doi.org/10.1177/0269215507085060.
Cui, X., Chen, W., Jin, X., Agrawal, S.K., 2017. Design of a 7-DOF cable-driven arm exoskeleton (CAREX-7) and a controller for dexterous motion training or assistance. IEEE/ASME Trans. Mechatron. 22 (1), 161–172. https://doi.org/10.1109/TMECH.2016.2618888.
Demircan, E., Yung, S., Choi, M., Baschshi, J., Nguyen, B., Rodriguez, J., 2020. Operational space analysis of human muscular effort in robot assisted reaching tasks. Robot. Auton. Syst. 125, 103429. https://doi.org/10.1016/j.robot.2020.103429.
Deng, L., Janabi-Sharifi, F., Wilson, W.J., 2002. Stability and robustness of visual servoing methods. In: IEEE International Conference on Robotics and Automation, 2002. Proceedings. ICRA'02, vol. 2. IEEE, pp. 1604–1609.
Espiau, B., Chaumette, F., Rives, P., 1992. A new approach to visual servoing in robotics. IEEE Trans. Robot. Autom. 8 (3), 313–326.
Gandolfi, M., Formaggio, E., Geroin, C., Storti, S.F., Boscolo Galazzo, I., Bortolami, M., Saltuari, L., Picelli, A., Waldner, A., Manganotti, P., Smania, N., 2018. Quantification of upper limb motor recovery and EEG power changes after robot-assisted bilateral arm training in chronic stroke patients: a prospective pilot study. Neural Plast. 2018, 15. https://doi.org/10.1155/2018/8105480.
Gans, N.R., Hutchinson, S.A., Corke, P.I., 2003. Performance tests for visual servo control systems, with application to partitioned approaches to visual servo control. Int. J. Robot. Res. 22 (10–11), 955–981.

Garrec, P., Friconneau, J.P., Measson, Y., Perrot, Y., 2008. ABLE, an innovative transparent exoskeleton for the upper-limb. In: 2008 IEEE/RSJ International Conference on Intelligent Robots and Systems, pp. 1483–1488.
Gopura, R.A.R.C., Kiguchi, K., Li, Y., 2009. SUEFUL-7: a 7DOF upper-limb exoskeleton robot with muscle-model-oriented EMG-based control. In: 2009 IEEE/RSJ International Conference on Intelligent Robots and Systems, pp. 1126–1131.
Guidali, M., Duschau-Wicke, A., Broggi, S., Klamroth-Marganska, V., Nef, T., Riener, R., 2011. A robotic system to train activities of daily living in a virtual environment. Med. Biol. Eng. Comput. 49 (10), 1213. https://doi.org/10.1007/s11517-011-0809-0.
Gunasekara, M., Gopura, R., Jayawardena, S., 2015. 6-REXOS: upper limb exoskeleton robot with improved pHRI. Int. J. Adv. Robot. Syst. 12 (4), 47. https://doi.org/10.5772/60440.
Halder, A.M., Itoi, E., An, K.-N., 2000. Anatomy and biomechanics of the shoulder. Orthop. Clin. 31 (2), 159–176. https://doi.org/10.1016/S0030-5898(05)70138-3.
Hogan, N., Krebs, H.I., Charnnarong, J., Srikrishna, P., Sharon, A., 1992. MIT-MANUS: a workstation for manual therapy and training. I. In: Proceedings IEEE International Workshop on Robot and Human Communication, pp. 161–165.
Huang, A.-C., Chien, M.-C., 2010. Adaptive Control of Robot Manipulators: A Unified Regressor-Free Approach. World Scientific, Singapore.
Hutchinson, S., Hager, G.D., Corke, P.I., 1996. A tutorial on visual servo control. IEEE Trans. Robot. Autom. 12 (5), 651–670.
Irshaidat, M., Soufian, M., Al-Ibadi, A., Nefti-Meziani, S., 2019. A novel elbow pneumatic muscle actuator for exoskeleton arm in post-stroke rehabilitation. In: 2019 2nd IEEE International Conference on Soft Robotics (RoboSoft), pp. 630–635.
Islam, M.R., Assad-Uz-Zaman, M., Rahman, M.H., 2019. Design and control of an ergonomic robotic shoulder for wearable exoskeleton robot for rehabilitation. Int. J. Dyn. Control. https://doi.org/10.1007/s40435-019-00548-3.
Islam, M.R., Rahmani, M., Rahman, M.H., 2020. A novel exoskeleton with fractional sliding mode control for upper limb rehabilitation. Robotica, 1–22. https://doi.org/10.1017/S0263574719001851.
Kazerooni, H., 2005. Exoskeletons for human power augmentation. In: 2005 IEEE/RSJ International Conference on Intelligent Robots and Systems, August, pp. 3459–3464. https://doi.org/10.1109/IROS.2005.1545451.
Khalil, H.K., Grizzle, J., 1996. Nonlinear Systems, vol. 3. Prentice Hall, Englewood Cliffs, NJ.
Khan, A.M., Usman, M., Ali, A., Khan, F., Yaqub, S., Han, C., 2016a. Muscle circumference sensor and model reference-based adaptive impedance control for upper limb assist exoskeleton robot. Adv. Robot. 30 (24), 1515–1529.
Khan, A.M., Yun, D.-W., Ali, M.A., Zuhaib, K.M., Yuan, C., Iqbal, J., Han, J., Shin, K., Han, C., 2016b. Passivity based adaptive control for upper extremity assist exoskeleton. Int. J. Control Autom. Syst. 14 (1), 291–300.
Khan, A.M., Yun, D.-W., Zuhaib, K.M., Iqbal, J., Yan, R.-J., Khan, F., Han, C., 2017. Estimation of desired motion intention and compliance control for upper limb assist exoskeleton. Int. J. Control Autom. Syst. 15 (2), 802–814.
Kiguchi, K., Esaki, R., Tsuruta, T., Watanabe, K., Fukuda, T., 2003. An exoskeleton system for elbow joint motion rehabilitation. In: Proceedings 2003 IEEE/ASME International Conference on Advanced Intelligent Mechatronics (AIM 2003), vol. 2, pp. 1228–1233.
Kiguchi, K., Rahman, M.H., Sasaki, M., Teramoto, K., 2008. Development of a 3DOF mobile exoskeleton robot for human upper-limb motion assist. Robot. Auton. Syst. 56 (8), 678–691. https://doi.org/10.1016/j.robot.2007.11.007.
Kim, B., Deshpande, A.D., 2015. Controls for the shoulder mechanism of an upper-body exoskeleton for promoting scapulohumeral rhythm. In: 2015 IEEE International Conference on Rehabilitation Robotics (ICORR), pp. 538–542.

Kim, B., Deshpande, A.D., 2017. An upper-body rehabilitation exoskeleton harmony with an anatomical shoulder mechanism: design, modeling, control, and performance evaluation. Int. J. Robot. Res. 36 (4), 414–435. https://doi.org/10.1177/0278364917706743.

Kim, G., Lim, S., Kim, H., Lee, B., Seo, S., Cho, K., Lee, W., 2017. Is robot-assisted therapy effective in upper extremity recovery in early stage stroke?—A systematic literature review. J. Phys. Ther. Sci. 29 (6), 1108–1112. https://doi.org/10.1589/jpts.29.1108.

Krebs, H.I., Volpe, B.T., Williams, D., Celestino, J., Charles, S.K., Lynch, D., Hogan, N., 2007. Robot-aided neurorehabilitation: a robot for wrist rehabilitation. IEEE Trans. Neural Syst. Rehabil. Eng. 15 (3), 327–335. https://doi.org/10.1109/TNSRE.2007.903899.

Krebs, H.I., Edwards, D., Hogan, N., 2016. Forging mens et manus: the MIT experience in upper extremity robotic therapy. In: Reinkensmeyer, D.J., Dietz, V. (Eds.), Neurorehabilitation Technology. Springer International Publishing, Cham, pp. 333–350. https://doi.org/10.1007/978-3-319-28603-7_16.

Krstic, M., Kanellakopoulos, I., Kokotovic, P.V., et al., 1995. Nonlinear and Adaptive Control Design, vol. 222. Wiley, New York, NY.

Laschi, C., Cianchetti, M., 2014. Soft robotics: new perspectives for robot bodyware and control. Front. Bioeng. Biotechnol. 2, 3. https://doi.org/10.3389/fbioe.2014.00003.

Lee, B.-K., Lee, H.-D., Lee, J.-Y., Shin, K., Han, J.-S., Han, C.-S., 2012. Development of dynamic model-based controller for upper limb exoskeleton robot. In: 2012 IEEE International Conference on Robotics and Automation (ICRA), IEEE, pp. 3173–3178.

Lee, K.W., Kim, S.B., Lee, J.H., Lee, S.J., Kim, J.W., 2017. Effect of robot-assisted game training on upper extremity function in stroke patients. Ann. Rehabil. Med. 41 (4), 539–546. https://doi.org/10.5535/arm.2017.41.4.539.

Li, Z., Su, C.-Y., Li, G., Su, H., 2015. Fuzzy approximation-based adaptive backstepping control of an exoskeleton for human upper limbs. IEEE Trans. Fuzzy Syst. 23 (3), 555–566.

Li, Z., Huang, Z., He, W., Su, C.Y., 2017. Adaptive impedance control for an upper limb robotic exoskeleton using biological signals. IEEE Trans. Ind. Electron. 64 (2), 1664–1674. https://doi.org/10.1109/TIE.2016.2538741.

Li, J., Cao, Q., Zhang, C., Tao, C., Ji, R., 2019. Position solution of a novel four-DOFs self-aligning exoskeleton mechanism for upper limb rehabilitation. Mech. Mach. Theory 141, 14–39. https://doi.org/10.1016/j.mechmachtheory.2019.06.020.

Liu, C., Cheah, C.C., Slotine, J.-J.E., 2006. Adaptive Jacobian tracking control of rigid-link electrically driven robots based on visual task-space information. Automatica 42 (9), 1491–1501.

Liu, L., Shi, Y.-Y., Xie, L., 2016. A novel multi-DOF exoskeleton robot for upper limb rehabilitation. J. Mech. Med. Biol. 16 (8), 1640023. https://doi.org/10.1142/s0219519416400236.

Liu, H., Tao, J., Lyu, P., Tian, F., 2020. Human-robot cooperative control based on SEMG for the upper limb exoskeleton robot. Robot. Auton. Syst. 125, 103350. https://doi.org/10.1016/j.robot.2019.103350.

Lo, H.S., Xie, S.Q., 2012. Exoskeleton robots for upper-limb rehabilitation: state of the art and future prospects. Med. Eng. Phys. 34 (3), 261–268. https://doi.org/10.1016/j.medengphy.2011.10.004.

Luna, C.O., Rahman, M.H., Saad, M., Archambault, P., Zhu, W.-H., 2016. Virtual decomposition control of an exoskeleton robot arm. Robotica 34 (7), 1587–1609.

Madani, T., Daachi, B., Djouani, K., 2017. Modular-controller-design-based fast terminal sliding mode for articulated exoskeleton systems. IEEE Trans. Control Syst. Technol. 25 (3), 1133–1140. https://doi.org/10.1109/TCST.2016.2579603.

Mahdavian, M., Toudeshki, A.G., Yousefi-Koma, A., 2015. Design and fabrication of a 3DoF upper limb exoskeleton. In: 2015 3rd RSI International Conference on Robotics and Mechatronics (ICROM), pp. 342–346. https://doi.org/10.1109/ICRoM.2015.7367808.

Malis, E., Chaumette, F., 2002. Theoretical improvements in the stability analysis of a new class of model-free visual servoing methods. IEEE Trans. Robot. Autom. 18 (2), 176–186.

Manna, S.K., Dubey, V.N., 2018. Comparative study of actuation systems for portable upper limb exoskeletons. Med. Eng. Phys. 60, 1–13. https://doi.org/10.1016/j.medengphy.2018.07.017.

Marcheschi, S., Salsedo, F., Fontana, M., Bergamasco, M., 2011. Body extender: whole body exoskeleton for human power augmentation. In: 2011 IEEE International Conference on Robotics and Automation, May, pp. 611–616. https://doi.org/10.1109/ICRA.2011.5980132.

Nef, T., Mihelj, M., Kiefer, G., Perndl, C., Muller, R., Riener, R., 2007. Armin-exoskeleton for arm therapy in stroke patients. In: IEEE 10th International Conference on Rehabilitation Robotics, 2007. ICORR 2007, IEEE, pp. 68–74.

Nef, T., Guidali, M., Klamroth-Marganska, V., Riener, R., 2009a. Armin—exoskeleton robot for stroke rehabilitation. In: Dössel, O., Schlegel, W.C. (Eds.), In: World Congress on Medical Physics and Biomedical Engineering, September 7–12, 2009, Munich, Germany, Springer, Berlin, Heidelberg, pp. 127–130.

Nef, T., Guidali, M., Riener, R., 2009b. ARMin III—arm therapy exoskeleton with an ergonomic shoulder actuation. Appl. Bionics Biomech. 6(2), https://doi.org/10.1080/11762320902840179.

Ochoa Luna, C., Habibur Rahman, M., Saad, M., Archambault, P.S., Bruce Ferrer, S., 2015. Admittance-based upper limb robotic active and active-assistive movements. Int. J. Adv. Robot. Syst. 12 (9), 117.

Otten, A., Voort, C., Stienen, A., Aarts, R., van Asseldonk, E., van der Kooij, H., 2015. LIMPACT: a hydraulically powered self-aligning upper limb exoskeleton. IEEE/ASME Trans. Mechatron. 20 (5), 2285–2298. https://doi.org/10.1109/TMECH. 2014.2375272.

Perry, J.C., Rosen, J., Burns, S., 2007. Upper-limb powered exoskeleton design. IEEE/ASME Trans. Mechatron. 12 (4), 408–417. https://doi.org/10.1109/TMECH.2007.901934.

Perry, J.C., Maura, R., Bitikofer, C.K., Wolbrecht, E.T., 2019. BLUE SABINO: development of a bilateral exoskeleton instrument for comprehensive upper-extremity neuromuscular assessment. In: Masia, L., Micera, S., Akay, M., Pons José, L. (Eds.), Converging Clinical and Engineering Research on Neurorehabilitation III. Springer International Publishing, Cham, pp. 493–497.

Peternel, L., Noda, T., Petrič, T., Ude, A., Morimoto, J., Babič, J., 2016. Adaptive control of exoskeleton robots for periodic assistive behaviours based on EMG feedback minimisation. PLoS ONE 11 (2), 1–26. https://doi.org/10.1371/journal.pone.0148942.

Poli, P., Morone, G., Rosati, G., Masiero, S., 2013. Robotic technologies and rehabilitation: new tools for stroke patients' therapy. BioMed. Res. Int. 2013, 8. https://doi.org/10.1155/2013/153872.

Pratt, G.A., Williamson, M.M., 1995. Series elastic actuators. In: Proceedings 1995 IEEE/RSJ International Conference on Intelligent Robots and Systems. Human Robot Interaction and Cooperative Robots, August, vol. 1, pp. 399–406.

Priyadarshini, R.G., Suryarajan, R., Prasad, J., 2018. Development of electromyogram based rehabilitation device for upper limb amputation using neural network. In: 2018 3rd International Conference on Communication and Electronics Systems (ICCES), pp. 826–830. https://doi.org/10.1109/CESYS.2018.8723958.

Rahman, M.H., Kittel-Ouimet, T., Saad, M., Kenné, J.-P., Archambault, P.S., 2012. Development and control of a robotic exoskeleton for shoulder, elbow and forearm movement assistance. Appl. Bionics Biomech. 9(3), https://doi.org/10.3233/ABB-2012-0061.

Rahman, M.H., Saad, M., Kenné, J.-P., Archambault, P.S., 2013a. Control of an exoskeleton robot arm with sliding mode exponential reaching law. Int. J. Control Autom. Syst. 11 (1), 92–104. https://doi.org/10.1007/s12555-011-0135-1.

Rahman, M.H., Saad, M., Kenné, J.-P., Archambault, P.S., 2013b. Control of an exoskeleton robot arm with sliding mode exponential reaching law. Int. J. Control Autom. Syst. 11 (1), 92–104.

Rahman, M.H., Rahman, M.J., Cristobal, O.L., Saad, M., Kenné, J.P., Archambault, P.S., 2014. Development of a whole arm wearable robotic exoskeleton for rehabilitation and to assist upper limb movements. Robotica 33 (1), 19–39. https://doi.org/10.1017/S0263574714000034.

Reinkensmeyer, D.J., Kahn, L.E., Averbuch, M., McKenna-Cole, A.N., Schmit, B., Rymer, W., 2000. Understanding and treating arm movement impairment after chronic brain injury: progress with the ARM guide. J. Rehabil. Res. Dev. 37 (6), 653–662.

Rocon, E., Ruiz, A.F., Raya, R., Schiele, A., Pons, J.L., Belda-Lois, J.M., Poveda, R., Vivas, M.J., Moreno, J.C., 2008. Human-robot physical interaction. In: Wearable Robots, John Wiley & Sons, Ltd, pp. 127–163. https://doi.org/10.1002/9780470987667.ch5.

Sale, P., Franceschini, M., Mazzoleni, S., Palma, E., Agosti, M., Posteraro, F., 2014. Effects of upper limb robot-assisted therapy on motor recovery in subacute stroke patients. J. Neuroeng. Rehabil. 11 (1), 104. https://doi.org/10.1186/1743-0003-11-104.

Sanjuan, J.D., Castillo, A.D., Padilla, M.A., Quintero, M.C., Gutierrez, E.E., Sampayo, I.P., Hernandez, J.R., Rahman, M.H., 2020. Cable driven exoskeleton for upper-limb rehabilitation: a design review. Robot. Auton. Syst. 103445. https://doi.org/10.1016/j.robot.2020.103445.

Schenkman, M., de Cartaya, V.R., 1987. Kinesiology of the shoulder complex. J. Orthop. Sports Phys. Ther. 8 (9), 438–450. https://doi.org/10.2519/jospt.1987.8.9.438.

Schiele, A., van der Helm, F.C.T., 2006. Kinematic design to improve ergonomics in human machine interaction. IEEE Trans. Neural Syst. Rehabil. Eng. 14 (4), 456–469. https://doi.org/10.1109/TNSRE.2006.881565.

Sidney, S., Rosamond, W.D., Howard, V.J., Luepker, R.V., 2013. The "Heart Disease and Stroke Statistics2013 Update" and the Need for a National Cardiovascular Surveillance System. American Heart Association, Dallas, TX, pp. 21–23.

Slotine, J.-J.E., Li, W., et al., 1991. Applied Nonlinear Control, vol. 199. Prentice Hall, Englewood Cliffs, NJ.

Song, R., Tong, K.Y., Hu, X.L., Zheng, X.J., 2007. Myoelectrically controlled robotic system that provide voluntary mechanical help for persons after stroke. In: 2007 IEEE 10th International Conference on Rehabilitation Robotics, pp. 246–249. https://doi.org/10.1109/ICORR.2007.4428434.

Stienen, A.H.A., Hekman, E.E.G., van der Helm, F.C.T., van der Kooij, H., 2009. Self-aligning exoskeleton axes through decoupling of joint rotations and translations. IEEE Trans. Robot. 25 (3), 628–633. https://doi.org/10.1109/TRO.2009.2019147.

Statistics, Stroke, 2019. Upperlimb. http://www.strokecenter.org/patients/about-stroke/stroke-statistics/.

Stroppa, F., Loconsole, C., Marcheschi, S., Frisoli, A., 2017. A robot-assisted neurorehabilitation system for post-stroke patients' motor skill evaluation with Alex exoskeleton. In: Ibáñez, J., González-Vargas, J., Azorín, J.M., Akay, M., Pons, J.L. (Eds.), Converging Clinical and Engineering Research on Neurorehabilitation II. Springer International Publishing, Cham, pp. 501–505.

Sutapun, A., Sangveraphunsiri, V., 2015. A 4-DOF upper limb exoskeleton for stroke rehabilitation: kinematics mechanics and control. Int. J. Mech. Eng. Robot. Res. 4 (3), 269–272. https://doi.org/10.18178/ijmerr.4.3.269-272.

Tang, Z., Zhang, K., Sun, S., Gao, Z., Zhang, L., Yang, Z., 2014. An upper-limb power-assist exoskeleton using proportional myoelectric control. Sensors (Basel, Switzerland) 14 (4), 6677–6694. https://doi.org/10.3390/s140406677.

Teasell, R.W., Kalra, L., 2004. What's new in stroke rehabilitation. Stroke 35 (2), 383–385. https://doi.org/10.1161/01.str.0000115937.94104.76.

Triwiyanto, Wahyunggoro, O., Nugroho, H.A., Herianto, 2016. String actuated upper limb exoskeleton based on surface electromyography control. In: 2016 6th International Annual Engineering Seminar (InAES), pp. 176–181. https://doi.org/10.1109/INAES.2016.7821929.

Tsagarakis, N.G., Caldwell, D.G., 2003. Development and control of a "soft-actuated" exoskeleton for use in physiotherapy and training. Auton. Robot. 15 (1), 21–33. https://doi.org/10.1023/A:1024484615192.

Tu, X., Zhou, X., Li, J., Su, C., Sun, X., Han, H., Jiang, X., He, J., 2017. Iterative learning control applied to a hybrid rehabilitation exoskeleton system powered by PAM and FES. Clust. Comput. 20 (4), 2855–2868. https://doi.org/10.1007/s10586-017-0880-x.

Ueda, J., Ming, D., Krishnamoorthy, V., Shinohara, M., Ogasawara, T., 2010. Individual muscle control using an exoskeleton robot for muscle function testing. IEEE Trans. Neural Syst. Rehabil. Eng. 18 (4), 339–350.

Veerbeek, J.M., Langbroek-Amersfoort, A.C., van Wegen, E.E.H., Meskers, C.G.M., Gert, K., 2016. Effects of robot-assisted therapy for the upper limb after stroke: a systematic review and meta-analysis. Neurorehabil. Neural Repair 31 (2), 107–121. https://doi.org/10.1177/1545968316666957.

Walha, L., Fakhfakh, T., Haddar, M., 2006. Backlash effect on dynamic analysis of a two-stage spur gear system. J. Fail. Anal. Prev. 6 (3), 60–68. https://doi.org/10.1007/BF02692330.

Walsh, C.J., Paluska, D., Pasch, K., Grand, W., Valiente, A., Herr, H., 2006. Development of a lightweight, underactuated exoskeleton for load-carrying augmentation. In: Proceedings 2006 IEEE International Conference on Robotics and Automation, 2006. ICRA 2006, May, pp. 3485–3491. https://doi.org/10.1109/ROBOT. 2006.1642234.

Xiao, F., 2019. Proportional myoelectric and compensating control of a cable-conduit mechanism-driven upper limb exoskeleton. ISA Trans. 89, 245–255. https://doi.org/10.1016/j.isatra.2018.12.028.

Xiao, F., Gao, Y., Wang, Y., Zhu, Y., Zhao, J., 2017. Design of a wearable cable-driven upper limb exoskeleton based on epicyclic gear trains structure. Technol. Health Care 25 (S1), 3–11. https://doi.org/10.3233/THC-171300.

Xiao, F., Gao, Y., Wang, Y., Zhu, Y., Zhao, J., 2018. Design and evaluation of a 7-DOF cable-driven upper limb exoskeleton. J. Mech. Sci. Technol. 32 (2), 855–864. https://doi.org/10.1007/s12206-018-0136-y.

Xie, S., et al., 2016. Advanced robotics for medical rehabilitation. Springer Tracts Adv. Robot. 108, 1–41.

Yao, B., 1996. Adaptive Robust Control of Nonlinear Systems With Application to Control of Mechanical Systems (Ph.D. thesis). University of California, Berkeley.

Yazarel, H., Cheah, C.-C., 2002. Task-space adaptive control of robotic manipulators with uncertainties in gravity regressor matrix and kinematics. IEEE Trans. Autom. Control 47 (9), 1580–1585.

Yoo, D.H., Kim, S.Y., 2015. Effects of upper limb robot-assisted therapy in the rehabilitation of stroke patients. J. Phys. Ther. Sci. 27 (3), 677–679. https://doi.org/10.1589/jpts.27.677.

Youcef-Toumi, K., Ito, O., 1990. A time delay controller for systems with unknown dynamics. J. Dyn. Syst. Meas. Control. 112 (1), 133–142.

Youcef-Toumi, K., Shortlidge, C.C., 1991. Control of robot manipulators using time delay. In: 1991 IEEE International Conference on Robotics and Automation, 1991, IEEE, pp. 2391–2398.

Zhang, L., Li, J., Cui, Y., Dong, M., Fang, B., Zhang, P., 2020. Design and performance analysis of a parallel wrist rehabilitation robot (PWRR). Robot. Auton. Syst. 125, 103390. https://doi.org/10.1016/j.robot.2019.103390.

Zhou, R., Zhao, N., Tan, Z., Zhang, M., Guo, H., 2019. Computerized generation and meshing simulation of a backlash-adjustable face-gear drive with a tapered involute pinion. Mech. Mach. Theory 140, 479–503. https://doi.org/10.1016/j.mechmachtheory.2019.06.017.

Zhu, C., Okada, Y., Yoshioka, M., Yamamoto, T., Yu, H., Yan, Y., Duan, F., 2014. Power augmentation of upper extremity by using agonist electromyography signals only for extended admittance control. IEEJ J. Ind. Appl. 3 (3), 260–269. https://doi.org/10.1541/ieejjia.3.260.

CHAPTER 10

A double pendulum model for human walking control on the treadmill and stride-to-stride fluctuations: Control of step length, time, velocity, and position on the treadmill

Alireza Bahramian[a], Farzad Towhidkhah[a], Sajad Jafari[a], Olfa Boubaker[b]
[a]Biomedical Engineering Department, Amirkabir University of Technology, Tehran, Iran
[b]University of Carthage, National Institute of Applied Sciences and Technology, Tunis, Tunisia

1 Introduction

Walking can be defined as a rhythmic movement of limbs that maintains animals and humans in constant forward displacement (Rao et al., 2010). It is generated by sophisticated interactions between motor centers in the brain, neural spinal circuits and reflexes, numerous muscles, different sensory systems, and the environment. The nervous system controls the body as a complex system via the integration of sensory feedbacks including visual (Salinas et al., 2017), vestibular (Larsson et al., 2016), and proprioceptive (Pearson, 2004), as well as a type of internal model(s) in the brain (Karimian et al., 2006) to maintain dynamic stability and effective walking. There are several internal (neurobiological) (Lewek et al., 2009; Faisal et al., 2008) as well as external (environment) (Su and Dingwell, 2007) sources of uncertainties (noises) that cause each step to become slightly different from the others (Hausdorff, 2007). These variabilities may contain valuable information about how the nervous system reacts to these noises to regulate limb movements. For example, the variabilities of movements increase with advancing age in older adults (Kang and Dingwell, 2008). These variations are correlated with the individuals' falls history (Toebes et al., 2012) and the likelihood of falls in the future (Verghese et al., 2009). Also, the increase of variabilities in movements facilitates motor learning during rehabilitation

Control Theory in Biomedical Engineering
https://doi.org/10.1016/B978-0-12-821350-6.00010-X

(Lewek et al., 2009; Ziegler et al., 2010). Besides motor and sensory sources of biological noises (Osborne et al., 2005), high degrees of freedom can be considered as another source of these variabilities (Todorov and Jordan, 2002; Scott, 2004). A great number of muscles and joints provide infinite choices for the nervous system to achieve the same goal (Engelbrecht, 2001), such as walking (Bohnsack-McLagan et al., 2016) or running (Agresta et al., 2019). There is a popular hypothesis that assumes the nervous system chooses an optimal solution to reach a task (Srinivasan and Ruina, 2006). This assumption is proper to express the average of attempts, however, it discusses less about movement variability (Kang and Dingwell, 2008; Dingwell and Marin, 2006).

Stride length (L_n), time (T_n), and velocity (V_n) during walking are global parameters that have largely been studied (Duncan, 2018; Terrier, 2016; Decker et al., 2016; Malcolm et al., 2018). Detrended fluctuation analysis (DFA) (Hu et al., 2001; Kantelhardt et al., 2001) is a method to determine statistical persistence/antipersistence of signals. This method has been widely used to analyze time series of these important parameters (Kirchner et al., 2014; Hausdorff et al., 1999; Roerdink et al., 2019). It has been revealed that sequences of L_n, T_n, and V_n during walking over the ground have high statistical persistence showing low control effort applied for each parameter (Terrier et al., 2005). Using metronome changes T_n to antipersistence time series, however, neither S_n nor L_n has been changed. This represents a higher control effort on T_n (Terrier et al., 2005). For another instance, the studies on walking on a fixed-speed treadmill showed an increase in S_n antipersistent characteristic (Terrier and Dériaz, 2012; Dingwell et al., 2010). Position on the treadmill after each stride (P_n), is another important parameter considered in the literature (Dingwell et al., 2010; Dingwell and Cusumano, 2015).

To answer the question of which strategy (position control or velocity control) the nervous system employs for walking on the treadmill, J. B. Dingwell and J. P. o. Cusumano designed four black-box models to generate the sequence of L_n and T_n, based on different control strategies (Dingwell and Cusumano, 2015). They revealed that L_n, T_n, V_n, and P_n of the model with a velocity control strategy could produce results similar to those of the experimental data (Dingwell and Cusumano, 2015). Although such black-box modeling is useful to simplify and clarify walking behavior, it produces no information about the bio-mechanics of the movement. In this case, the application of simple and conceptual dynamical robotic models can be helpful. These models benefit both the advantages of simplicity as well as gait mechanical dynamics. As a typical model in this category, we can point to the simple and well-known mass-spring model suggested to re-generate the pattern of ground-reaction force of walking (Aloulou and Boubaker, 2011, 2015). This model explains locomotion stability (Aloulou and Boubaker, 2010). The pendulum can be also mentioned as another effective and simple model (Boubaker and Iriarte, 2017). It has been used to study walking stability on a ramp (Nourian Zavareh et al., 2018; Nazarimehr et al., 2017) as well as the optimality of step size and velocity of walking over the ground (Kuo, 2001, 2002). To the best of the authors' knowledge, there is no study based on the pendulum model to investigate T_n, L_n, V_n, and P_n variabilities during walking on the treadmill.

In this chapter, simple models using a double pendulum model for control of walking on the treadmill are developed. These models focus on important parameters of walking, that is, T_n, L_n, V_n, and P_n, and their variabilities. Initially, four controllers are designed for each of these parameters during walking on the treadmill. Afterward, models using different combinations of these controllers are proposed. Then, by adding noises to model parameters, variabilities are included. Finally, the results of a comparison between these different models are presented.

2 Material and method

2.1 Double pendulum model

In this chapter, a double pendulum is used as a human walking model (Kuo, 2002; Garcia et al., 1998) as shown in Fig. 1. In this model two major phases of walking, single support and double support, are considered. g is gravitational constant, L is leg length, m is foot mass, and M is hip mass. ϕ and θ are defined

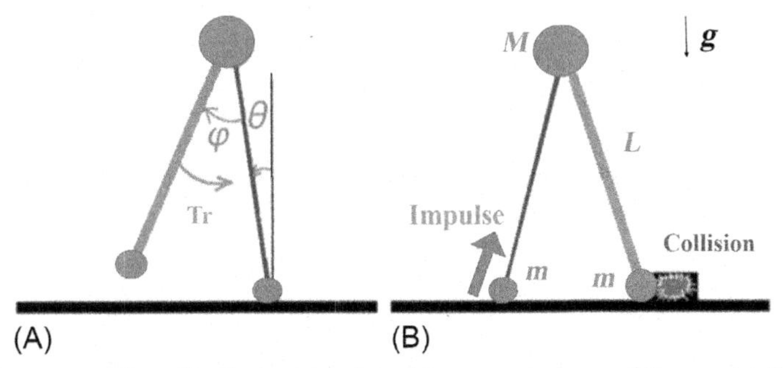

Fig. 1 Representation of both single and double support phase of the model. (A) Single support phase. (B) Double support phase.

as swing and stance leg angles, respectively. The mathematical equations of this model are presented as follows.

Eqs. (1), (2) are for the single support phase.

$$\ddot{\theta} - g/L\sin(\theta) = 0 \tag{1}$$

$$m/M\ddot{\phi} - m/M(1 - \cos(\phi))\ddot{\theta} - m/M\dot{\theta}^2\sin(\phi) - m/M\sin(\theta - \phi) = Tr \tag{2}$$

Tr is the swing hip torque assumed to be generated through a rotational spring with stiffness, k, and equilibrium angle, ϕ_0 (Eq. 3).

$$Tr = k(\phi_0 - \phi) \tag{3}$$

The single support phase is terminated when:

$$\phi - 2\theta = 0 \tag{4}$$

Double support phase is also represented by Eq. (5) as an event:

$$\begin{bmatrix} \theta \\ \dot{\theta} \\ \phi \\ \dot{\phi} \end{bmatrix}^{+} = \begin{bmatrix} -1 & 0 & 0 & 0 \\ 0 & \cos(2\theta) & 0 & 0 \\ -2 & 0 & 0 & 0 \\ 0 & \cos(2\theta)(1-\cos(2\theta)) & 0 & 0 \end{bmatrix} \begin{bmatrix} \theta \\ \dot{\theta} \\ \phi \\ \dot{\phi} \end{bmatrix}^{-} + \begin{bmatrix} 0 \\ \sin 2\theta \\ 0 \\ 1-\cos(2\theta) \end{bmatrix} Impulse \tag{5}$$

In Eq. (5), (−) and (+) indicate states values before and after foot-contact condition. Also, *Impulse* is considered as the push-off impact amplitude.

In this model, the sum of two consecutive steps is considered as a stride.

2.2 Controller design

Impulse and k were considered as control parameters of the model (Bahramian et al., 2019) (other model parameters values are mentioned in Table 1). To clarify the effect of each parameter, first, values of $Impulse = 0.4$ and $k = 4$ were set. Then, after the system reached a steady state, the k value

Table 1 Model parameters values.

Parameter	Value
M	50 kg
m	5 kg
L	1 m
g	9.8 m/s^2
Φ_0	0.04 rad

was fixed and the *Impulse* value was set by other values. After three strides, stride time and stride length were measured to reveal the effect of *Impulse* on these parameters. The same method was used for exhibiting the effect of *k*. Figs. 2 and 3 exhibit the effect of these parameters on step length and step time, respectively.

As shown in Figs. 2 and 3, step duration is much more affected by *k* and step length by *Impulse* in the same way. So *k* and *Impulse* are considered as control parameters for step length and step time, respectively.

A control method, so-called neural signals controllers (Huang et al., 2017), was used for controlling step length and time. The step length controller formulations are explained in Eqs. (6), (7).

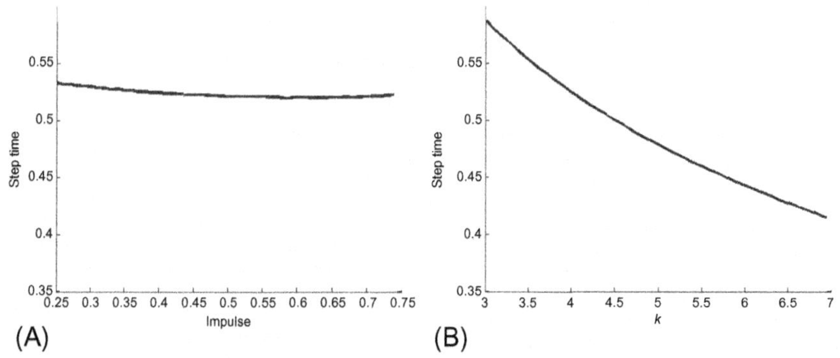

Fig. 2 Effect of *Impulse* and *k* as control parameters on stride time are shown (A) As it can be seen, changing of *Impulse* value has a negligible effect on stride time (B) Step time is affected by *k* and it decreases when *k* increases.

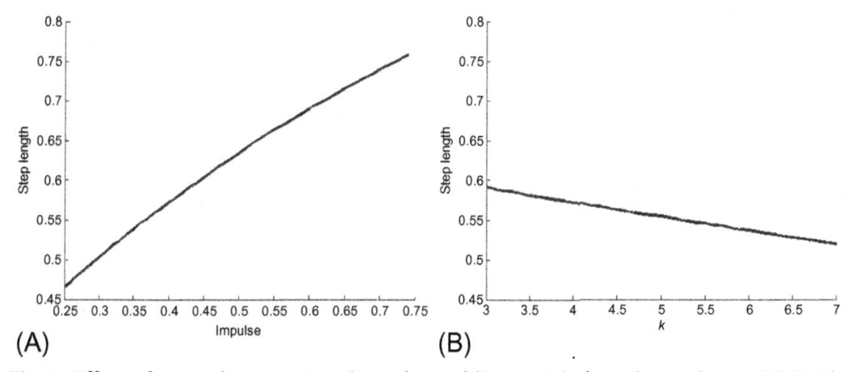

Fig. 3 Effect of control parameters (*Impulse* and *k*) on stride length are shown (A) Stride length is increased by increasing *Impulse* value (B) Effect of *k* on stride length is much less than *Impulse*.

$$C\_pu\_L(n) = \boldsymbol{P_L} \times (\text{step length (desired)} - \text{step length } (n)) + \boldsymbol{I_L} \times (C\_pu\_L(n-1)) + \boldsymbol{D_L} \times (\text{step length } (n) - \text{step length } (n-1)) \tag{6}$$

$$Impulse(n+1) = Impulse(n) + C\_pu\_L(n) \tag{7}$$

In these equations, $\boldsymbol{P_L}$, $\boldsymbol{I_L}$, and $\boldsymbol{D_L}$ are the step length controller's coefficients, and C_pu_L is control variable to change *Impulse* result in step length regulation. It is important to note that the traditional "neural signals controller" method includes only P and D coefficients. In this study, the I coefficient was added to improve the controller performance.

Similarly, the step time controller formulations are explained in Eqs. (8), (9).

$$C\_k\_T(n) = \boldsymbol{P_T} \times (\text{step time (desired)} - \text{step time } (n)) + \boldsymbol{I_T} \times (C\_k\_T(n-1)) + \boldsymbol{D_T} \times (\text{step time } (n) - \text{step time } (n-1)) \tag{8}$$

$$k(n+1) = k(n) + C\_k\_T(n) \tag{9}$$

In these equations, $\boldsymbol{P_T}$, $\boldsymbol{I_T}$, and $\boldsymbol{D_T}$ are step time controller's coefficients, and C_k_T is control variable to change k result in step time regulation. Depending on the walking parameters to be controlled, we designed four control system models. The first control system model contains controllers for step length and time referred to as c_TL.

The step velocity can be regulated by changes in step time and/or step length. In order to increase/decrease step velocity, it is required to increase/decrease step length and/or decrease/increase step time. Thus, in the second control system model, a controller for velocity was designed and added to c_TL by Eqs. (10)–(13). This control system model, which controls step length, time, and velocity, is named as c_TLV.

$$C\_pu\_V(n) = \boldsymbol{P_{VL}} \times (\text{treadmill velocity (desired velocity)} - \text{step velocity } (n)) + \boldsymbol{I_{VL}} \times (C\_pu\_V(n-1)) + \boldsymbol{D_{VL}} \times (\text{step velocity } (n) - \text{step velocity } (n-1)) \tag{10}$$

$$Impulse(n+1) = Impulse(n) + C\_pu\_L(n) + C\_pu\_V(n) \tag{11}$$

$$C\_k\_V(n) = \boldsymbol{P_{VT}} \times (\text{treadmill velocity (desired velocity)} - \text{step velocity } (n)) + \boldsymbol{I_{VT}} \times (C\_k\_V(n-1)) + \boldsymbol{D_{VT}} \times (\text{step velocity } (n) - \text{step velocity } (n-1)) \tag{12}$$

$$k(n+1) = k(n) + C\_k\_T(n) + C\_k\_V(n) \tag{13}$$

In these equations, $\boldsymbol{P_{VL}}$, $\boldsymbol{I_{VL}}$, and $\boldsymbol{D_{VL}}$ are step velocity controller coefficients and *C_pu_V* is control variable to change *Impulse* and step length result in step velocity regulation. $\boldsymbol{P_{VT}}$, $\boldsymbol{I_{VT}}$, and $\boldsymbol{D_{VT}}$ are step velocity controller coefficients and *C_k_V* is a control variable to change *k* and step time result in step velocity regulation.

In the third control system model, the position of human walking on the treadmill, P_n, was considered in the control system design. P_n is calculated by Eq. (14) (Dingwell and Cusumano, 2015).

$$P_n = P_{(n-1)} + [(\text{step length}) - (\text{treadmill velocity}) \times (\text{step duration})] \quad (14)$$

The initial value of P_n is set to 0, representing the middle of the treadmill position. Similar to step velocity control, the position controller was designed and added to c_TLV using Eqs. (15)–(18). This control system model is referred to as c_TLVP.

$$\begin{aligned} C\_pu\_P(n) = &\boldsymbol{P_{PL}} \times (\text{treadmill desired position } (0) - \text{step position } (n)) \\ &+\boldsymbol{I_{PL}} \times (C\_pu\_P(n-1)) \\ &+\boldsymbol{D_{PL}} \times (\text{step position } (n) - \text{step position } (n-1)) \end{aligned} \quad (15)$$

$$Impulse(n+1) = Impulse(n) + C\_pu\_L(n) + C\_pu\_V(n) + C\_pu\_P(n) \quad (16)$$

$$\begin{aligned} C\_k\_P(n) = &\boldsymbol{P_{PL}} \times (\text{treadmill desired position } (0) - \text{step velocity } (n)) \\ &+\boldsymbol{I_{PL}} \times (C\_k\_P(n-1)) \\ &+\boldsymbol{D_{PL}} \times (\text{step position } (n) - \text{step position } (n-1)) \end{aligned} \quad (17)$$

$$k(n+1) = k(n) + C\_k\_T(n) + C\_k\_V(n) + C\_k\_P(n) \quad (18)$$

In these equations, $\boldsymbol{P_{PL}}$, $\boldsymbol{I_{PL}}$, and $\boldsymbol{D_{PL}}$ are position controller coefficients and *C_pu_P* is control variable to change *Impulse* and step length result in step position regulation. $\boldsymbol{P_{PT}}$, $\boldsymbol{I_{PT}}$, and $\boldsymbol{D_{PT}}$ are step velocity controller coefficients and *C_k_P* is a control variable to change *k* and step time result in step position regulation.

The fourth control system model contains only position and velocity controllers (i.e., only *C_k_V(n)*, *C_pu_V(n)*, *C_k_P(n)*, and *C_pu_P(n)*) and is referred to as c_VP.

2.3 Experimental data

The experimental data (Dingwell et al., 2010) was used for comparing the outcomes of the different control system models and are available at:

https://doi.org/10.5061/dryad.sk55m. These experimental results include data of walking on the treadmill for 17 young adults. Each subject was asked to walk on the treadmill for 5 min with self-selected speed (for more information see Dingwell et al., 2010). As an example of this data, the stride time, length, velocity, and position of the first trial of subject 13 are shown in Fig. 4.

2.4 Adding uncertainty to the model

To model uncertainty and the subject's variabilities, we added random noises to the four control system models as shown in Eqs. (19), (20).

$$Impulse(n+1) = Impulse(n) + C\_pu\_L(n) + C\_pu\_V(n) + C\_pu\_P(n) + \eta \times rand \tag{19}$$

$$k(n+1) = k(n) + C\_k\_T(n) + C\_k\_V(n) + C\_k\_P(n) + \gamma \times rand \tag{20}$$

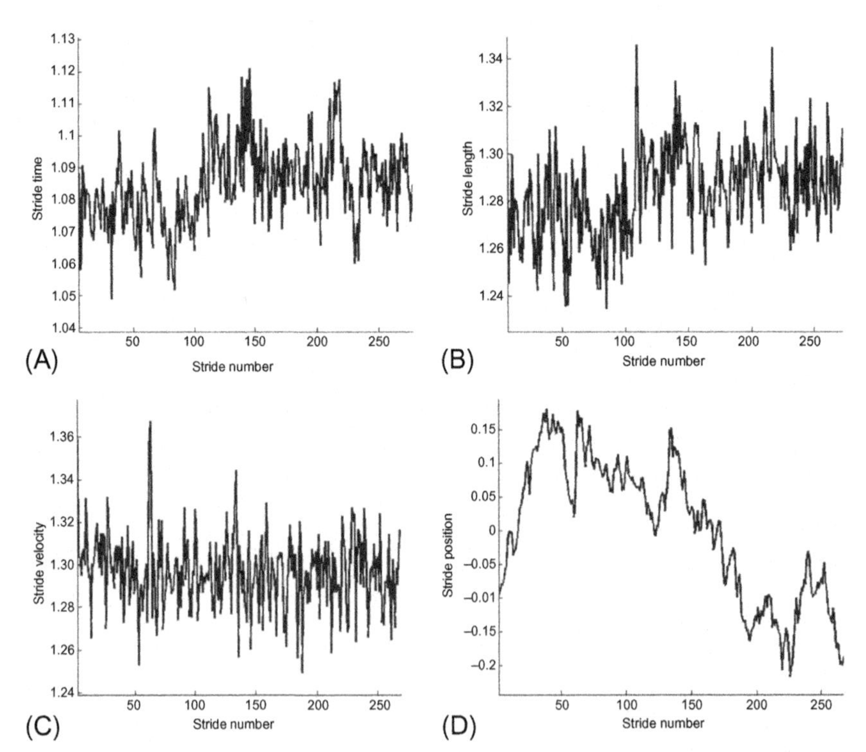

Fig. 4 An example of time series about walking on the treadmill. (A), (B), (C), and (D) represent stride length, duration time, velocity, and position, respectively.

Table 2 Controller parameters for step length and time.

Step length controller	Values	Step time controller	Values
P_L	−2	P_L	0.03
I_L	0.02	I_L	0.03
D_L	0.03	D_L	0.002

Table 3 Controller parameters for step length and time.

Step velocity controller	Values	Step position controller	Values
P_{VL}	0.02	P_{PL}	0.001
I_{VL}	0.01	I_{PL}	0.02
D_{VL}	0.001	D_{PL}	0.001
P_{VT}	1	P_{PT}	0.001
I_{VT}	0.02	I_{PT}	0.02
D_{VT}	0.05	D_{PT}	0

In these equations, η and γ are coefficients of added noises and *rand* is the equivalence of probability density functions that generate a random number between −0.5 and 0.5. The controllers' coefficients, shown in Tables 2 and 3, and η and γ values were set manually in the c_TLVP model in order to produce similar standard deviations (SD) of stride length and time in both model and experimental data. In this way, the η and γ were set to 0.0008 and 0.0037, respectively.

In the next step and to evaluate the effect of each controller, the control variables were removed step by step and the other control system models (i.e., c_TLV by deletion of position controller, and then c_TL by deletion of velocity controller) were simulated and the results were compared to those of experiments. The same approach was used for c_VP by deleting stride length and time controllers.

3 Results

First, we set length (desired) to 0.55, step time (desired) to 0.5, and treadmill velocity to 1.1, and transient responses of the four models were computed.

The stride time transient responses of four control system models are plotted in Fig. 5. As shown, stride time of c_TL reached its desired value more quickly compared to the others.

The stride length transient responses of four control system models are plotted in Fig. 6. The stride length of c_TL reached its desired value more quickly compared to the others. In addition, c_VP did not approach its desired value.

The stride velocity transient responses of four control system models are plotted in Fig. 7. Stride velocity c_TLVP reached its desired value less quickly than c_TLV.

The stride position transient responses of four control system models are plotted in Fig. 8. Only c_TLVP and c_VP closed to the proper value of the position. c_TLVP closed to the desired value much more quickly than c_VP.

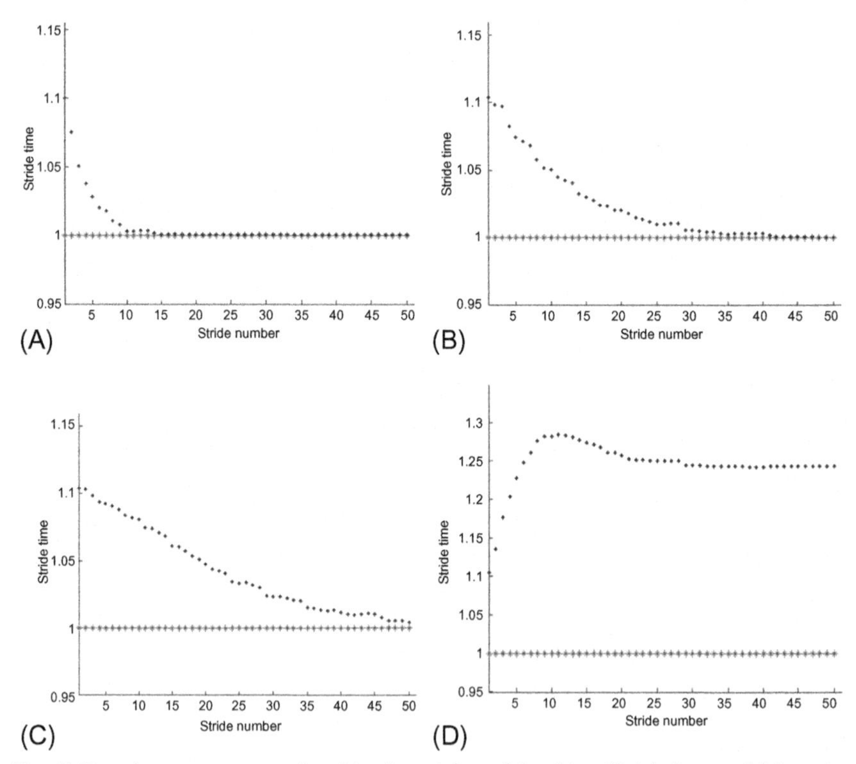

Fig. 5 Transient responses of stride time (plotted by *blue* "." (*dark gray* "." in print version)) and its desired values (plotted by *red* "*" (*light gray* "*" in print version)) are shown in (A), (B), (C), and (D) for c_TL, c_TLV, c_TLVP, and c_VP, respectively.

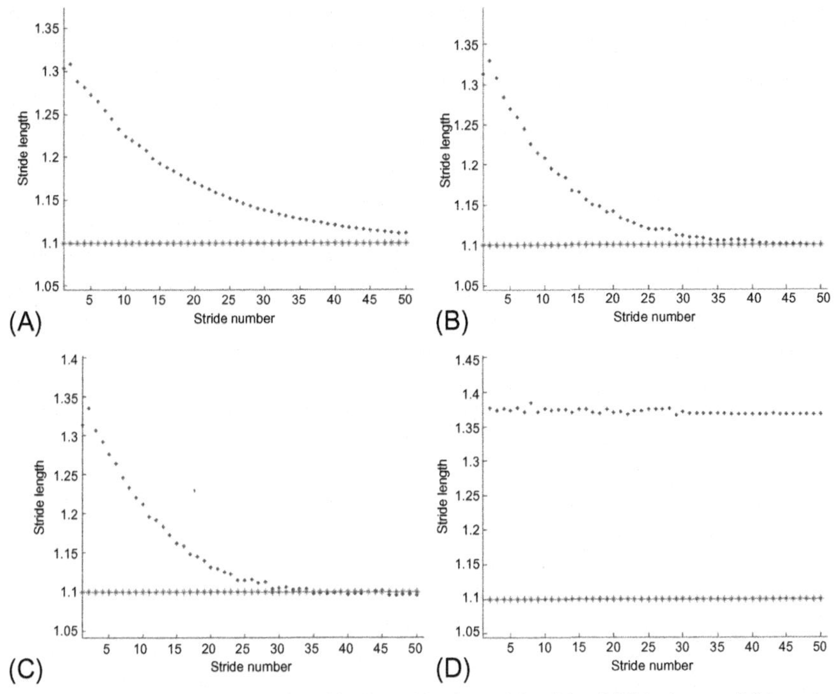

Fig. 6 Transient responses of stride length (plotted by *blue* "." (*dark gray* "." in print version)) and its desired values (plotted by *red* "*" (*light gray* "*" in print version)) are shown in (A), (B), (C), and (D) for c_TL, c_TLV, c_TLVP, and c_VP, respectively.

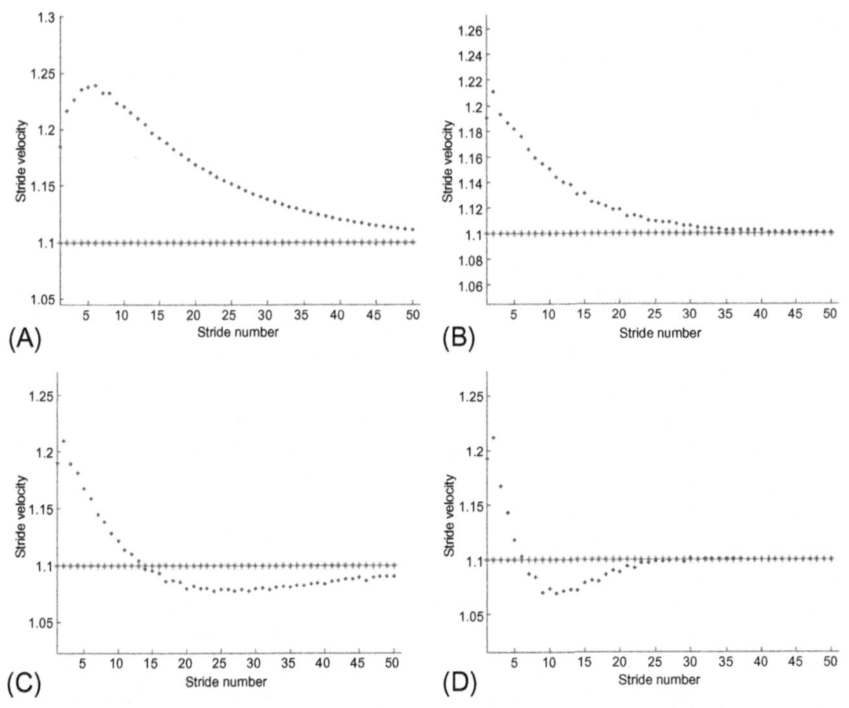

Fig. 7 Transient responses of stride velocity (plotted by *blue* "." (*dark gray* "." in print version)) and its desired values (plotted by *red* "*" (*light gray* "*" in print version)) are shown in (A), (B), (C), and (D) for c_TL, c_TLV, c_TLVP, and c_VP, respectively.

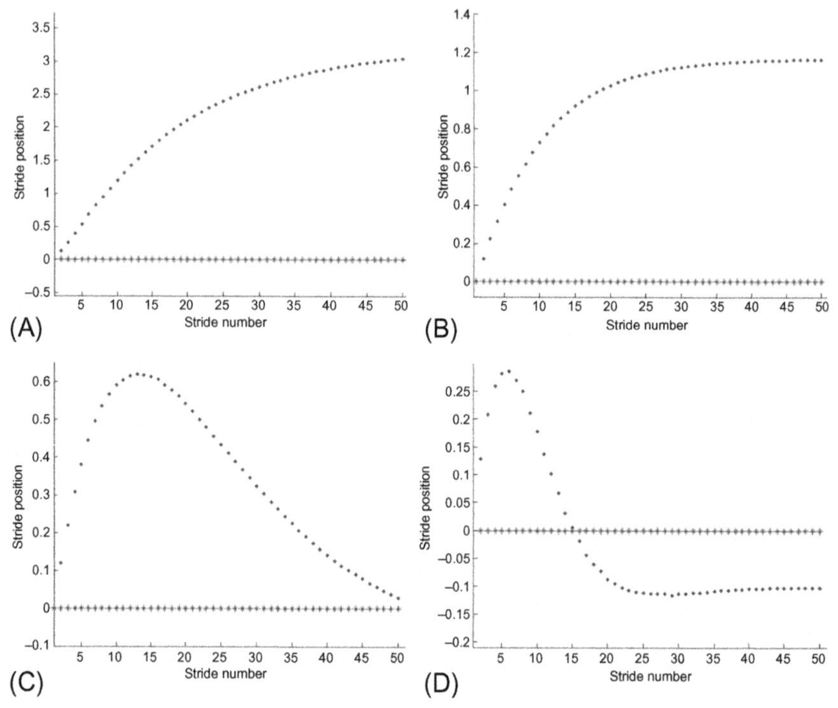

Fig. 8 Transient responses of stride position (plotted by *blue* "." (*dark gray* "." in print version)) and its desired values (plotted by *red* "*" (*light gray* "*" in print version)) are shown in (A), (B), (C), and (D) for c_TL, c_TLV, c_TLVP, and c_VP, respectively.

Next, the means of step length, time, position, and velocity in the experimental data were used as their desired values in the controllers. Then, the values of means, SD, and DFA scaling exponents (α) were calculated for these parameters.

Means, SD and DFA scaling exponents (α) for stride times of four control system models are plotted in Fig. 9. As shown, α is between 0.5 and 1 for all models. It is the same as that of experimental data.

Means, SD, and DFA scaling exponents (α) for stride length of four control system models are plotted in Fig. 10. Except c_VP, α of other models is between 0.5 and 1, as it is in the experimental data.

Means, SD, and DFA scaling exponents (α) for stride velocity of four control system models are plotted in Fig. 11. Only c_TLVP value of (α) is like that of the experimental data.

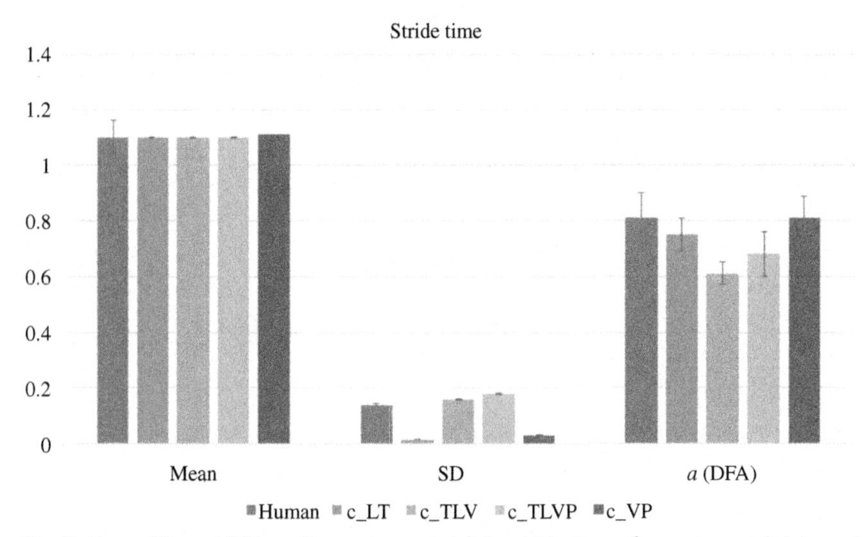

Fig. 9 Mean, SD, and DFA scaling exponents (α) for stride time of experimental data and c_TL, c_TLV, c_TLVP, c_VP are demonstrated.

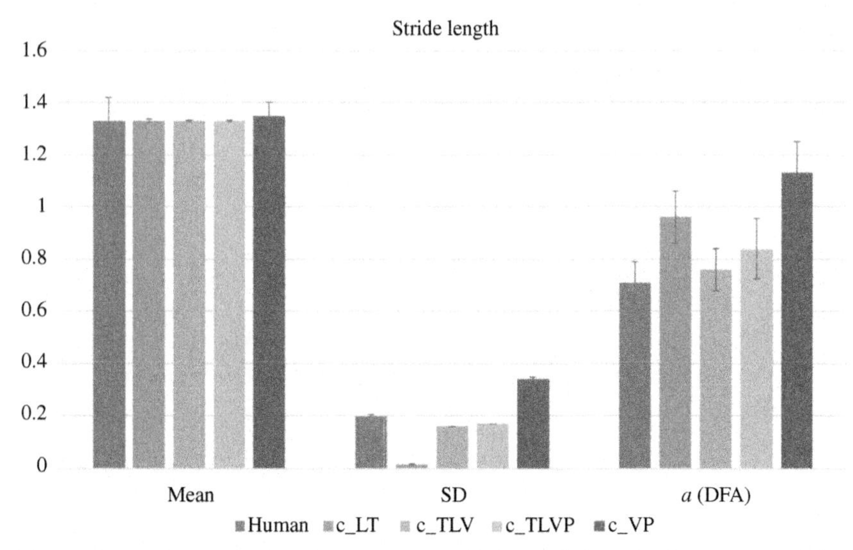

Fig. 10 Mean, SD, and DFA scaling exponents (α) for stride length of experimental data and c_TL, c_TLV, c_TLVP, c_VP are demonstrated.

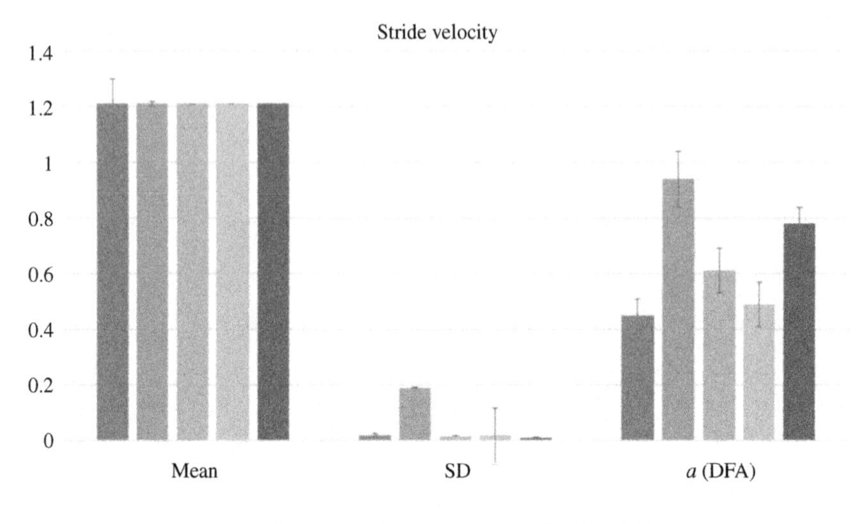

Fig. 11 Mean, SD, and DFA scaling exponents (α) for c_TL, c_TLV, c_TLVP, c_VP and experimental data of human are exhibited.

Means, SD, and DFA scaling exponents (α) for stride position of four control system models are plotted in Fig. 12. Only c_TLVP value of (α) is like that of the experimental data. SD of all models are approximately similar to human data except c_TL. All models' value of (α) is more than 1, just as in the experimental data.

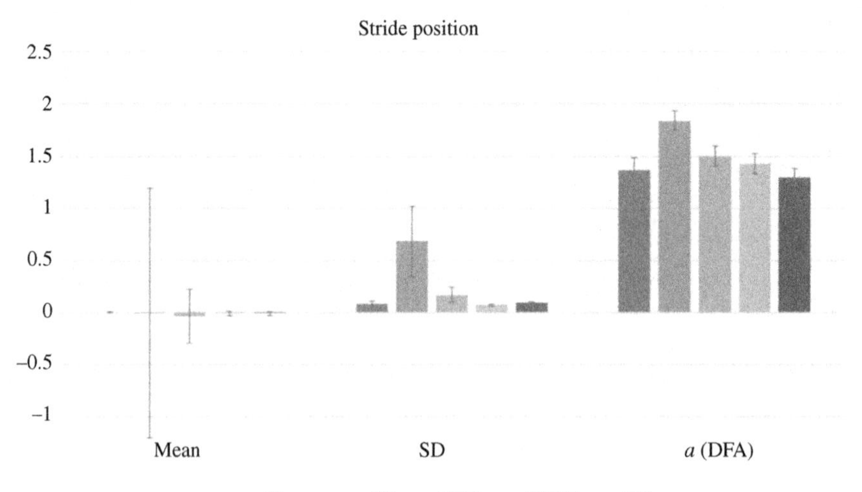

Fig. 12 Mean, SD, and DFA scaling exponents (α) for c_TL, c_TLV, c_TLVP, c_VP, and experimental data of human for the position are shown.

4 Discussion

Human movement variability, its origins, and valuable information are hot topics in the field of human motor control and rehabilitation. In this work, we presented some biped-based models for human walking. The focus of the models is the variabilities of stride-to-stride walking parameters. The models are built by different combinations of four controllers (for stride length, time, velocity, and position). This approach lends itself to application in modeling of human walking in different environmental conditions (walking overground or on a treadmill), subjects with different ages, or dual tasks studies. In this research, the model that contains all controllers (for stride length, time, velocity, and position) showed acceptable similarity of α values to experimental data for walking on the treadmill.

It is believed that the average human movement is approximately optimal, for doing the same trials of a task (Arechavaleta et al., 2008). According to this assumption, for each self-selected speed, stride length is chosen in such a way as to minimize energy consumption (Kuo, 2001). Thus the optimal values of stride length and time can be chosen as their desired values. As shown in Fig. 6, the model with no control on steps length and time (c_VP) cannot reach the desired optimal values. In addition, the stride length SD value in this model is much higher than those of other models and experimental data. Therefore, it seems that to save metabolic energy on the treadmill, besides controlling velocity and position, step length and time also need to be controlled by the central nervous system.

Two theories have been suggested for human position control during walking on the treadmill. One of these theories suggests that the position on the treadmill is only corrected when the body is near to the treadmill edges (Dingwell et al., 2010). The other theory suggests a lazy controller for this case (Dingwell and Cusumano, 2015). The presented models are in the direction of the second theory by choosing small values for coefficients of position controller.

Several studies discussed the differences (Rosenblatt and Grabiner, 2010; Lazzarini and Kataras, 2016) and similarities (Riley et al., 2007; Bollens et al., 2010) between treadmill and overground walking. One of these differences is statistical persistence of velocity that exists during walking overground, but not on the treadmill (Terrier et al., 2005). The result of the c_TL model showed similarity to walking overground w.r.t α (DFA). Consequently, this suggests that in self-selected speed during walking overground, only step length and time are likely controlled by the central nervous system and no (or lazy) velocity control is required.

As shown in Figs. 5 and 7, by adding velocity control to the model, transient time of stride duration increased and velocity decreased. Also, the SD value for the stride time of c_TLV was much more than that of c_TL. It was shown that the more the specific parameter receives control, the more SD value the other one gets. These results might be consistent with the concept of the uncontrolled manifold hypothesis (Scholz and Schöner, 1999; Monaco et al., 2018). More research is suggested to investigate this hypothesis.

5 Conclusion

This chapter presented a simple bio-mechanic model for human walking on the treadmill. This study showed that in addition to velocity and position control, control of stride length and time is also necessary to keep walking at its optimal mode. Moreover, it was also shown that this framework of modeling has the capability to explain differences between α (DFA) in time series of stride velocity of walking overground and on a treadmill. Further research is required to explore the relationship between α (DFA) of stride length, time, velocity, and position as well as propose controllers in different motor and mental tasks.

References

Agresta, C.E., Goulet, G.C., Peacock, J., Housner, J., Zernicke, R.F., Zendler, J.D., 2019. Years of running experience influences stride-to-stride fluctuations and adaptive response during step frequency perturbations in healthy distance runners. Gait Posture 70, 376–382.

Aloulou, A., Boubaker, O., 2010. Modeling and controlling a humanoid robot in the three dimensional space. In: Proceeding of the IEEE International Symposium on Robotics and Intelligent Sensors, March 2010, pp. 8–11.

Aloulou, A., Boubaker, O., 2011. Minimum jerk-based control for a three dimensional bipedal robot. In: Jeschke, S., Liu, H., Schilberg, D. (Eds.), Intelligent Robotics and Applications. ICIRA 2011. In: Lecture Notes in Computer Science, vol. 7102. Springer, Berlin, Heidelberg.

Aloulou, A., Boubaker, O., 2015. A minimum jerk-impedance controller for planning stable and safe walking patterns of biped robots. In: Carbone, G., Gomez-Bravo, F. (Eds.), Motion and Operation Planning of Robotic Systems. Mechanisms and Machine Science. In: vol. 29. Springer, Cham.

Arechavaleta, G., Laumond, J.-P., Hicheur, H., Berthoz, A., 2008. An optimality principle governing human walking. IEEE Trans. Robot. 24 (1), 5–14.

Bahramian, A., Shamsi, E., Towhidkhah, F., Jafari, S., 2019. Hip retraction enhances walking stability on a ramp: an equilibrium point hypothesis-based study. biorxiv. 638635.

Bohnsack-McLagan, N.K., Cusumano, J.P., Dingwell, J.B., 2016. Adaptability of stride-to-stride control of stepping movements in human walking. J. Biomech. 49 (2), 229–237.
Bollens, B., Crevecoeur, F., Nguyen, V., Detrembleur, C., Lejeune, T., 2010. Does human gait exhibit comparable and reproducible long-range autocorrelations on level ground and on treadmill? Gait Posture 32 (3), 369–373.
Boubaker, O., Iriarte, R. (Eds.), 2017. The Inverted Pendulum in Control Theory and Robotics: From Theory to New Innovations. In: vol. 111. IET.
Decker, L.M., Cignetti, F., Hunt, N., Potter, J.F., Stergiou, N., Studenski, S.A., 2016. Effects of aging on the relationship between cognitive demand and step variability during dual-task walking. Age 38 (4), 363–375.
Dingwell, J.B., Cusumano, J.P., 2015. Identifying stride-to-stride control strategies in human treadmill walking. PLoS One 10(4), e0124879.
Dingwell, J.B., Marin, L.C., 2006. Kinematic variability and local dynamic stability of upper body motions when walking at different speeds. J. Biomech. 39 (3), 444–452.
Dingwell, J.B., John, J., Cusumano, J.P., 2010. Do humans optimally exploit redundancy to control step variability in walking? PLoS Comput. Biol. 6(7), e1000856.
Duncan, A.W., 2018. Bridging the Gap: Individual Relationships between Long Range Correlations and Dexterity in Walking. University of Nebraska at Omaha.
Engelbrecht, S.E., 2001. Minimum principles in motor control. J. Math. Psychol. 45 (3), 497–542.
Faisal, A.A., Selen, L.P., Wolpert, D.M., 2008. Noise in the nervous system. Nat. Rev. Neurosci. 9 (4), 292.
Garcia, M., Chatterjee, A., Ruina, A., Coleman, M., 1998. The simplest walking model: stability, complexity, and scaling. J. Biomech. Eng. 120 (2), 281–288.
Hausdorff, J.M., 2007. Gait dynamics, fractals and falls: finding meaning in the stride-to-stride fluctuations of human walking. Hum. Mov. Sci. 26 (4), 555–589.
Hausdorff, J., Zemany, L., Peng, C.-K., Goldberger, A., 1999. Maturation of gait dynamics: stride-to-stride variability and its temporal organization in children. J. Appl. Physiol. 86 (3), 1040–1047.
Hu, K., Ivanov, P.C., Chen, Z., Carpena, P., Stanley, H.E., 2001. Effect of trends on detrended fluctuation analysis. Phys. Rev. E 64 (1), 011114.
Huang, Y., Huang, Q., Wang, Q., 2017. Chaos and bifurcation control of torque-stiffness-controlled dynamic bipedal walking. IEEE Trans. Syst. Man Cybern. Syst. Hum. 47 (7), 1229–1240.
Kang, H.G., Dingwell, J.B., 2008. Separating the effects of age and walking speed on gait variability. Gait Posture 27 (4), 572–577.
Kantelhardt, J.W., Koscielny-Bunde, E., Rego, H.H., Havlin, S., Bunde, A., 2001. Detecting long-range correlations with detrended fluctuation analysis. Physica A 295 (3–4), 441–454.
Karimian, M., Towhidkhah, F., Rostami, M., 2006. Application of model predictive impedance control (MPIC) in analysis of human walking on rough terrains. Int. J. Appl. Electromagn. Mech. 24 (3–4), 147–162.
Kirchner, M., Schubert, P., Liebherr, M., Haas, C.T., 2014. Detrended fluctuation analysis and adaptive fractal analysis of stride time data in Parkinson's disease: stitching together short gait trials. PLoS One 9(1), e85787.
Kuo, A.D., 2001. A simple model of bipedal walking predicts the preferred speed–step length relationship. J. Biomech. Eng. 123 (3), 264–269.
Kuo, A.D., 2002. Energetics of actively powered locomotion using the simplest walking model. J. Biomech. Eng. 124 (1), 113–120.

Larsson, J., Miller, M., Hansson, E.E., 2016. Vestibular asymmetry increases double support time variability in a counter-balanced study on elderly fallers. Gait Posture 45, 31–34.
Lazzarini, B.S.R., Kataras, T.J.J.G., 2016. Treadmill walking is not equivalent to overground walking for the study of walking smoothness and rhythmicity in older adults. Gait Posture 46, 42–46.
Lewek, M.D., Cruz, T.H., Moore, J.L., Roth, H.R., Dhaher, Y.Y., Hornby, T.G., 2009. Allowing intralimb kinematic variability during locomotor training poststroke improves kinematic consistency: a subgroup analysis from a randomized clinical trial. Phys. Ther. 89 (8), 829–839.
Malcolm, B.R., Foxe, J.J., Butler, J.S., Molholm, S., De Sanctis, P., 2018. Cognitive load reduces the effects of optic flow on gait and electrocortical dynamics during treadmill walking. J. Neurophysiol. 120 (5), 2246–2259.
Monaco, V., Tropea, P., Rinaldi, L.A., Micera, S., 2018. Uncontrolled manifold hypothesis: organization of leg joint variance in humans while walking in a wide range of speeds. Hum. Mov. Sci. 57, 227–235.
Nazarimehr, F., Jafari, S., Golpayegani, S.M.R.H., Kauffman, L.H., 2017. Investigation of bifurcations in the behavior of the process equation. Preprint arXiv: 00419.
Nourian Zavareh, M., Nazarimehr, F., Rajagopal, K., Jafari, S., 2018. Hidden attractor in a passive motion model of compass-gait robot. Int. J. Bifurcation Chaos 28 (14), 1850171.
Osborne, L.C., Lisberger, S.G., Bialek, W., 2005. A sensory source for motor variation. Nature 437 (7057), 412.
Pearson, K.G., 2004. Generating the walking gait: role of sensory feedback. In: Progress in Brain Research. vol. 143. Elsevier, pp. 123–129.
Rao, P.G.K., Subramanyam, M., Reddy, K.M., Sarma, G., Rupa, P., 2010. Un-obstructive capturing of power parasitically from normal human walking. J. Instrum. Soc. India 40 (3), 231–234.
Riley, P.O., Paolini, G., Della Croce, U., Paylo, K.W., Kerrigan, D.C., 2007. A kinematic and kinetic comparison of overground and treadmill walking in healthy subjects. Gait Posture 26 (1), 17–24.
Roerdink, M., de Jonge, C.P., Smid, L.M., Daffertshofer, A., 2019. Tightening up the control of treadmill walking: effects of maneuverability range and acoustic pacing on stride-to-stride fluctuations. Front. Physiol. 10.
Rosenblatt, N.J., Grabiner, M.D., 2010. Measures of frontal plane stability during treadmill and overground walking. Gait Posture 31 (3), 380–384.
Salinas, M.M., Wilken, J.M., Dingwell, J.B., 2017. How humans use visual optic flow to regulate stepping during walking. Gait Posture 57, 15–20.
Scholz, J.P., Schöner, G., 1999. The uncontrolled manifold concept: identifying control variables for a functional task. Exp. Brain Res. 126 (3), 289–306.
Scott, S.H., 2004. Optimal feedback control and the neural basis of volitional motor control. Nat. Rev. Neurosci. 5 (7), 532.
Srinivasan, M., Ruina, A., 2006. Computer optimization of a minimal biped model discovers walking and running. Nature 439 (7072), 72.
Su, J.L.-S., Dingwell, J.B., 2007. Dynamic stability of passive dynamic walking on an irregular surface. J. Biomech. Eng. 129 (6), 802–810.
Terrier, P., 2016. Fractal fluctuations in human walking: comparison between auditory and visually guided stepping. Ann. Biomed. Eng. 44 (9), 2785–2793.
Terrier, P., Dériaz, O., 2012. Persistent and anti-persistent pattern in stride-to-stride variability of treadmill walking: influence of rhythmic auditory cueing. Hum. Mov. Sci. 31 (6), 1585–1597.

Terrier, P., Turner, V., Schutz, Y., 2005. GPS analysis of human locomotion: further evidence for long-range correlations in stride-to-stride fluctuations of gait parameters. Hum. Mov. Sci. 24 (1), 97–115.

Todorov, E., Jordan, M.I., 2002. Optimal feedback control as a theory of motor coordination. Nat. Neurosci. 5 (11), 1226.

Toebes, M.J., Hoozemans, M.J., Furrer, R., Dekker, J., van Dieën, J.H., 2012. Local dynamic stability and variability of gait are associated with fall history in elderly subjects. Gait Posture 36 (3), 527–531.

Verghese, J., Holtzer, R., Lipton, R.B., Wang, C., 2009. Quantitative gait markers and incident fall risk in older adults. J. Gerontol. Ser. A Biol. Med. Sci. 64 (8), 896–901.

Ziegler, M.D., Zhong, H., Roy, R.R., Edgerton, V.R., 2010. Why variability facilitates spinal learning. J. Neurosci. 30 (32), 10720–10726.

CHAPTER 11

Continuum NasoXplorer manipulator with shape memory actuators for transnasal exploration

Phoebe Lim, Leoni Goh Yi Ting, Ong Kwok Chin Douglas, Honglin An, Chwee Ming Lim, Hongliang Ren
Department of Biomedical Engineering, Faculty of Engineering, National University of Singapore, Singapore, Singapore

1 Clinical needs and intended engineering design objectives

According to the Singapore Cancer Registry Annual Registry Report in 2015, nasopharyngeal cancer (NPC) was found to be the seventh most frequent cancer in males in the period 2011–15 (National Registry of Diseases Office, 2017; Albery, 1995). NPC cancer is proved to be one of the most lethal cancers in Singapore. Hence it is crucial for patients to seek treatment as early as possible. However, NPC does not display pronounced symptoms in the early stages. As a result, patients often only get diagnosed with NPC in the later stages of the disease. At the same time, it is also crucial to note that the later the patient gets diagnosed with NPC, the lower the chance of survival (Taylor et al., 2019). In addition, the rate of recurrence after primary treatment of NPC ranges from 15% to 58% (Kalairaj et al., 2019). The interval between the initial treatment and recurrence ranges between 1 month and 10 years (Kalairaj et al., 2019). As a result, there is a need for the monitoring of the nasopharynx, especially for patients that have received treatment for NPC, to enable early detection of recurrence, thus allowing for immediate medical attention. Patients with a family medical history of NPC can also benefit from a monitoring device for the nasopharynx, enabling them to assess the region and the surrounding areas.

There are two methods for diagnosing NPC: direct nasopharyngoscopy and indirect nasopharyngoscopy by a skilled physician. For indirect nasopharyngoscopy, the physician employs the use of small unique mirrors

Control Theory in Biomedical Engineering
https://doi.org/10.1016/B978-0-12-821350-6.00011-1

and a bright light source, which are inserted into the patient's mouth for the physician to observe the nasopharynx region. While this method is not as invasive as direct nasopharyngoscopy, it is unable to give a clear view, especially if the patient has a narrow, constricted throat or tends to exhibit a vomiting reflex. As such, the diagnosis will not be as conclusive. Direct nasopharyngoscopy, on the other hand, employs a nasopharyngoscope that is inserted through the nostril of the patient to allow the physician to view the nasopharynx region clearly. Local anesthesia is usually applied to the nasal cavity before carrying out the procedure. This method can provide more conclusive diagnosing of NPC in patients as compared to indirect nasopharyngoscopy. However, the disadvantage of this method is that its invasive nature may tend to cause discomfort to the subject. Both procedures need a skilled physician to perform them. This makes it more tedious for the subject as they have to make an appointment at a hospital or clinic before being able to undergo the procedure.

We developed a home-based kit that allows patients to maneuver the nasopharyngoscope with ease. The following are the key aims:

- A nasopharyngoscope can be self-directed, without the presence of a clinician, as well as affordable for the public. This will increase the frequency of observations and reduce the healthcare burden on professionals.
- To have a good bending range to allow smooth motion in the nasopharynx and to increase the viewing range of the nasopharyngoscope. This will enable users to be able to detect problem areas in the nasopharynx easily.
- To reduce discomfort for patients that will be self-administering the device through the nasopharynx region. This is done by minimizing coincidence with the surrounding tissue.
- To manufacture a device that is compact and lightweight for increased convenience. This will increase the portability of the device, which will help reduce the healthcare burden and increase the frequency of observations.
- To develop an economical device that will increase the affordability and accessibility of health care to all, regardless of economic status.

2 Methods

A new nasopharyngoscope, **NasoXplorer**, allows for self-administered nasopharyngoscopy at home instead of having the procedure done at hospitals or clinics. As the diameter of the nostril opening averages 10 mm (Schriever et al., 2013), our device aims to have a total diameter of less than 5 mm. This not only allows the device to move smoothly within the region, but also reduces the likelihood of the device touching any sensitive tissue. The final prototype of NasoXplorer has a width of 3–3.84 mm at the working end. The size of the nostrils' opening also restricts the number of tubes and wires we can use. The NasoXplorer consists of a single tube, with a single working lane. Materials used in the NasoXplorer include an acrylic tube and a shape memory alloy (SMA) spring.

Component	Material
Body	Acrylic tube
Body	Aluminum tube
Spring	Flexinol actuator spring CH 337
Loose wires	Enamel coated copper wire
Bendy tube	FLX-9070

The NasoXplorer has 2 degrees of freedom—one is characterized using Arduino, and the other is by manually rotating the working device. NasoXplorer also aims to minimize patient discomfort. This was crucial in our manufacturing process since the NasoXplorer is to be self-administered. Conventional nasal endoscope treatments may use a topical anesthetic (Atar, 2014). However, it is worth noting that the use of the anesthetic does not guarantee any lack of discomfort and will only serve to alleviate patient discomfort. In a study, patients rated the discomfort in sedated nasopharyngoscopy as about 2–3 on a scale of 1–10 with 10 being the most painful (Gaviola et al., 2013; Huang et al., 2001; Morgan et al., 1995). Hence, the NasoXplorer aims to minimize patient discomfort by minimizing the diameter of its working end and creating a simple maneuvering system to allow patients to navigate the nasopharynx easily.

Subsequent prototypes will aim at containing the control system as well as reducing its size to reduce cost and allow the device to be more portable for the convenience of patients.

As the design specification consultant, the requirements and guidelines mention in this document will ensure that the information stated is met and adhered to.

This section discusses the specifications and design features of the device as well as restrictions imposed on its design. The NasoXplorer has no intention of curing NPC and will not be sufficient to provide treatment to patients with NPC. The NasoXplorer aims to monitor and increase the frequency of observations of NPC while reducing the burden on healthcare professionals. We focus on the mechanics of the NasoXplorer and will not discuss further considerations for camera models and electronic transfer or quality of captured images. For the final prototype, we used optical fibers to test the viewing angle of the NasoXplorer. Although the NasoXplorer has only 2 degrees of freedom, it has a wide viewing range of up to 90 degrees.

2.1 Device specifications from anatomical considerations

Besides user needs and intended use of a device, other design considerations had to be made to consider the different anatomical variations of different people. We observed that several characteristics played a part in anatomical differences.

2.1.1 Anatomical variations in shape

The NasoXplorer aims at minimizing end-user discomfort, especially for a self-administered device. As the distal segment of the device will seek to work as a probe, a flexible and soft material was deemed most suitable for this function. The flexibility of the NasoXplorer allows the camera to have a broader optical viewing range, while the softness of the device prevents significant user discomfort. By minimizing the size of the NasoXplorer, patients with different nose shapes are easily be able to navigate the nasopharynx region. Hence, the functionality of NasoXplorer is maximized in the observation of the nasal cavity and nasopharynx region.

2.1.2 Anatomical variations in size

Anatomical differences exist between individuals. For the NasoXplorer to cater to different individuals, anthropometric data had to be studied. As the NasoXplorer is designed to be self-administered, the reduction of end-user discomfort was of importance. This meant that we had to cater to all patients regardless of anatomical differences. The rigidity of the NasoXplorer provides a fuss-free, easy-to-use solution for patients. A softer, flexible tip was employed to allow monitoring of the more in-depth part in the nasopharynx region. However, our device may not be suitable for viewing the deeper regions in the nasopharynx region. This can be improved in the future by using a more extended tip. However, adjustments may have to be made to the length of the spring as a more significant force will be required to actuate the longer tip. Hence, we studied various randomized trials and reviewed the anthropometric measurements.

2.1.3 Anatomical variations based on age and gender

In a 2013 study, among 69 healthy volunteers, it was observed that males tend to have larger nostrils than females (Schriever et al., 2013). Males had an average nostril opening cross-sectional area of 407 mm^2, whereas females had an average cross-sectional area of 310 mm^2. This suggests a diameter of 10–11 mm. This study did not consider different age groups. However, it is good to note that the average age of all 69 subjects is 35 years old. In another study done in 2005, boys had an average nostril opening cross-sectional area of 72.3 mm^2, while girls had an average nostril opening cross-sectional area of 66.0 mm^2, suggesting a diameter to ~5 mm^2, which is half that of an adult (Mori et al., 2005).

2.1.4 Estimation of the distance between nasal inlet to the channel

The distance from the nasal inlet to the nasopharynx region can extend up to 14 cm (Taylor et al., 2019). Although the NasoXplorer is 22.7 cm long, only 8.5 cm was intended for use within the human body. Other parts of the NasoXplorer are not intended to be inside the human body. However, these are average measurements for Caucasian males only, and it should be noted that Asians may have a smaller distance between nasal inlet and channel. Since our target audience is Asians, the NasoXplorer is sufficient for viewing most of the nasopharynx region.

2.1.5 Estimation of area of narrowest path in the nasopharynx region

One cross-sectional study examined the differences in nasopharyngeal area in 96 individuals aged 6–59 years old without facial abnormalities (De Araújo et al., 2016). Subjects were split into four age groups, but no differences were found in their nasopharyngeal cross-sectional areas, and all were slightly more prominent than 1 cm^2. Since the segment of the NasoXplorer entering the nose has a diameter of a maximum of 4 mm, we have met the requirements that we set out to achieve.

2.1.6 Device design specifications

Following the preceding findings, our own design specifications aim to cater to people of various anatomical differences.

Parameters	Anatomical restrictions	Design specifications	Comments
Diameter (outer)	Should ideally be less than 10mm Data collected was based on Caucasians, should be adjusted to 8mm	3.92mm	Parameter met with 2mm of space between device and lumen on each end
Diameter (inner)		3.9mm	The diameter of the camera is 3.9mm, too large for the aluminum tube. Further improvements can be made to use a smaller version of the IntroSpicio camera instead
Insertion length Length of workspace in region (sagittal)	Minimum of 100mm	8.5mm	Parameter not met, but still sufficient to capture more than 80% of the nasal cavity Improvements can be made to lengthen the flexible tip
Bending angle (degrees)		145	Parameters met
Control system			Further improvements will be made to contain the control system and improve the portability of the device

2.2 Overall design

Our device in Fig. 1 encompasses the traits of a flexible nasopharyngoscope, which includes a small diameter of 3mm and a high deflection angle of 145 degrees (Table 1). Both the diameter and deflection angles are comparable to those that are commercially available.

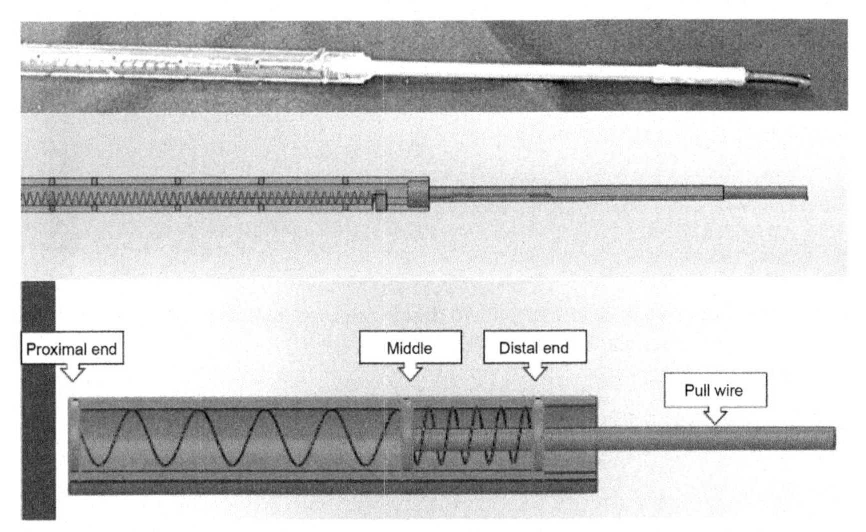

Fig. 1 The prototype, optical image (top), CAD model (middle) and illustrations (bottom).

Table 1 NasoXplorer specifications.

NasoXplorer	
Diameter (tip)/mm	3
Diameter (proximal end)/mm	8.4
Angulation range (deflection)/degrees	145
Working length (active)/cm	2
Working length (passive)/cm	9.8

2.3 Design components and design rationale

2.3.1 Optical zooming segment and camera

The camera we used is the IntroSpicio 110 micro CCD camera. It has a diameter of 3.9 mm and is threaded through the TangoBlack flexible segment, as well as the sole SMA spring in the NasoXplorer (Fig. 2). The distal end of the NasoXplorer is the flexible tip, bending up to an angle of ~100 degrees. This is supported by the User and Technical Specification Report. The NasoXplorer works using an SMA spring that extends and contracts with the application of a small current. Fig. 3 shows part of the NasoXplorer when relaxed and contracted, respectively. The spring used in the NasoXplorer is 0.1 in. in diameter and 0.6 in. in length. Further details are provided in Fig. 3. When the NasoXplorer enters the nasal inlet of the

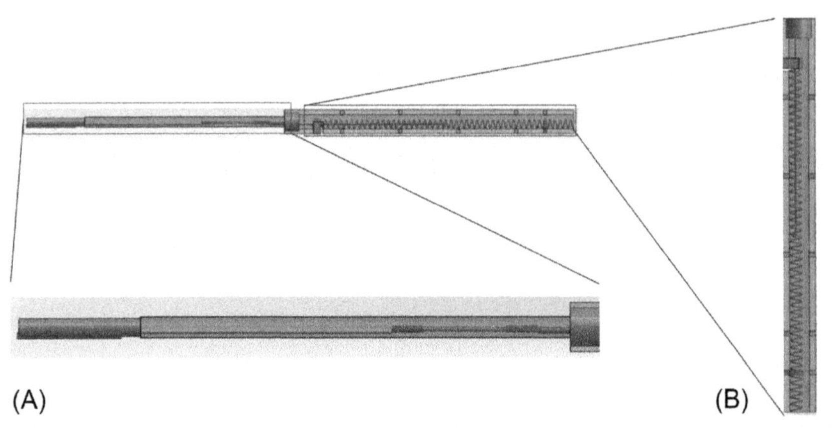

Fig. 2 Blown-up images featuring the two design features (hybrid pull wire (A) and the spring casing with ventilation outlets (B)).

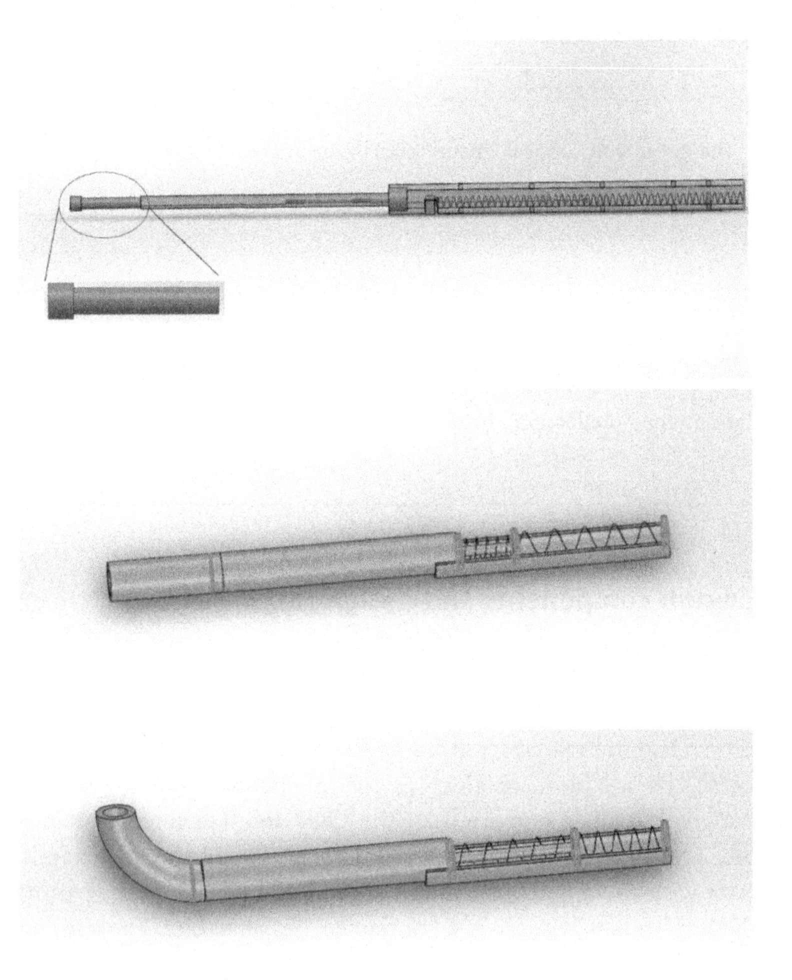

Fig. 3 (Top and middle) Relaxed state; (bottom) during actuation; more animations is in the support Video 1 in the online version at https://doi.org/10.1016/B978-0-12-821350-6.00011-1.

patient, the flexible tip is in a relaxed state (Fig. 3). Upon entering the curved region, the patient is able to actuate the device and begin bending of the flexible tip through passing a small current (Fig. 3). This allows end users to navigate easily the curved region of the nasopharynx, as well as increases the optical viewing range of the device. With the NasoXplorer there is a reduced risk of interaction and contact with the surrounding tissue of the nasopharynx region, thereby reducing patient discomfort.

2.3.2 Actuation and control of the bending segment

The bending segment refers to the proximal end, which is responsible for bending the flexible tip on the distal end. This segment will not enter the lumen, but instead is held by the user. The central working part of the bending segment is the SMA spring (Kalairaj et al., 2019), the Flexinol Actuator Spring with the following specifications: wire diameter of 0.015 in., outer diameter of 0.100 in., compressed length of 0.6 in., 40 coils, maximum temperature of 90°C. The 2 degrees of freedom in the NasoXplorer allows a proper viewing angle inside the nasopharynx region. The bending segment is actuated using one single spring, which can carry out two actions. This is done by using three electrodes on a single SMA spring, at the proximal, middle, and distal, as represented in Figs. 1 and 4.

The SMA spring is heated using an electrical current transmitted by enamel-coated copper wires with stripped ends. We chose copper wires due to their thinness and low heat dissipation during electricity conduction. When the circuit is completed at the proximal and middle segments (as shown in Fig. 5), the SMA spring in that region contracts to generate a force sufficient to pull the wire backward, hence creating a bending moment at the flexible tip. Meanwhile, the SMA spring in the other region is stretched. The pull wire, a copper-leather hybrid, prevents the transmission of electricity and heat. When the circuit is completed at the distal and middle segment, the SMA spring in that region contracts and allows for slack in the pull wire, thereby allowing the already bent flexible tip to straighten. It should be noted that the proximal and distal ends of the SMA spring are fixed with regards to the spring casing and does not move. Only the middle electrode

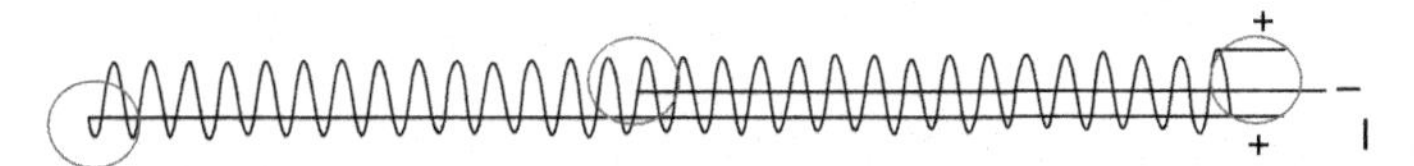

Fig. 4 Diagram showing three electrodes on one SMA spring.

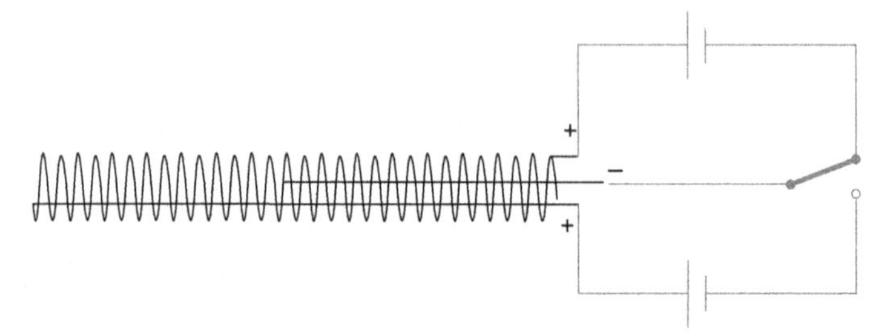

Fig. 5 Diagram showing closed-circuit connections. Attachment of spring and electrode to the distal end, attachment of electrode, and pull wire to middle portion and attachment of spring with an electrode to a proximal end. Right: illustration of the switching circuit.

moves during actuation. The electrodes are run out from the proximal end of the device to the switching circuit as shown in Fig. 5.

The NasoXplorer includes the following parts: (1) spring casing, (2) SMA spring, (3) pull wire (consisting of leather and copper, inside the extension tube), and (4) extension tube. For Part 2, we used a Flexinol Actuator Spring, chosen due to its SMA properties. Using SMA, we could implement the three electrodes to act as pseudo-two springs. This meant that additional springs would not be required to help straighten the flexible tip once the SMA spring was compressed. The NasoXplorer could then be reduced to having one working lane instead of two, which helps to reduce cost and diameter.

For Part 3, we chose a copper wire for increased durability and conductivity. For Part 4, we chose aluminum. This segment encases only the pull wire. Hence, the thicker acrylic that encased the actuation mechanism could be replaced by a thinner alternative. Since the pull wire, now a copper-leather hybrid, does not conduct both electricity and heat, a variety of materials could be considered. Eventually, we chose the aluminum tube as it is cheap and has a diameter of 3.97 mm.

2.3.3 Stiffness modulation

After the warping of the initial prototype due to heat, another opportunity to utilize the material for another purpose presented itself. The glass-transition temperature of the material could be used to modulate the stiffness instead. With the ability to change stiffness when required, it potentially allows us to vary the bending configurations, despite having a simple actuator. Hence, to explore the potential of this as an extra novelty of the design, we printed tubes of similar dimensions (OD 3 mm ID 2 mm) with four different materials that are a mixture of VeroClear and TangoBlack+ in different ratios.

After stiffness tests, it can be observed that there is stiffness change when the material is heating. There is also some evidence of returning stiffness when the heating is removed. While comparing across materials, we found that heating does indeed increase the flexibility of all materials if it does not break apart. However, constructs consisting of purely one material seem to yield unstable results. Instead, materials that contain a mix of rigid and flexible materials can yield consistent increases in flexibility while increasing the heating effect.

This finding allowed us to enhance our device further. By fabricating the tube in an optimal mix of materials and heat supply, the device is able to deform into irregular shapes. By doing so, it increases user comfort, as the device is then be able to account for physiological differences between patients.

3 Design verification

3.1 Bending capability: Determine the bending angle

A camera was fixed in a position to capture all the videos of the bending while carrying voltage and pulse-width modulation (PWM). These videos were passed through an online application of Physlet's called "Tracker," which provided coordinates that could be translated into displacement using Microsoft Excel, and then converted into a bending angle. A uniform white background was used in all the videos so that the bending would be more precise. The device was placed horizontally, as shown in Fig. 6. A small blue tack was used to lift the NasoXplorer up slightly so that the flexible tip was not in contact with the paper.

We conducted a bending angle test to determine the highest bending angle possible for the NasoXplorer (Fig. 7). At 100% PWM, a maximum angle of close to 80 degrees is attainable at 7.5 V. However, the large bending angle of 80 degrees over long periods of time may cause fractures to the flexible portion of the NasoXplorer due to the fragile properties of TangoBlack. Contrary to our expectations, the bending angle for 90% PWM exceeded that of 100% PWM (a bending angle of 66 degrees for NasoXplorer at 100% PWM at 7 V). We attributed this to the loosening of the pull wire in the NasoXplorer, which prevents the efficient transfer of momentum and force generated by the actuation of the SMA coils. To evaluate the effectiveness of our prototype, we ran various tests to show the optimal

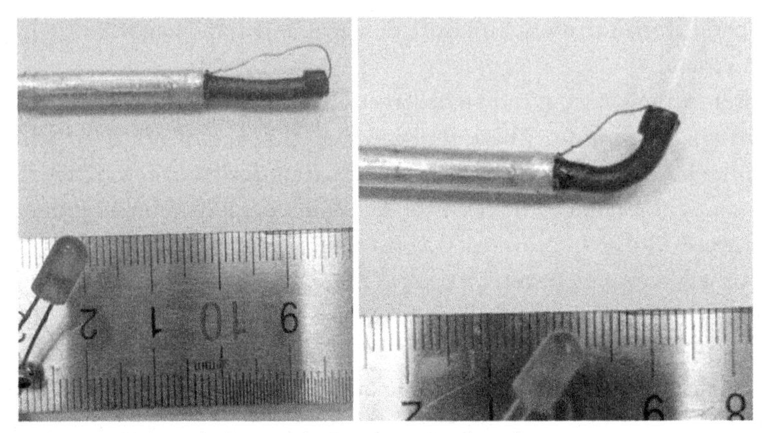

Fig. 6 The NasoXplorer before and after actuation/bending.

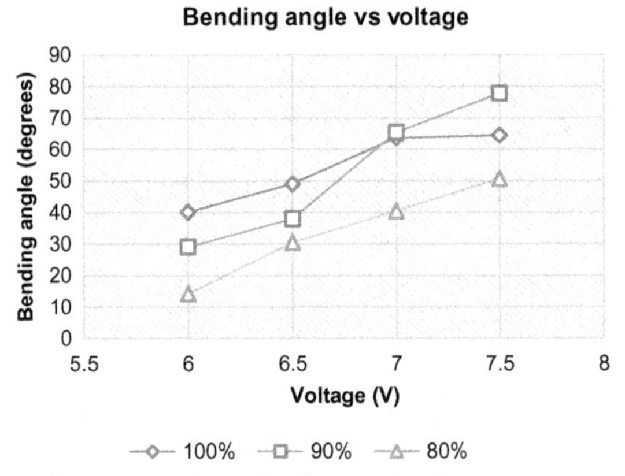

Fig. 7 Bending angle against voltage feeding to the SMA.

performance of the NasoXplorer. PWM along with voltage was varied to show how NasoXplorer performed under the varying conditions. It was concluded that at 100% PWM and 7V, the NasoXplorer was able to perform best as it displayed an optimal bending angle of 63 degrees. Although a bending angle of 66 degrees can be achieved at 7.5V, due to the properties of TangoBlack, it is possible for to optimize the fracture at the flexible portion of the bending over long periods of time.

3.2 Temperature monitoring during both actuation and retraction

We used a Seek thermal camera to monitor the temperature of the device during about 150 seconds of actuation and retraction. The camera was plugged into an iPhone to capture the temperature of the device at each time frame. We used an Arduino program such that the circuit would turn on for 25 seconds, and cool for the next 25 seconds.

One drawback of the design of the NasoXplorer is that the acrylic tube prevented cooling. Even when the circuit is off, the temperature of the device remains at a constant, with few reductions observed. This was the case even after we drilled holes into the working end to prevent the device from overheating. Hence, we shifted the working end ex vivo.

None of the test results at all four voltages reported temperatures close to 80°C, the temperature for acrylic warping (Figs. 8 and 9). However, these are the results for just one cycle of the actuation. When running the circuit for 120 seconds, temperatures stabilized around 53°C. Hence, it is concluded that the acrylic tube is safe to use in the NasoXplorer. Another thing

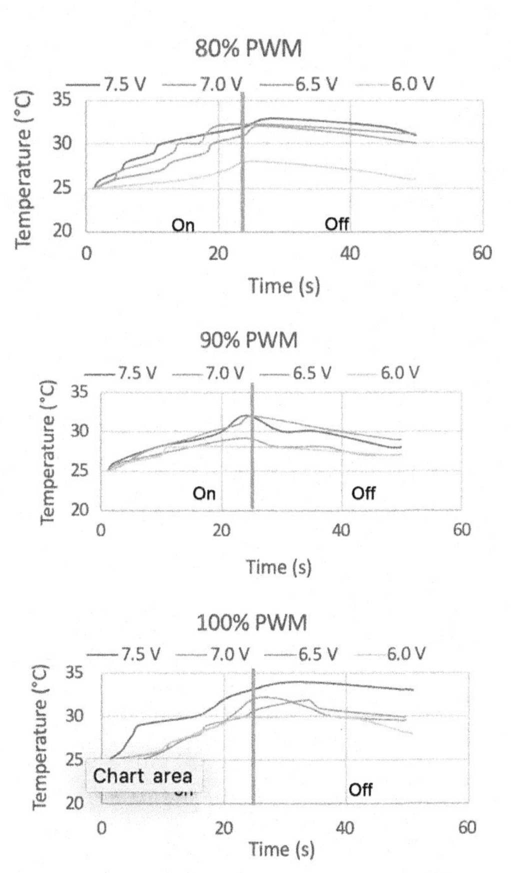

Fig. 8 Temperature-time relationship at 80%, 90%, 100% PWM.

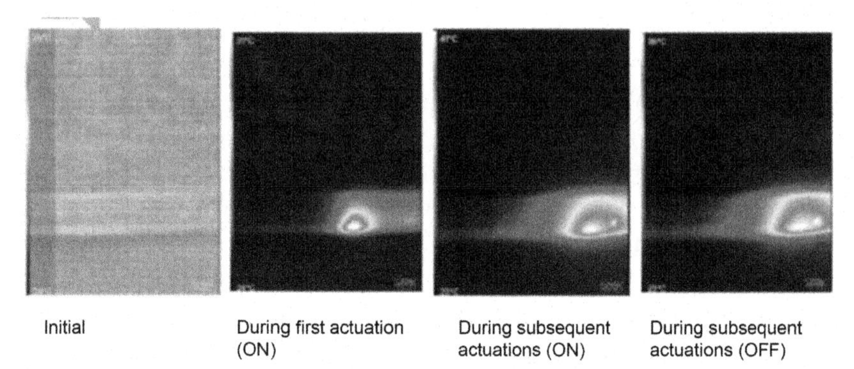

Fig. 9 Example of Seek thermal application used during actuation.

to note is the stability in 90% and 80% PWM. As shown in the graphs, temperatures in the 80% and 90% PWM rise steadily in a staircase manner while cooling down almost instantaneously. However, the 100% PWM produced results that are more haphazard and less predictable. Hence, these two PWMs are favored for use in the NasoXplorer.

3.3 Dynamic force test with changes in temperature

The objective is to test out varying stiffness of each material (Figs. 10–14) and assess suitability for our project.

Larger currents produce more significant amounts of heat. This translates to more significant flexibility changes and can be seen in all the graphs where values for 0.5 A were consistently higher than that of 0.3 A, except for RGD 8705 during the early parts of the actuation. For RGD 8730 (pure VeroClear), the sample broke apart during testing. Hence, through this test, we recommend a mixture of rigid and flexible polymers.

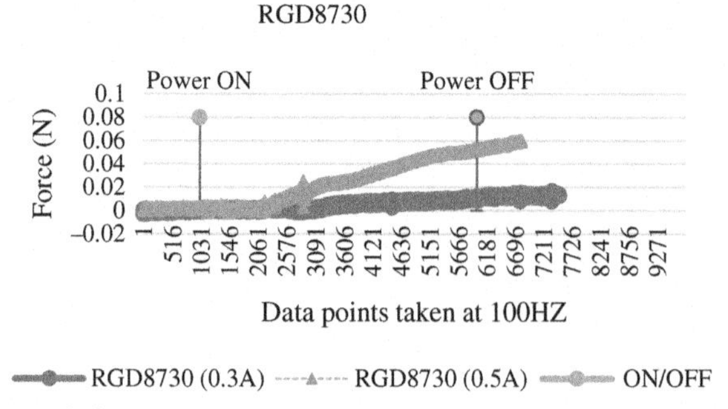

Fig. 10 Force graph for material RGD-8730.

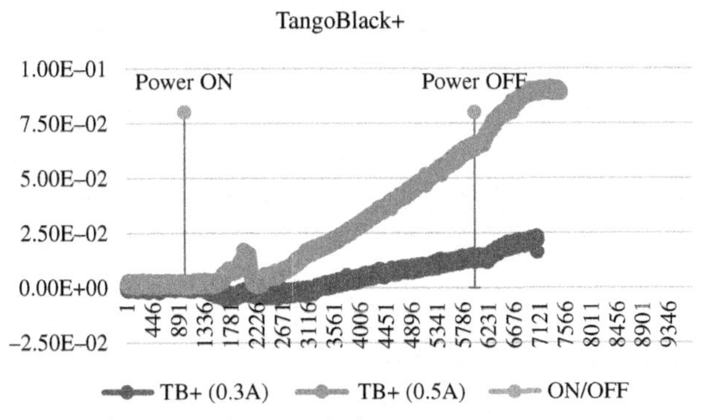

Fig. 11 Force graph for material TangoBlack+.

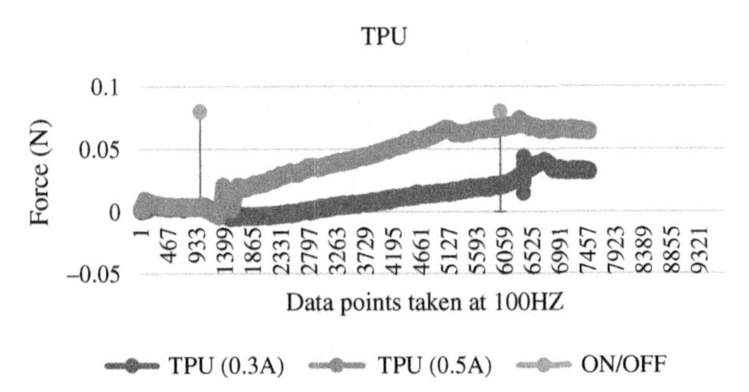

Fig. 12 Force graph for material TPU.

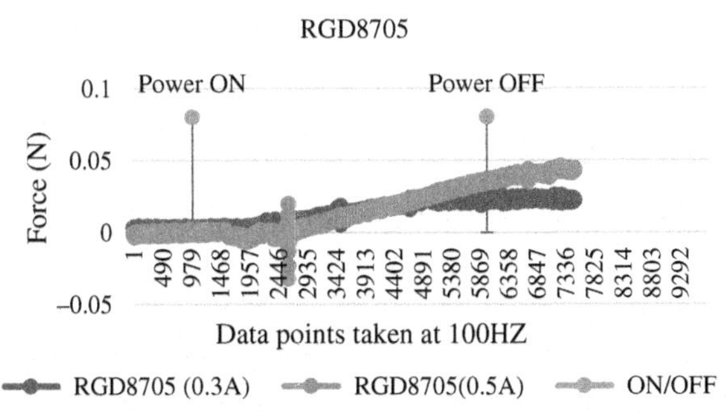

Fig. 13 Force graph for material RGD-8705.

As shown in Fig. 15, there is a consensus that a more considerable heating amount leads to a larger deformation response. This can be seen from the larger forces produced, collected by the force sensor.

Through the displacement tests in Fig. 16, 90% PWM showed the most displacement. We also noticed that at both 7.5 V and 7.0 V for 90% and 80%

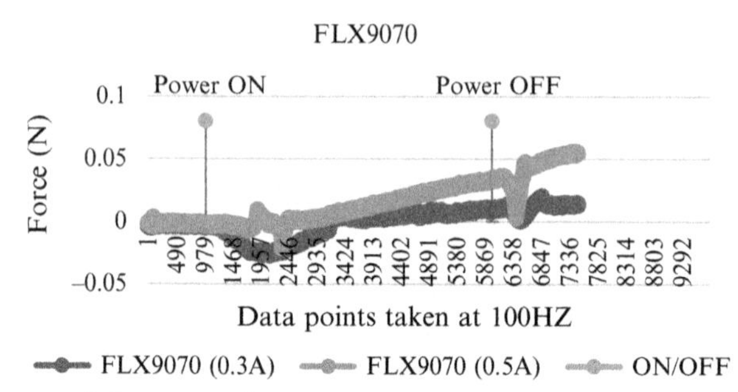

Fig. 14 Force graph for material FLX-9070.

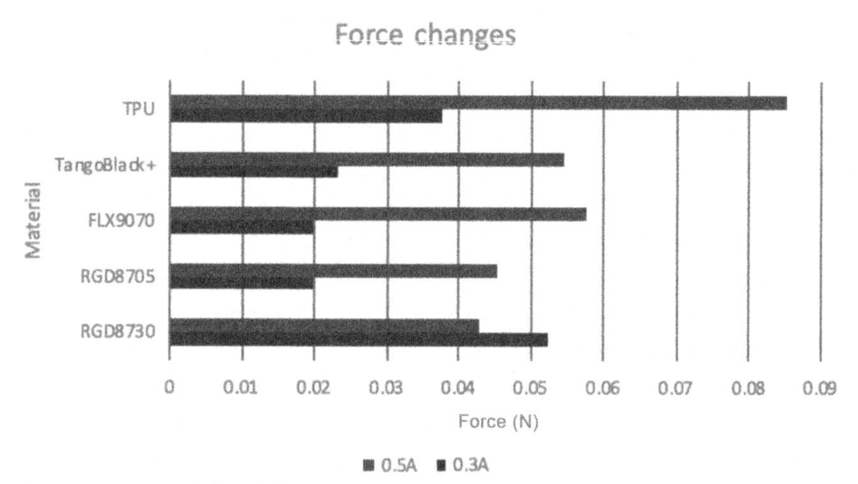

Fig. 15 Force graph for different materials under 0.5 and 0.3 A currents.

PWM, the curve was in a steady and relatively consistent staircase shape, hence showing potential in terms of control.

3.4 Insertion test

By using a plastic skull to model a human nasal passage (Fig. 17, support Video 1 in the online version at https://doi.org/10.1016/B978-0-12-821350-6.00011-1), our team conducted multiple experiments to deliver the endoscope from the nostril to the nasopharyngeal region. Points were indicated on the inside of the skull. An optic fiber was taped to our device, and a laser was used to shine light through the end of the fiber. The

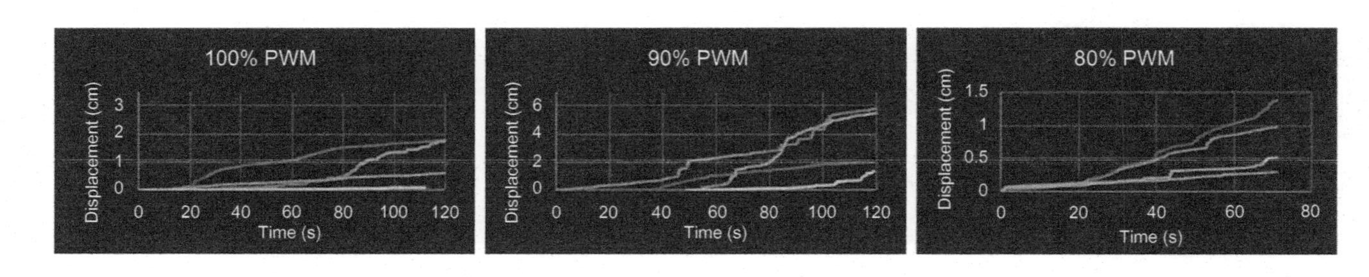

Fig. 16 Displacement tests under various PWM.

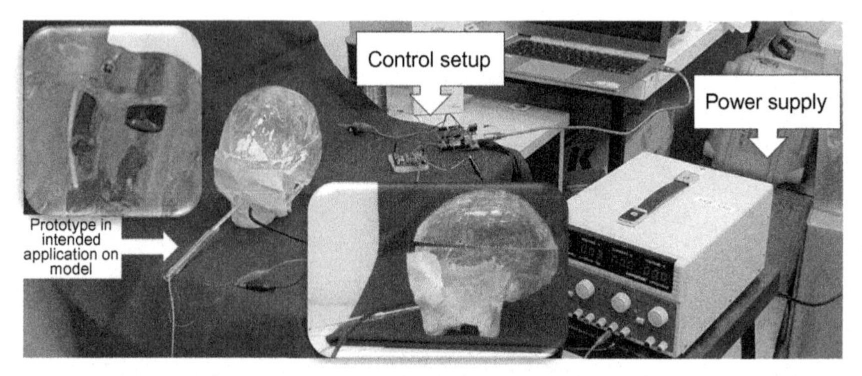

Fig. 17 Illustration of the intended usage of the prototype in a skeleton model. (Inset bottom) Set-up for insertion test (side view); inset top-left: set-up for insertion test (top view).

illumination was used to confirm if the NasoXplorer could capture images where the points were if it had a camera inside.

Using this set-up, the NasoXplorer successfully pointed at all the points inside the skull. Hence, it can capture all parts of the nasopharyngeal region.

Summary of design verification tests and associated acceptance criteria.

Testing subjects	Related methods of testing	Acceptance criteria	Results after testing
Breakage of flexible tip during endoscopy (due to cyclic bending)	ASTM F2942-13	No breakage of the tip. Breakage could potentially cause harm as it may cause the device to be in contact with sensitive tissue	Due to the reinforced tip, the flexible tip was much sturdier, and no breakage was observed
Wire breakage during endoscopy	ASTM A931-18 ASTM B1-13 (2018)	It does not break after repeating cycles	No breakage was observed

4 Design review

We conducted a Design Review to study design results, serve as feedback control, and assess the progress of the project. The NasoXplorer provides a larger bending angle and has the potential to reduce the diameter and to improve end-user control. The large bending angle allows better imaging of the nasopharyngeal region, and the soft robotic flexible tip possibly reduces patient discomfort. These aspects of the NasoXplorer aim to encourage more patients to conduct home-based nasopharyngoscopy.

4.1 Failure mode analysis

We conducted failure mode analysis to assess various situations in which the NasoXplorer may fail. This allowed our team to identify components that have a higher risk of failure and provide adjustments as well as mitigation methods to further ensure the safety of the NasoXplorer.

4.2 Remarks on the prior comparative art

Endoscopes have been around for many years. Instruments that are assumed to be prototypes were uncovered in the ruins of Pompeii (Olympus museum). The first recorded attempt to use the endoscope was in 1805 by Philipp Bozzini. Since then, there have more than 1000 different designs of endoscopic devices and more than five different designs for actuators that involve SMAs. The first generations of endoscopes are designed with only the flexibility of the tube in mind as well as the manipulation of the flexure of the tube. A notable mention would be the patent filed in 1967 by Olympus Corp (US3610231A), where the flexure of the tube is controlled via a set of cables to maneuver the narrow paths of varying curvature in the bronchi. From then on, there were also generations of endoscopes that are rigid. These types of endoscopes forgo flexibility in exchange for the increased image fidelity output via solid-state imaging devices. Due to the focus on the device being flexible, further elaboration on these types of endoscopes is not relevant to this review.

With the unique shape-shifting properties of SMAs, endoscopes that incorporate these technologies were developed. There were also inventions that allowed the torsional and rotational displacement of the tip of the device, such as US6672338B1, that created a tendon used to guide catheters past torturous terrain. With many actuators, a precise movement consisting of both translational and pivotal motion was achieved with a network of SMAs placed along the length of the device and between the coils of the spring. The sheer amount of SMAs required has the potential to produce a massive amount of heat, which is extremely detrimental to surrounding tissue, especially tissue that is laden with nerve endings in a sensing organ such as the nose.

Failure mode analysis

Failure mode	Cause of failure	Harm	Initial risk (with 1 = least, 10 = most)			Risk controls and mitigation measures	Verification and validation
			Likelihood	Severity	Risk		
Breakage of endoscope tip during endoscopy	Repeated bending of the tip	The flexible tip is trapped in a patient's body	2	4	4	Benchmarking of a current device	Test the force required to detach the camera/tip ASTM F2942-13 ASTM D6412/D6412M-99(2012)
Breakage of main body design	Exceeding principal stress of the material	Part of endoscope trapped in a patient's body	1	3	6	Ensure that the acrylic material is strong enough to withstand the stress of copper wire	Three-point bending test ASTM F2606-08 Wires do not break the main body while bending at 90 degrees
Wire breakage during endoscopy	Exceeding tensile strength of the enamel coated copper wire	Part of endoscope trapped in a patient's body (unable to retract the endoscope)	1	3	3	Benchmarking of a current device	Tensile strength test for enamel coated copper wire. The force required to bend the endoscope does not exceed the tensile strength of nylon wire—ASTM A931-18
Biocompatibility of materials used (surface	Inadequate biocompatibility of the material	Foreign body response, inflammation	1	3	3	FDA standards	Biocompatibility test ISO-10993

Failure mode analysis—cont'd

Failure mode	Cause of failure	Harm	Initial risk (with 1 = least, 10 = most)			Risk controls and mitigation measures	Verification and validation
			Likelihood	Severity	Risk		
contact, wear particles, etc.)							
Inability to maneuver/ bend the tubes	Wires entanglement in the device Diamond-cuts damaged	Inability to perform endoscopy/ money and time lost	1	3	3	Revision of design Motorization to ensure efficient cable driven mechanism	Using applications to measure the angle of bending Entangled wires prevent efficient power transfer
Poor guidance of tube	User competence	Endothelial tear	3	4	6	Doctor instructions Revision of design to increase stability and ease of patient self-administration	Testing using a plastic skull
Poor vision/ imaging	Optical lens (light transmission)	Poor results obtained from endoscopy	1	1	1	Change the camera with better resolution	In vivo clinical testing and cadaveric testing
Bad viewing angle for camera lens	Cable-driven mechanism	Poor results obtained from endoscopy	1	1	1	Ensure that the cables are not entangled Motorization of the device to ensure efficient and effective cable-driven mechanism	Functionality test (testing using the plastic skull) Testing using applications to measure the bending angle

Since an elevated temperature is a requirement before the actuation of the SMA to return to its original shape, there have been some methods developed to achieve this aim. The most popular is the usage of current to increase the temperature via heat due to the resistance of the wire. In addition to allowing the SMA to keep its thin profile, it also facilitates even heating of the alloy. Some other methods of actuation also include using fluid, as per US5645520A, where an insert regulates the temperature via the ratio of hot and cold water as well as installed heating elements. The active endoscope invented by Ikuta et al. makes uses of a fluid cooling system as well. The inclusion of such mechanisms is bound to increase the girth of the tube significantly, which is not desirable considering that girth is a luxury when designed for use in the respiratory tract.

In conventional bending applications, the most used mechanic of bending is having the number of bending actuators be equal to that of the number of directions that the device can pull in. However, this hinders ability to reduce the diameter of the tube. A method to reduce diameter is to reduce the total amount of actuators needed while retaining the same amount of directions in which bending can occur. One such example is patent US5897488A, where the number of actuators was reduced from four to three. One notable mention is the mask that was designed to facilitate the endoscopy process involving entry via orifices in the upper half of the body (US 0100368). The mask, however, is directed at subjects that are required to be sedated, with ports allowing the insertion of an oxygen tube and to test for tidal carbon dioxide. This design incorporates features that are redundant in a procedure that is to be done without anesthesia. The proximity of the mask wall to the nostril is also too long for accurate insertion of the device. Nasal dilators have also been developed, for example patent 5895509. While they have potential in terms of reducing the proximity of the scope to the orifice and the widening of the airway, the lack of self-positioning mechanism may lead to imprecise positioning of the scope. This leads to discomfort, which is an unwanted outcome.

While there are many prior arts regarding endoscopes that have been developed for medical use, there is little mention of a design that is catered to home use. In addition, the insertion of the various control mechanisms means that there is still a baseline on the minimum thickness achievable, which makes it unfeasible for self-administering of the endoscopy process because the girth of the tube might cause severe trauma to the surrounding tissues if navigation is poorly done.

In Table 2, we compare the following devices: (1) Schölly—ENT fiberscope 311003s, (2) Optim—ENTity SDXL (small diameter) LED NasoView Pharyngoscope, (3) Huger Endoscopy System—FN-32A Portable Fiber Nasopharyngoscope, (4) Olympus Rhino-Laryngol Fiberscope ENF-XP, and (5) Fuji Film ENT-Serie ER-530S2.

Table 2 shows that these representative commercial nasopharyngoscopes have a small diameter ranging from 2.2 to 3.2 mm. The slim designs of these nasopharyngoscopes enable them to be used in pediatrics, as they can pass nasal cavities with smaller diameters. Despite the slim designs of these nasopharyngoscopes, their field of view is not compromised. In addition, these are flexible nasopharyngoscopes, which explains the high deflection angles (110–140 degrees) of these devices. The high deflection angles also suggest that these nasopharyngoscopes can capture a broader range of views, allowing for easy observation of areas of interest.

While they are much more flexible than conventional rigid nasopharyngoscopes, these commercially available nasopharyngoscopes are still not suitable for home use.

4.3 Satisfaction benchmarking in clinical needs

Table 3 shows the satisfaction benchmarking in clinical needs.

4.4 Target metrics

Table 4 describes the target specifications that we aimed to achieve, and the marginal values describe the current specifications of the NasoXplorer.

4.5 Needs-metrics mapping matrix

Table 5 maps the needs and metrics in a matrix format.

4.6 Metrics benchmarking

Table 6 compares between NasoXplorer and five commercially available nasopharyngoscopes.

5 Conclusion and future work

The NasoXplorer is an innovative yet highly functional device that aims to provide a home-based kit for nasopharyngoscopy. With functional capabilities that are able to compete with most existing nasopharyngoscopes, the NasoXplorer allows patients to conduct nasopharyngoscopy in the comfort

Table 2 Specifications for five commercially available nasopharyngoscopes.

	Diameter (insertion tube)/mm	Field of view/ degrees	Angle of view (deflection)/degrees	Working length/mm	Cost
ENT fiberscope 311003s	3.2	85	140	300	~$6000[a]
ENTity SDXL LED Naso View	2.7	85	110	300	$6202.87[b]
FN-32A portable fiber nasopharyngoscope	3.0	90	130	300	~$6000[c]
ENF-XP	2.2	75	130	300	~$3500[d]
ENT-Serie ER-530S2	3.2	90	130	300	~$3500[e]

[a] Cost estimation of endoscope retrieved from https://www.global-medical-solutions.com.
[b] Cost of endoscope retrieved from https://www.gsaadvantage.gov/ref_text/V797D70047/V797D70047_online.htm.
[c] Cost estimation of endoscope retrieved from Huger Shanghai Medical.Co.
[d] Cost estimation of endoscope retrieved from DOTmed Listing.
[e] Cost estimation of endoscope retrieved from Green Medical Company Limited.

Table 3 Table of satisfaction benchmarking based on the needs of the users, with one ✓ representing the least satisfactory/important and five ✓ representing the most satisfactory/important.

Need #	Clinical needs	Importance (1–5)	NasoXplorer	ENT fiberscope 311003s	ENTity SDXL Nasoview	ENF-XP	ENT-Serie ER-530S2
1	Wide viewing angle	4	✓✓✓✓✓	✓✓✓✓	✓✓✓✓	✓✓✓✓	✓✓✓✓
2	High steerability	3	✓✓✓	✓✓✓✓	✓✓✓✓	✓✓✓✓	✓✓✓✓
3	Clear images	3	✓✓	✓✓✓✓✓	✓✓✓✓	✓✓✓✓	✓✓✓✓
4	User-friendly	4	✓✓✓✓	✓✓✓	✓✓✓	✓✓✓	✓✓✓
5	Safe for prolonged usage	5	✓✓✓✓	✓✓✓✓	✓✓✓✓	✓✓✓✓	✓✓✓✓
6	Home-based	5	✓✓✓✓✓	✓	✓	✓	✓
7	Patient self-administered	4	✓✓✓✓	✓	✓	✓	✓
8	Comfortable	5	✓✓✓✓	✓✓✓✓	✓✓✓✓	✓✓✓✓	✓✓✓✓

Table 4 Target specifications for developers to refer to for further developments and improvements of NasoXplorer.

Target specification of metrics						
Metric #	**Needs #**	**Metric**	**Importance (1–5)**	**Units**	**Marginal value**	**Ideal value**
1	1	Articulation range	5	degrees	±145	±145
2	2,3	Working length	5	mm	118	100
3	3	Stability	4	Yes/No	Yes	Yes
4	4,7	Automatic mechanism	4	Manual/semi/full	Semi	Semi
5	5	Resistance to deformation	2	Yes/No	Yes	Yes
6	8	Distal tip diameter	5	mm	3.0	3.0
7	4,8	Speed of insertion	3	Fast/slow	Slow	Slow
8	6	Cost of production	3	–	$270	$270

Table 5 Needs-metrics table for comparison of the requirements for each metric.

Needs-metrics matrix									
		Metrics #							
Needs #		1 Articulation range	2 Working length	3 Stability	4 Automatic mechanism	5 Resistance to deformation	6 Distal tip diameter	7 Speed of insertion	8 Cost of production
1	Viewing angle	×							
2	High steerability		×						
3	Clear images		×	×					
4	User-friendly				×			×	
5	Safe for prolonged usage					×			
6	Home-based		×						×
7	Patient self-administered				×				
8	Comfortable						×	×	

Table 6 Metrics for comparison between NasoXplorer and the five commercially available nasopharyngoscopes.

Metric #	Need #	Metric	Importance (1–5)	Units	NasoXplorer	ENT fiberscope 311003s	ENF-XP	ENT-Serie ER-530S2
1	1	Articulation range	5	degrees	±145	±140	±130	±130
2	4,6	Working length	5	mm	118	300	300	300
3	3	Stability	4	Yes/No	Yes	Yes	Yes	Yes
4	4,7	Automatic mechanism	4	Manual/semi/full	Semi	Semi	Semi	Semi
5	5	Resistance to deformation	2	Yes/No	Yes	Yes	Yes	Yes
6	8	Distal tip diameter	5	mm	3.0	3.2	2.2	3.2
7	4,8	Speed of insertion	3	Fast/slow	Slow	Slow	Slow	Slow
8	6	Cost of production	3	–	$270	~$6000	~$3500	~$3500

of their homes, at ease. While the device is slim and can be handled via one-handed operation, there are still some features that we would like to improve to ensure it is optimized for personal and home use. The NasoXplorer consists of features that allow patients to undergo smooth and easy self-administered endoscopy. These include the use of soft materials, the sizeable bending angle of the device, and external controls to increase patient comfort and enhance confidence in using the NasoXplorer. Future add-ons include the optimization of the Arduino code and the provision of a box for the control system. The inclusion of automation will reduce the time required for a full endoscopy procedure. Since the NasoXplorer is to be self-administered, training and demonstrations will be carried out to educate the patient on proper usage of the device. Usage of a home-based kit for nasopharyngoscopy is a more convenient, less time-consuming, and cheaper alternative for patients who require consistent nasopharyngoscopy in clinics and hospitals.

Appendix: Supplementary material

Supplementary material related to this chapter can be found on the accompanying CD or online at https://doi.org/10.1016/B978-0-12-821350-6.00011-1.

References

Albery, L., 1995. The efficacy of speech and language therapy for cleft palate speech disorders. Int. J. Lang. Commun. Disord. 30 (S1), 237–241. https://doi.org/10.1111/j.1460-6984.1995.tb01678.x.

Atar, M., 2014. Transnasal endoscopy: technical considerations, advantages, and limitations. World J. Gastrointest. Endosc. 6 (2), 41. https://doi.org/10.4253/wjge.v6.i2.41.

De Araújo, L.L., Da Silva, A.S.C., Araújo, B.M.A.M., Yamashita, R.P., Trindade, I.E.K., Fukushiro, A.P., 2016. Nasopharyngeal dimensions in normal individuals: normative data. CoDAS 28 (4), 403–408. https://doi.org/10.1590/2317-1782/20162015020.

Gaviola, G.C., Chen, V., Chia, S.H., 2013. A prospective, randomized, double-blind study comparing the efficacy of topical anesthetics in nasal endoscopy. Laryngoscope 123 (4), 852–858.

Huang, Z.-L., Wang, D.-Y., Zhang, P.-C., 2001. Evaluation of nasal cavity by acoustic rhinometry in Chinese, Malay and Indian ethnic groups. Acta Otolaryngol. 121 (7), 844–848.

Kalairaj, M.S., Banerjee, H., Lim, C.M., Chen, P.Y., Ren, H., 2019. Hydrogel-matrix encapsulated nitinol actuation with self-cooling mechanism. RSC Adv. 9 (59), 34244–34255.

Morgan, N.J., MacGregor, F.B., Birchall, M.A., Lund, V.J., Sittampalam, Y., 1995. Racial differences in nasal fossa dimensions by acoustic rhinometry. Rhinology 33, 224–228.

Mori, A., Nakajima, T., Kaneko, T., Sakuma, H., Aoki, Y., 2005. Analysis of 109 Japanese Children's lip and nose shapes using 3-dimensional digitizer. Br. J. Plast. Surg. 58 (3), 318–329. https://doi.org/10.1016/j.bjps.2004.11.019.

National Registry of Diseases Office, June 19, 2017. Singapore Cancer Registry Annual Registry Report 2015. .

Schriever, V.A., Hummel, T., Lundström, J.N., Freiherr, J., 2013. Size of nostril opening as a measure of intranasal volume. Physiol. Behav. 110–111, 3–5. https://doi.org/10.1016/j.physbeh.2012.12.007.

Taylor, A.J., Slutzky, T., Feuerman, L., Ren, H., Tokuda, J., Nilsson, K., Tse, Z.T.H., 2019. MR-conditional SMA-based origami joint. IEEE/ASME Trans. Mechatron. 24 (2), 883–888.

CHAPTER 12

Tunable stiffness using negative Poisson's ratio toward load-bearing continuum tubular mechanisms in medical robotics

Krishna Ramachandra[a], Catherine Jiayi Cai[a], Seenivasan Lalithkumar[a], Xinchen Cai[a], Zion Tszho Tse[b], Hongliang Ren[a]
[a]Department of Biomedical Engineering, Faculty of Engineering, National University of Singapore, Singapore, Singapore
[b]University of Georgia, Athens, GA, United States

1 Background

While there exists a small number of soft, flexible, and minimally invasive surgical robotic manipulators for minimally invasive surgery (MIS), the lack of dynamic load bearing capacity in these instruments has limited their use in applications requiring exertion of forces. Equipping these MIS devices with variable stiffness capabilities allows them to be used for isolating and operating procedures in surgical manipulations that require force applications, such as tumor isolation, manipulation, and safe extraction, without contaminating surrounding tissues. A variable stiffness module adds an element of a dynamic haptic feedback and additional load-bearing capacity, which makes the device more intuitive for surgeons. In this work, we take the design-centric approach and develop a novel method to achieve a tunable stiffness module aiming to widen the capabilities of MIS devices.

We aimed to create a new actuation mechanism with variable stiffness (Huan et al., 2016; Li et al., 2016, 2017). A flexible and compliant spring backbone can navigate tight and narrow spaces (Li et al., 2015b; Wu et al., 2017). However, in order to be fully implementable as part of a surgical system, our device requires additional fast stiffness tuning features. While a fully flexible spring body provides the necessary conformability for safer maneuverability inside the body, it is not well suited for applying force to the tissue. The lack of rigidity also makes it difficult for the instruments to resist external forces within the body. Ideally, surgical robots and

Control Theory in Biomedical Engineering
https://doi.org/10.1016/B978-0-12-821350-6.00012-3

instruments should be flexible when traversing long and narrow paths, but able to become rigid upon reaching the surgical site for force application.

Structures that allow for variable stiffness result in many desirable properties, such as curvilinear navigation, dynamic force feedback, and shape deformation capabilities (Chen et al., 2016; Li et al., 2014, 2015a; Tan and Ren, 2017). Having a module with tunable stiffness is imperative for our robotic device as it allows for load-bearing capability, which is essential for isolating, cutting, and extracting tumors in a safe manner without contaminating the surrounding healthy tissues. Currently explored variable stiffness mechanisms often involve at least one of the following: expensive and lengthy fabrication processes, bulky setups, and unsuitability for fast-changing applications. More details with regards to these limitations can be found in the literature review below.

We focus on creating a novel, tunable stiffness module using auxetic materials to allow us to vary the rigidity of the surgical instruments and the backbone as required. We created a method that integrates actuation and stiffness modulation that allows for easy implementation and low complexity. Additionally, we use simple 3D printing, paper cutting, and folding methods to ensure that our approach is economically feasible and easy to fabricate compared to other variable stiffness methods.

In the following section, we evaluate different existing methods and analyze their suitability for surgical robots.

2 Literature review/concept evaluation

Currently, there are a few different methods that have been explored as potential mechanisms for achieving tunable stiffness. The theories behind these methods and their limitations are highlighted.

2.1 Electro/magneto-rheological fluids

Electro/magneto-rheological (ER/MR) fluids can be used to achieve tunable stiffness.

In this method, the fluid is usually composed of a primary liquid, which is mostly silicone oil suspended particles of a polymer. This fluid undergoes a change in its rheological properties, such as viscosity, when subjected to an electric or magnetic field (Fig. 1A). As such, the fluid transforms from a liquid phase to a solid gel phase as the field density is increased. On a microscopic level, the particles in the fluid align with the electric or the magnetic field lines, resulting in a change in its yield stress, viscosity, and other fluid

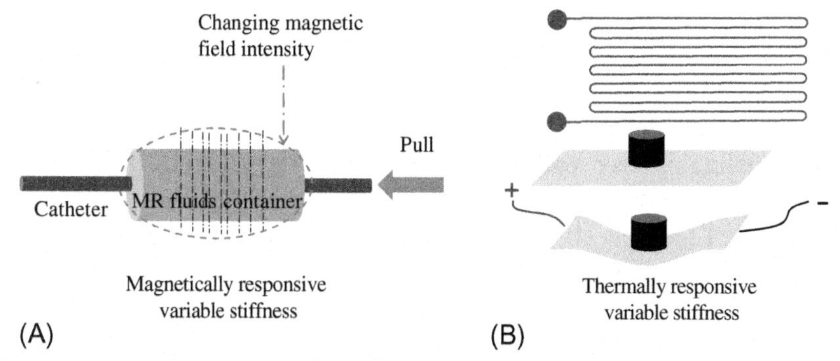

Fig. 1 Two typical mechanisms for stiffness tuning: (A) Variable stiffness principle with magnetically responsive material in action; (B) Construction of a phase change variable stiffness material with microchannels filled with a low melting point.

properties, which can be exploited for use in variable stiffness (Chen et al., 2007; Liu et al., 2001; Wu et al., 2010; Zhou, 2003). The advantage of using such fluids is that the orientation of particles can be changed on the order of milliseconds, thus allowing for a quick way to achieve stiffness tunability (Lindler and Wereley, 1999). There is a wide range of uses for such tunable stiffness fluids, and their engineering applications have been explored in various studies (Ditaranto, 1964; Nakamura et al., 2002; Parthasarathy and Klingenberg, 1996; Wei et al., 2007; Yeh and Shih, 2005). There is also literature available that details the process of preparing such ER/MR fluids and the integration of these fluids with devices as a variable stiffness module (Behbahani and Tan, 2017; Carlson and Jolly, 2000; Hao, 2002; Parthasarathy and Klingenberg, 1996; See et al., 2001; Tangboriboon et al., 2009; Varga et al., 2006), and the integration of this fluid into a device to be used as a variable stiffness module. However, the methods used to prepare ER/MR fluids involve complex setups and multiple treatment processes. Additionally, these methods require highly controlled environments, where even the slightest impurities can render the fluids unusable. Furthermore, integration of the ER/MR fluid into a modular device requires complex fabrication methods. Lastly, the application of an electric or a magnetic field typically requires a bulky setup.

2.2 Phase change materials

Phase-changing materials comprise another field of tunable stiffness materials. The stiffness of the material is altered by either atmospheric or direct joule heating, which leads to a reversible transition from solid to liquid phase

(Fig. 1B) (Taghavi et al., 2018). Many experiments have been conducted with the thermal activation properties of nonconductive shape memory polymers (SMP) (Balasubramanian et al., 2014; Clark et al., 2010; McKnight et al., 2010; Shan et al., 2013), coiled fibers (Haines et al., 2014), thermoplastics (including wax soaked) (Cheng et al., 2014; McEvoy and Correll, 2015), and low-melting point alloys (Schubert and Floreano, 2013; Shan et al., 2013; Taghavi et al., 2018) through the means of external heating elements or through self-joule heating. Based on the material properties and stiffness-varying mechanisms, these materials can be classified into either phase change materials (melting) or glass transition behavior (softening). The process of thermal activation is accelerated when the composite is embedded with an external heating element, which can remain conductive even under high levels of deformation. This is possible using microfluidic channels of liquid-phase metals, which can also be sewn into fabrics to create thin and modular variable stiffness structures (Taghavi et al., 2018). While this method of phase change materials seems attractive, there are challenges associated with it, especially in the context of surgical robotics. Using such microfluidic channels and the need for high heat inside the human body introduces sealing and insulation problems, requiring one to develop a means for effective separation of the thermally responsive material. Methods of heat dissipation must also be included, which increases the bulkiness of the device (Shan et al., 2015). Furthermore, phase change materials have low bandwidth and need relatively larger amounts of time to input and output thermal energy into the system (Taghavi et al., 2018). Thus, they cannot be used for fast-changing applications.

2.3 Jamming methods

Another class of tunable stiffness device involves jamming flexible actuation methods using friction or negative pressure, resulting in a change from a compliant to a rigid state. Jamming is broadly classified into particle/granular jamming and layer jamming methods. In granular jamming, many tiny solid grains/particles act as "fragile matter." In the absence of external stress, the bulk matter is free to move, and the granule system can act as a fluid-like compliant structure. Upon application of a vacuum, the granules jam together and transition into a solid-like state, thereby increasing its stiffness (Jiang et al., 2014). There already exists numerous applications of granular jamming, especially in the context of industrial robotic arms and manipulators, among others (Brown et al., 2010; Loeve et al., 2010). However, these applications make use of jamming for devices that are intended for direct human manipulation, and, as such, they are typically on the larger scale of a few centimeters in diameter (Jiang et al., 2014). Our robotic manipulator for MIS has a dimensional constraint of about 5 mm, so granular jamming is challenging given the complexities required for miniaturization of the technique. Additionally, there is a limit to the amount of achievable rigidity, as the internal pressure cannot be lower than absolute vacuum (Jiang et al., 2012). Furthermore, the bulk material properties of such materials are not well understood, making mathematical modeling of the process a challenge.

2.4 Negative pressure jamming

For a preliminary study, we explored the method of using negative pressure to create a tunable stiffness module. In this method, a thin film of PET was encapsulated around a flexible continuum robot with 3 degrees of rotational freedom. Fig. 2 shows the continuum robot with the encapsulation made by 3D-printed vertebrates supported by Ni-Ti wires. Pulling on one or two wires simultaneously causes an unbalanced load and a bending moment in the desired direction.

In order to change the configuration from a flexible to a rigid robot, a negative pressure was created through the means of a vacuum pump. When the pressure inside the membrane was lower than the atmospheric pressure, there was a net inward force, which caused the encapsulation membrane to clamp onto the. This normal contact between the membrane and the vertebrates of our robot created a frictional force that resisted bending motion.

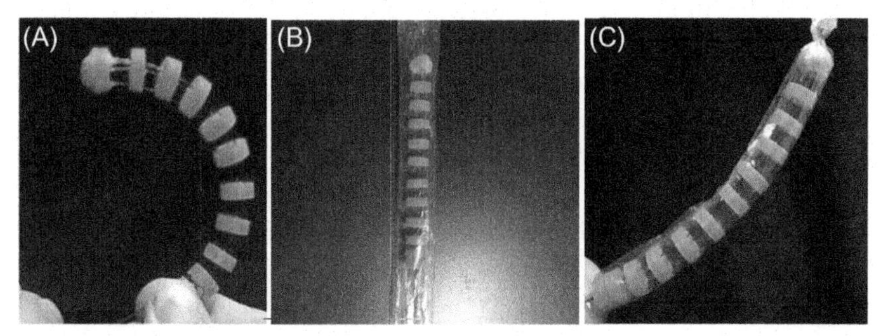

Fig. 2 (A) Flexible motion of the continuum robot through the means of Ni-Ti wires; (B) Continuum robot with bag without negative pressure. (C) Continuum robot with bag with negative pressure.

Through this mechanism, we theoretically analyzed and tested our design for a variable stiffness device.

From the tests, we observed that the created vacuum was not sufficient to hold the robot in place. This is because there is a physical limit to the amount of resistance by the membrane as the internal pressure cannot go below zero. Hence, the negative pressure is limited to the atmospheric pressure. Additionally, the coefficient of static friction between the membrane and the robot vertebrate is not large enough to resist motion even when an almost complete vacuum is created. As shown in Fig. 2, without negative pressure (A) and (B), the robot mostly returned to its original shape and was not stiff enough for load-bearing applications. We also encountered many sealing problems, as it was almost impossible to get a perfect vacuum and there were many microscopic holes near the end junctions where leakage was evident. For these reasons, we did not continue pursuing this method.

Based on our concept evaluation, we find that most of the existing methods cannot be readily applied as solutions for a variable stiffness module of tubular surgical instruments because they do not meet our design criteria or because they require complex modifications to achieve them. Hence, there is a need for us to come up with a fast mechanism that can better meet our specific needs of tunable stiffness of tubular flexible manipulators in medical robotics.

3 Concept combining jamming and continuum metamaterials with negative Poisson's ratio materials (auxetics)

While the method of negative pressure did not yield satisfactory results, the method of jamming using friction to resist motion is a promising direction for exploration. We are exploring tunable stiffness approaches through the means of auxetic continuum materials (also known as negative Poisson's ratio (NPR) materials), folding/unfolding and jamming mechanisms.

Poisson's ratio indicates transverse strain relative to longitudinal strain. Most common materials display a positive Poisson's ratio (PPR) as they get shorter in the transverse direction when elongated in the longitudinal direction, and vice versa. An auxetic material displays the opposite characteristics in Fig. 3, and they elongate in the transverse direction upon being elongated in the longitudinal direction.

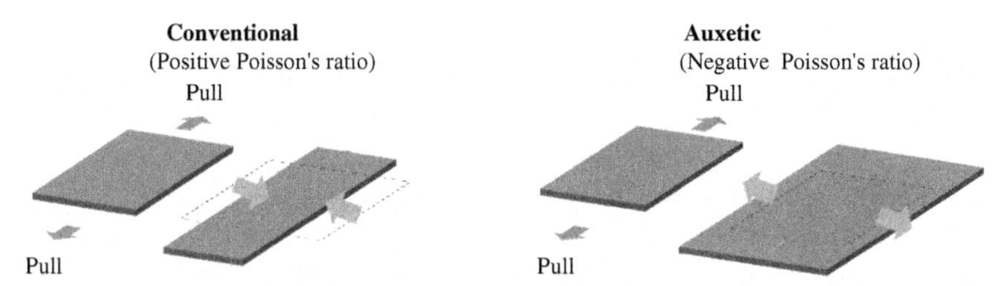

Fig. 3 Comparison of a positive Poisson's ratio material (left) vs a negative Poisson's ratio material (right).

These types of materials display interesting properties, such as light weight, indentation resistance, impact resistance, shear resistance, and better energy absorption capabilities as compared to conventional materials. While there has been previous work done in the field of purely auxetic materials, they have focused more on theoretical modeling than exploring the practical applications of such materials in the field of variable stiffness devices. In this study, we develop a stiffness tuning framework using continuum tubular auxetic materials.

For a proof-of-concept, we restricted ourselves to fabricating these materials using commonly available and cheap processes, thereby improving the speed and economic feasibility of manufacture. We employed structures that could be fabricated by simple 3D printing as well as kirigami and origami folding structures as simple, fast, and low-cost methods of fabrication to prove the concepts with the following design requirements in mind:

- Geometric constraint: 10–12mm maximum outer diameter
- Compatible with the tendon driven actuation method
- Shift from flexible to stiff and vice versa without using temperature regulations

4 Concentric continuum metastructures

Our primary hypothesis for a conceptualized mechanism is to vary the stiffness of our tubular surgical device by using the concept of jamming. However, unlike layer or space jamming, we use the inherent mechanism of auxetic materials, which expands upon elongation. This is more advantageous as compared to using a vacuum as there is no need for sealing, which is often one of the biggest challenges associated with jamming. By restricting the space available for the expansion of the auxetic material, it is possible to achieve jamming due to the contact force between the auxetic material and the restrictive material. A two-layered tubular system can be designed where the outer layer has fixed dimensions and is semirigid (i.e., it is only allowed to bend). The inner tube is made of an auxetic material, which expands upon elongation. However, since the available space is restricted, the inner tube will jam against the outer tube, creating friction which resists further motion. The mechanism and proposed design are shown in Figs. 4 and 5.

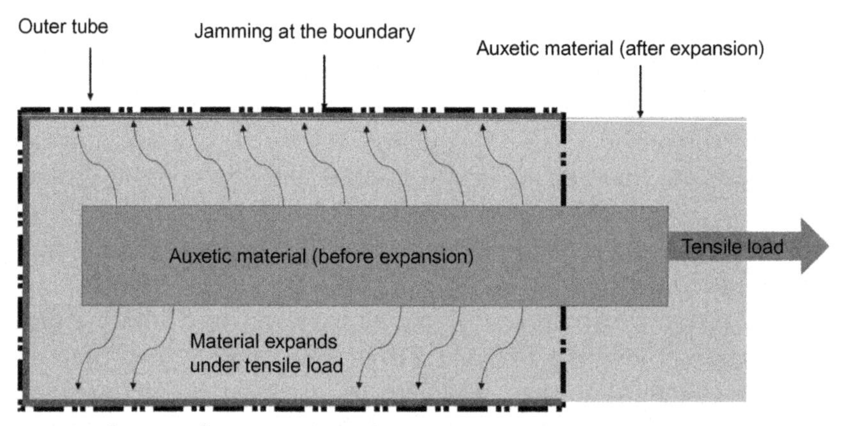

Fig. 4 Mechanism of jamming through auxetic materials.

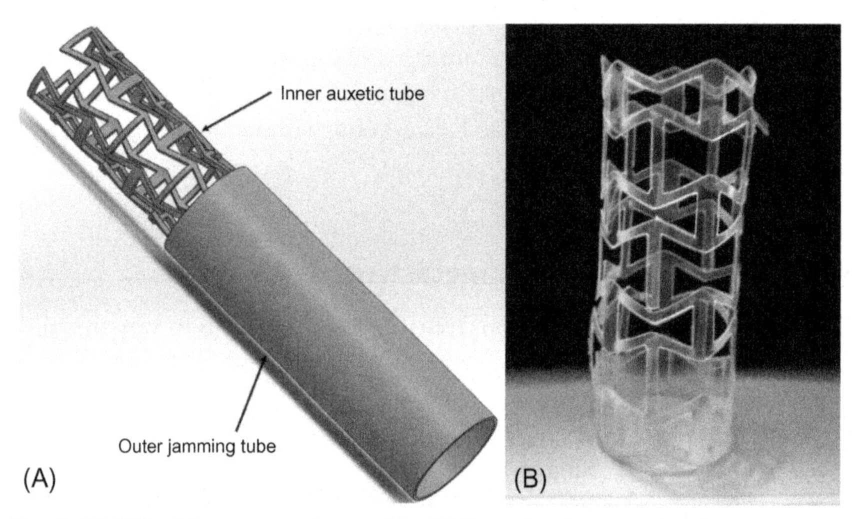

Fig. 5 (A) CAD of the proposed assembly. (B) Example of an auxetic tube.

We reasoned that we could further improve this design to increase the amount of jamming at the surface by having a counter force at the boundary pushing in the other direction. This would effectively double the normal force and, hence, double the friction. To implement this process, we used a combination of an auxetic tube material surrounded by a normal PPR material. The inner auxetic tube had a diameter of 11 mm while the outer tube had a diameter of 12 mm such that they fit snugly into each other with a space of 1 mm as shown in Fig. 6. We theorized that when the same tensile force is applied to both the PPR and NPR materials, the NPR material will expand while the PPR material will contract. Thus, the radius of the inner tube will increase, and the radius of the outer tube will decrease until the free space is occupied. As we continue to apply the tensile load, jamming will occur, and the resulting friction between the tubes will resist any bending forces, leading to a stiff structure. The load-bearing capacity can be increased by exerting a higher tensile force, which leads to a larger frictional force between the tubes.

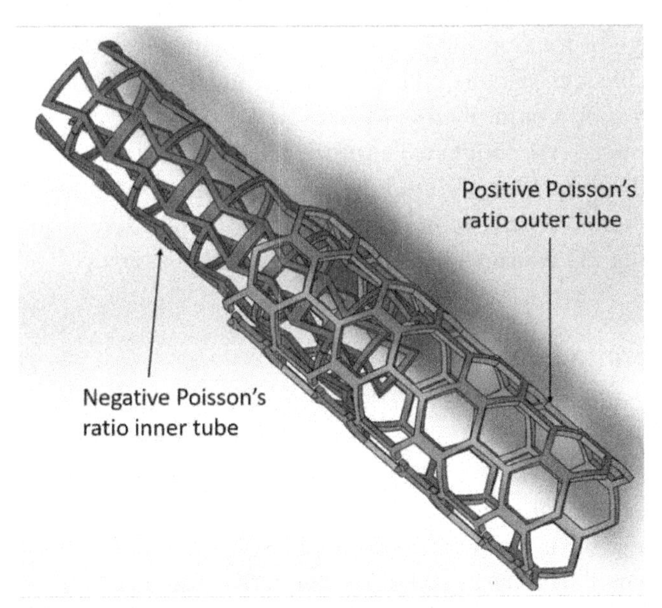

Fig. 6 CAD of the improved concept.

5 Fabrication methodology

Since our design will be subjected to tensile and compressive loads and undergo deformation, we require our material to be compliant and flexible. We used two different methods of fabrication to achieve our tubular structures. The first method involves cutting a two-dimensional sheet of soft material and rolling it into a tube. The second method involves direct 3D printing of the tubular structures. Each of the methods are detailed in the sections that follow.

5.1 Rolling two-dimensional sheets

As it is possible to create tubes by rolling materials on two-dimensional sheets, we use a standard re-entrant planar design on sheets of silicone and laser cut the pattern. We use Dragon Skin 30 Silicone rubber to create the sheets. We then spread out the mixture over a flat surface for a sheet thickness of 1 mm. After curing and cutting the sheets, they are rolled into a cylinder. When stretched, we did not observe much of an expansion or auxetic activity because of the fully compliant nature of silicone. Auxetic materials display their properties best when expansion is only constrained to the hinge of rotation while the other parts remain relatively rigid. This is not possible with silicone rubber as all parts undergo equal stretch, thereby giving a Poisson's ratio of close to zero.

Similar tests were conducted using the following varieties of silicone rubber with different Young's modulus and thickness. Upon observation, all these tests result in little to no auxetic activity. Hence, we abandoned the idea of using 2D planar sheets to fabricate auxetic materials.

5.2 3D printed auxetics

We took a different approach to the fabrication of auxetic materials than in the previous section. In this method, we first created a computer aided design (CAD) model of a sheet metal and superimposed the required sketch of the auxetic design. We then extruded the design and joined the bends of the fold to get our required structure. This part was then directly printed using different flexible materials. We used five different auxetic designs and three different materials for each design to conduct our tests. The materials and designs are described below.

5.2.1 Material filaments

As it is required for the part to be flexible, we used thermoplastic polyurethane (TPU), polyvinyl alcohol (PVA), and Pythonflex as three different materials to test our designs. All these materials are easily available and compatible with a standard FDM 3D printer.

Thermoplastic polyurethane (TPU): We use a 1.75 mm filament of TPU, with high elasticity and resistance to grease, oil, and abrasion.

Polyvinyl alcohol (PVA): PVA is a water-soluble material and is commonly used as a sacrificial support material for complex geometrical parts as they can be dissolved in water. Although their primary use is in making molds and supports, the material displays excellent elasticity when extruded at a slightly lower temperature (~10–20° lower than the rated temperature), the material has excellent elasticity and springback and hence can be used as a flexible spring-like material. However, care should be taken, as slight deviations from this range of temperature can quickly make the material brittle and hard.

Pythonflex: Pythonflex is a high-performance variant made from specially formulated TPU material.

5.3 Auxetic material designs

As mentioned earlier, we used five different auxetic designs to cover a wide range of concepts of auxetic behavior and to explore the robustness of our results. Each of our five designs were printed with the three materials mentioned in the previous section and subjected to mechanical tests. The conceptual theory behind the auxetic behavior of each design is elaborated below.

5.3.1 Hexagonal re-entrant honeycomb structure

Fig. 7 shows a traditional hexagonal re-entrant structure. When a force is applied in either direction, the diagonal ribs move and rotate in a way to produce an auxetic effect in the other direction. The auxetic effect is observed as the diagonal ribs aligned along the horizontal direction move apart in the vertical direction under tension. Tests have shown that most of the structures involving re-entrant honeycombs undergo deformation.

5.3.2 Chiral structure

A chiral formation is defined as a nonsuperimposable mirror image. Such structures exhibit a Poisson's ratio of close to −1 (Saxena et al., 2016). When

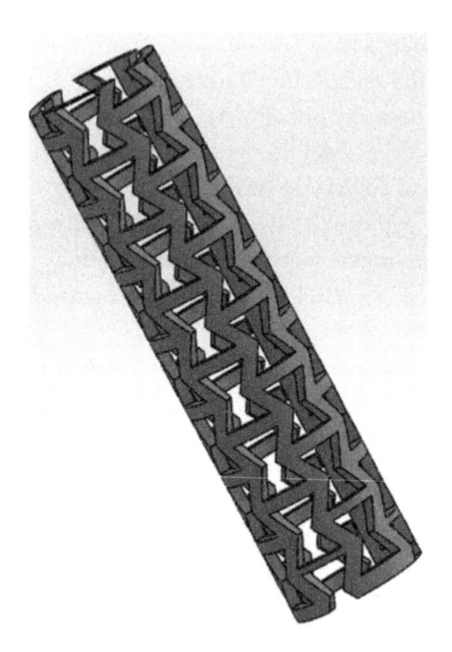

Fig. 7 Re-entrant structure CAD.

the ligaments are pulled (i.e., a longitudinal strain is applied), the rigid ring rotates, causing unwinding of the ligaments, which leads to its auxetic behavior.

As shown in Fig. 8, we designed and fabricated a tetra-chiral structure with each central node connected to four other nodes. While we wanted to construct a hexachiral structure, the geometric constraints set by our design specifications did not allow us to construct such designs as the tube diameter would have had to be increased to 20 mm.

5.3.3 Star honeycomb structure

A star honeycomb structure is a subset of re-entrant structures and a variant of the traditional hexagonal re-entrant design. As is common with re-entrancy, the deformation of the star structure is through the hinging of the walls of the cells, as shown in Fig. 9. These star-shaped building blocks have a rotational symmetry on the order $n=3$, 4, and 6. We have designed a Star 4 structure, which is said to display a Poisson's ratio of -0.845 (Kolken and Zadpoor, 2017).

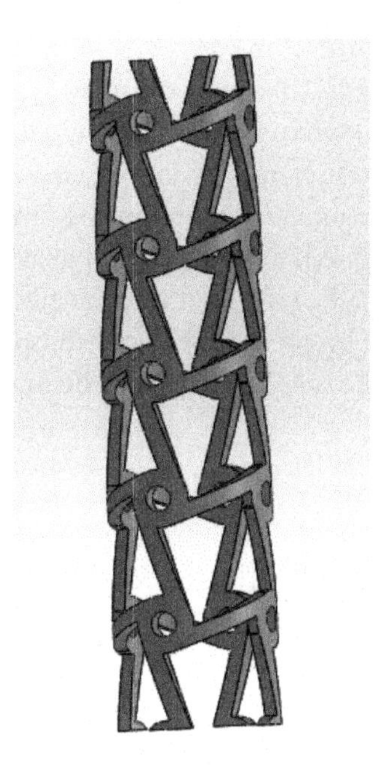

Fig. 8 Chiral structure CAD.

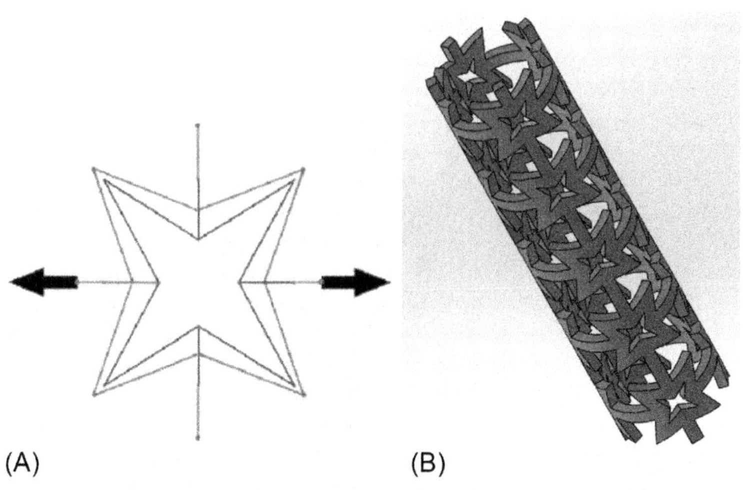

Fig. 9 (A) Auxetic behavior of Star 4. (B) CAD drawing of the Star 4 CAD design.

5.3.4 Missing rib structure

The missing rib pattern (Fig. 10) can be classified as a hybrid of the chiral and re-entrant designs. This structure is formed by taking the base hexagonal honeycomb re-entrant design and removing two opposite ribs out of the four. This is done because unit cells with four ribs are much stiffer than those with two ribs, and, as such, can display auxetic behavior at much lower loads. The missing rib model has broadly two types of auxetic geometries, namely the Lozenge grid and the square grid. For our testing, we used the Lozenge grid design as shown in the CAD in Fig. 11. The in-plane Poisson's ratio for

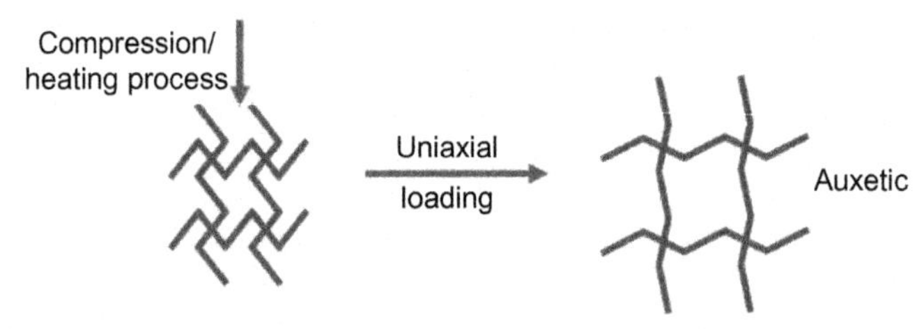

Fig. 10 Auxetic behavior of missing rib pattern.

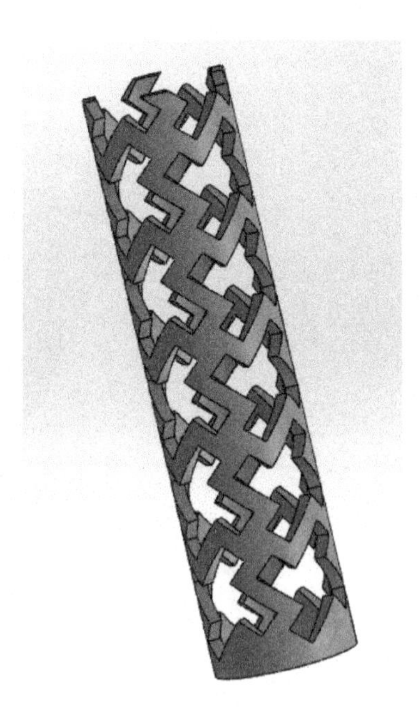

Fig. 11 CAD drawing of the missing rib design.

this geometry was found to be 0.43 (Kolken and Zadpoor, 2017). The detailed kinematic model of the missing rib configuration was presented by Zhai (Zhai et al., 2018). The auxetic effect of such a design is obtained by the concurrent out-folding of re-entrant cells and the missing ribs' rotation mechanism.

5.3.5 Double arrowhead structure

The double arrowhead design is another variant of the re-entrant honeycomb, which was first founded through the numerical topology optimization method. Based on the actual configuration of the arrowhead, any extension will cause the triangles to expand in the transverse direction while compressions will cause them to collapse. The auxetic behavior of such geometries is shown in Fig. 12. The CAD drawing of the double arrowhead structure is shown in Fig. 13. These structures are designed to exhibit a

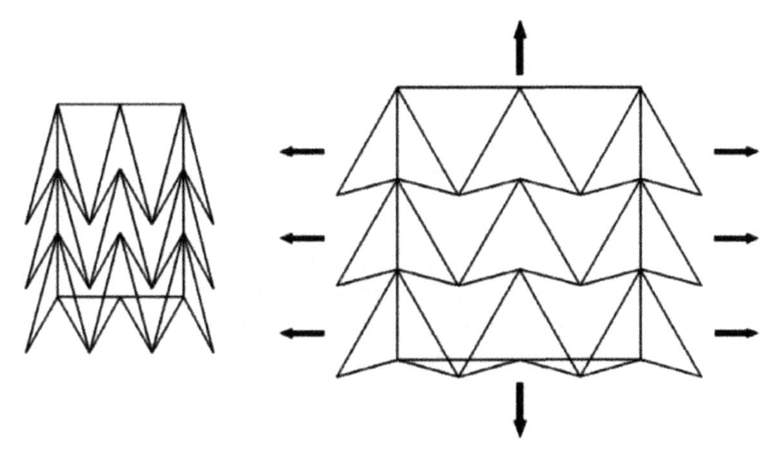

Fig. 12 Auxetic behavior of double arrowhead structure.

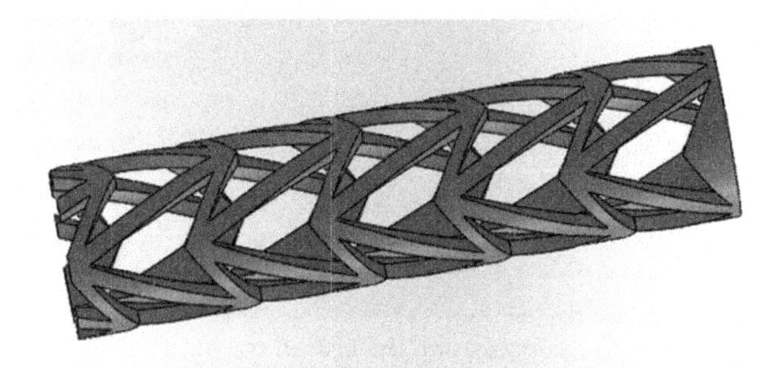

Fig. 13 CAD of the double arrowhead structure.

Poisson's ratio of 0.8. However, they have been measured to exhibit a Poisson's ratio of 0.92 at smaller strains (Kolken and Zadpoor, 2017).

6 Continuum metastructural test

6.1 Mechanical test specifications

As mentioned in the previous section, the five designs were all fabricated using the three different materials and assembled into a concentric tube structure with the auxetic material as the inner tube and the normal PPR tube as the outer tube. For each of the assemblies, we performed a three-point bending test (Instron universal testing machine) as it is the most typical representation of the type of movement that our surgical device undergoes during its course of operation. Each sample assembly was subjected to three test cases.

The first test case is a simple three-point bending with no horizontal extensions or loads (Fig. 14). This case represented the control case where there is no jamming between the layers. The second and the third test case involve applying a 3 mm and 6 mm horizontal elongation, respectively, to the sample and then conducting the three-point bending tests with the stretched samples. These cases were meant to represent the situations when

Fig. 14 Bending test of the sample using the Instron machine.

the elongation of the sample causes the auxetic material to expand to create jamming at the boundary. We chose 6 mm as it was the maximum elongation before cracks appeared on the surface of the sample. We also wanted to see the effects with half the maximum possible elongation, so we chose 3 mm as another test case. The test was conducted with a 5 mm per minute rate of vertical loading at 5 mm per minute and an end test criterion of 10 mm total vertical flexural elongation. Measurements were done using a 50 N load cell.

We hypothesize that with increasing horizontal elongation, the load vs vertical extension curve should shift upwards. This is because jamming resulting from the horizontal elongation will increase the stiffness of the sample causing it to resist bending. Hence, for the same vertical extension, a higher load will be required. Between 3 and 6 mm horizontal elongation, we expected the load vs extension graph to be higher for the 6 mm test case because a larger horizontal elongation should lead to a larger expansion of the auxetic material and hence increase jamming and the associated stiffness. The test results and the associated discussions are presented in the next section.

6.2 Continuum metastructural tests

6.2.1 Re-entrant honeycomb structure tests

We used the standard hexagonal re-entrant design that is snugly fitted into the hexagonal PPR outer tube and subjected it to the bending test. As shown in Fig. 15, among the three materials, only TPU showed a negligible increase in stiffness after the effects of jamming following the application of a horizontal elongation. Furthermore, this increase in stiffness was load-dependent and not consistent over all the load values. The other two materials did not show such an increase, but rather showed a decrease in the stiffness, especially at higher levels of load.

6.2.2 Missing rib structure tests

Like in the case of the re-entrant design, the missing rib design only showed an increase in stiffness upon elongation when using the TPU material. The increase in stiffness was consistent across different loads. We saw a wide variance when using PVA, suggesting that the material was not suited for such test applications. We also observe a small increase in stiffness when using Pythonflex, which is consistent across the loads. However, there was not much of a difference between the 3 and 6 mm elongations, and the differences in stiffness between the base case and the elongated case were minute (Fig. 16).

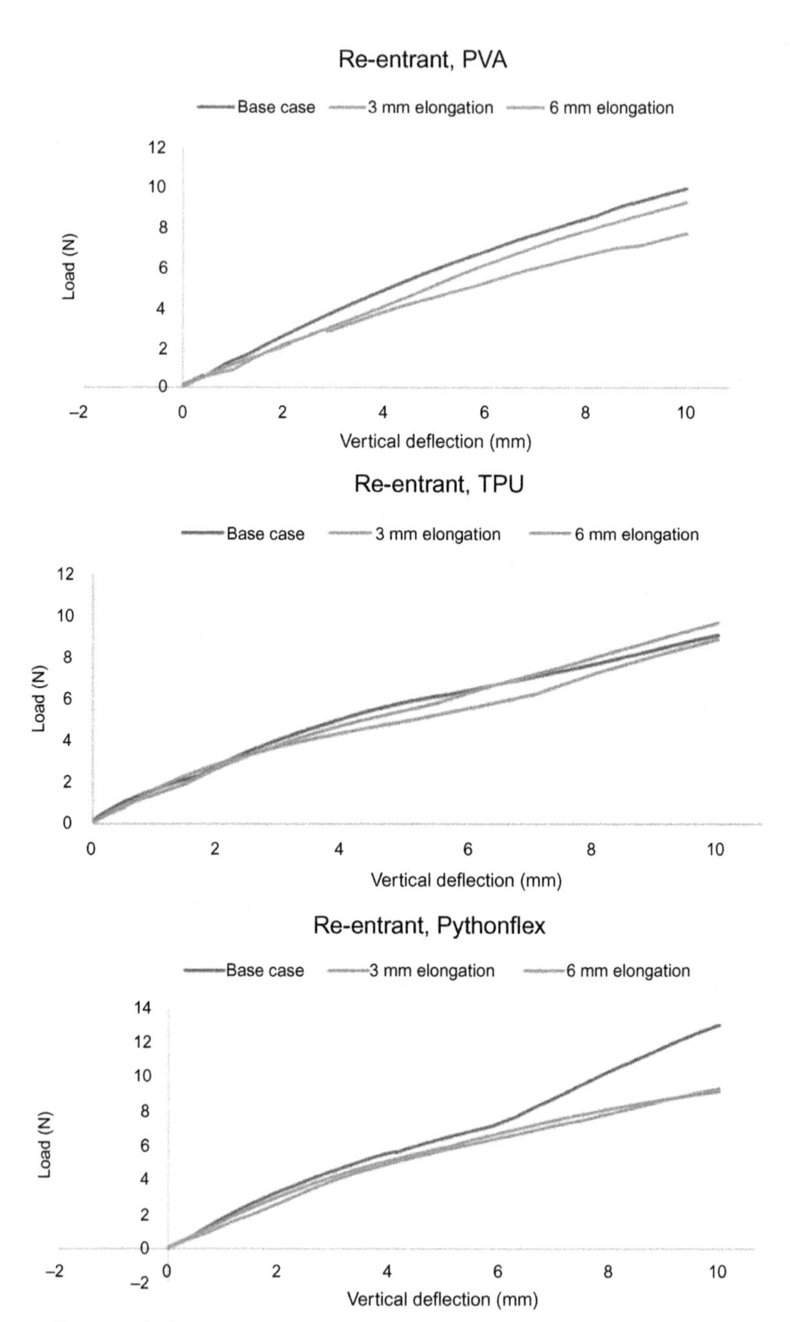

Fig. 15 Test results for re-entrant structures.

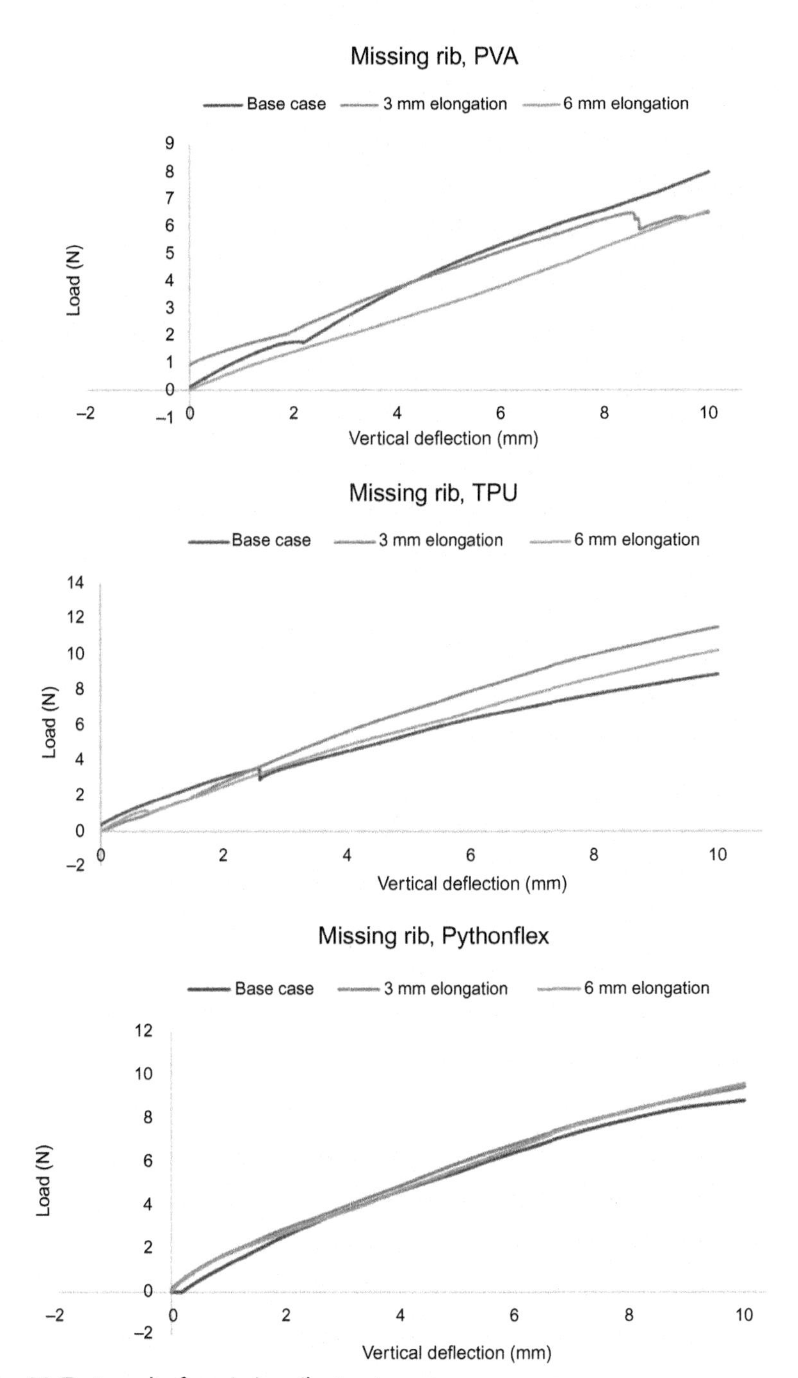

Fig. 16 Test results for missing rib structures.

6.2.3 Double arrow honeycomb structure tests

Like the previous two designs, the TPU material exhibited an increase in stiffness compared to the case with no elongation (Fig. 17). Like the missing rib design, this increase was consistent across different loads. Tests with Pythonflex again showed no significant change in stiffness. We observed that the variation in stiffness when using PVA was very high. It is unclear at this point if such a difference was actually present or if tiny human errors in loading geometry might have led to such a difference. Repeated tests are required to confirm this. Furthermore, using a 6 mm elongation in PVA material is not recommended, as after a certain load the material starts to rupture or develop a crack as shown by the sharp drop in the load in the graph.

6.2.4 Chiral structure tests

The chiral design was the only design that showed an increase in stiffness upon elongation with more than one material (Fig. 18). Both TPU and PVA data showed that the chiral design had higher stiffness, especially at higher loads. Like the previous two cases, Pythonflex did not show much variation among the three cases. The increase in stiffness was consistent across the loads when using PVA. However, it was inconsistent when using TPU.

6.2.5 Star structure tests

This star design showed least capability of stiffness tuning out of all the five designs across the three materials (Fig. 19). We saw a relatively wider variance among the base case and elongation cases, and in all three materials, the base case had higher stiffness. This means that upon elongation, the samples displayed a lower stiffness, which opposed our hypothesized concept of increasing stiffness through jamming.

6.3 Results discussions

In the previous section, we tested the possibility of using the mechanism of jamming created by the expansions of auxetic materials to achieve variable stiffness. We tested five designs and three flexible materials with 0, 3, and 6 mm horizontal elongations. Our results showed that across the designs and materials, there was little evidence to show that there was a significant difference in the change in stiffness. In some cases, the change in stiffness was observable as the load vs vertical extension curve shifted upwards under horizontal elongation. However, this change was not consistent and not observable in the test cases. Among all the different auxetic designs, the change in

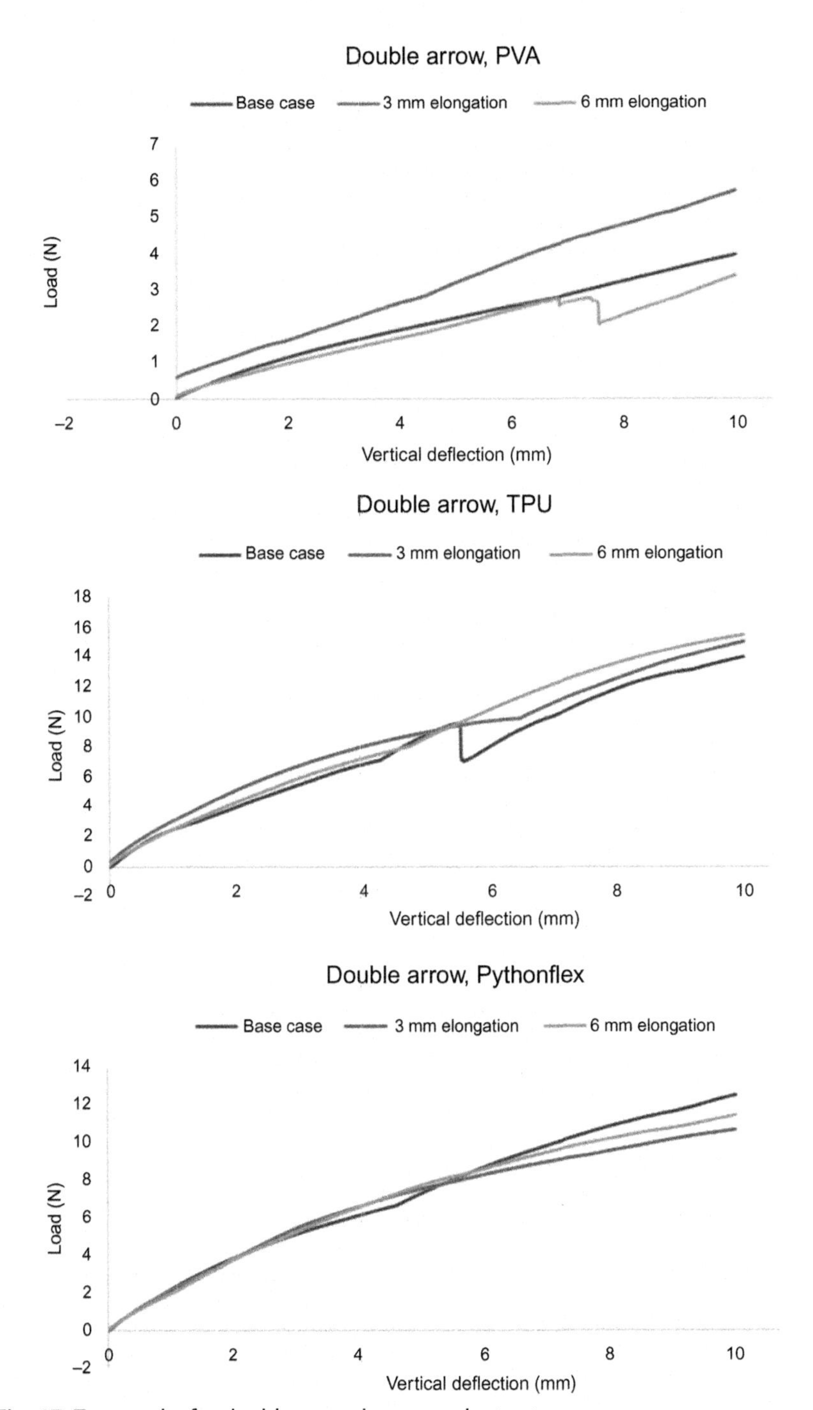

Fig. 17 Test results for double arrow honeycomb structure.

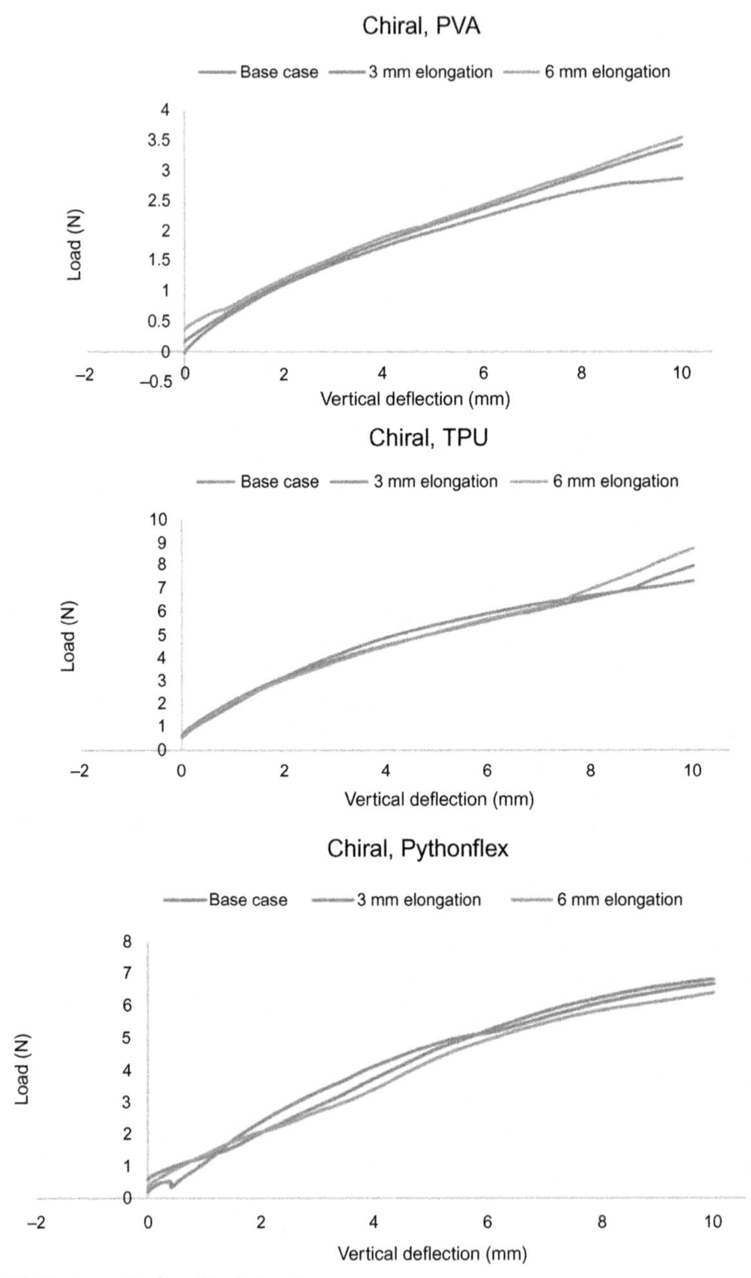

Fig. 18 Test results for chiral structures.

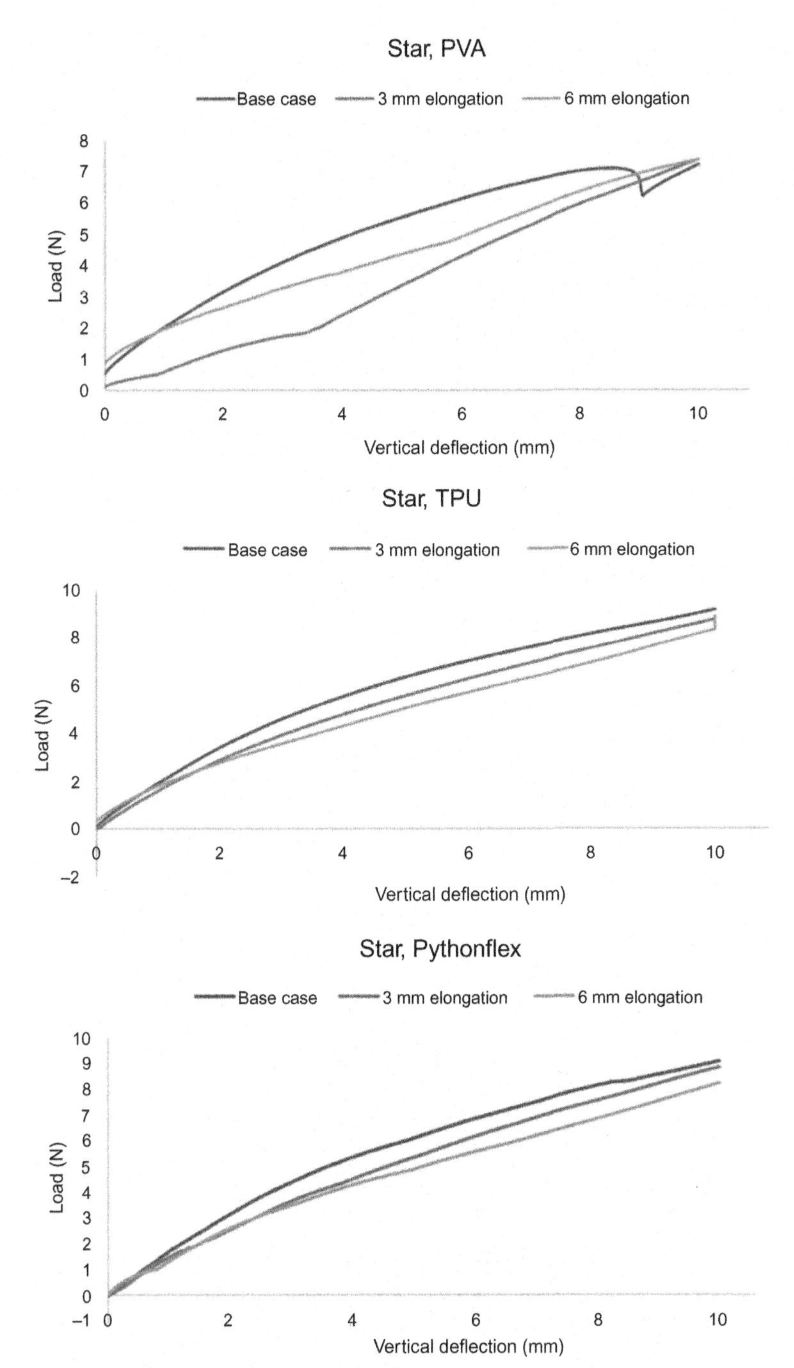

Fig. 19 Test results for Star structures.

stiffness was consistently observed in the chiral design. We inferred that this lack of evidence of jamming was because the test was very sensitive to the loading geometry and fabrication precision, which was less controlled in the preliminary study. While care was taken to ensure that the test piece was as central as possible, it is possible that changes in the position might have occurred while applying the elongation. Furthermore, we observed that the PVA material showed a wide variance for test cases with 6 mm elongation. This might due to the PVA had passed its elastic phase and entered the plastic phase at 6 mm, resulting in cracks.

NPR materials display interesting and unique characteristics that are inherent to their metamaterial structure. These characteristics can be utilized by precisely designing the structure and carefully altering it to suit the stiffness tuning needs.

7 Kirigami and origami methods

7.1 Kirigami methods

Recently, a new form of fabrication process for 2D materials called kirigami, the art of paper cutting, has gained interest. This method of fabrication is quick and economical and in this study was done through the help of Silhouette Curio, a DIY cutting machine. Auxetics structures such as rotating polygons are best suited for fabrication through this process, and their behavior is obtained from the rotation of rigid polygons about their hinges. Fig. 20 shows the mechanism of the opening of the slots when a longitudinal strain is

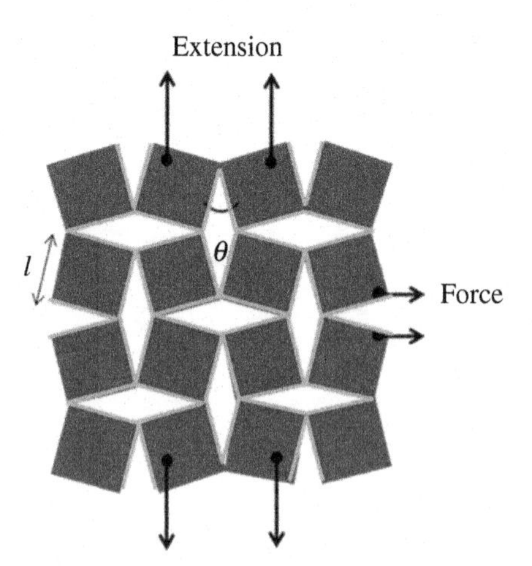

Fig. 20 2D rotating squares.

applied on both ends. The rotation of the rigid squares about the connecting hinges leads to the auxetic behavior. We explored this property by testing different base materials and folding designs and considered the potential applications. For this experiment, three different base materials were used, namely, silicon rubber sheets, high-density foam sheets, and cardboard paper sheets. The behavior of each of the materials is described in the sections that follow.

7.1.1 Cardboard paper

We used a cardboard paper as the individual squares, as it possesses greater structural integrity compared to normal paper, which helps in load-bearing capacity. A 4×12 rectangular matrix comprised of 15 mm squares was cut out and its behavior is shown below. The Poisson's ratio is -1 in both directions of stretching and the young's modulus for a unit thickness of squares is given to be:

$$E_1 = E_2 = K_h \frac{8}{l^2} * \frac{1}{(1 - \sin\theta)}, \tag{1}$$

where K_h is the stiffness constant of the hinges, θ is the angle between the squares, and l is the length of the squares.

Fig. 20 shows the auxetic behavior of such a structure. To apply this property for variable stiffness devices, we fold the rectangular matrix into a tube-like structure as shown in Fig. 21. Pulling the tube from both ends causes the squares to rotate and consequently the structure to expand.

In order to simplify the understanding of such structures, a similar cut-based on rectangular matrix was folded into a triangular prism-like structure. We observe that as the squares rotate and open, spaces are created between each element where bending is possible along the cutting hinges (Fig. 22A). However, when the shape of the tube is restored, the spaces are closed up

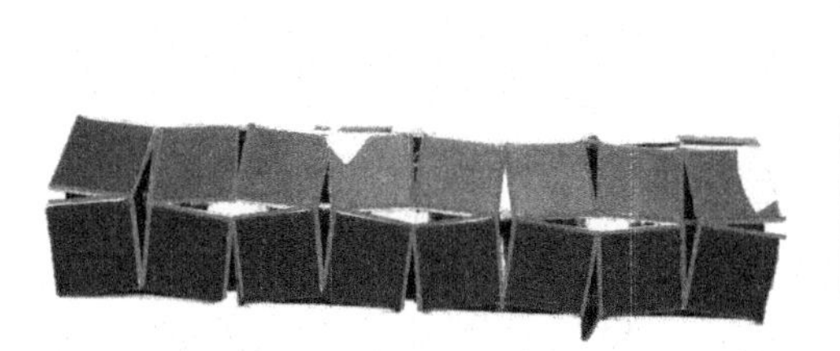

Fig. 21 Rotating squares design folded into tubular structures.

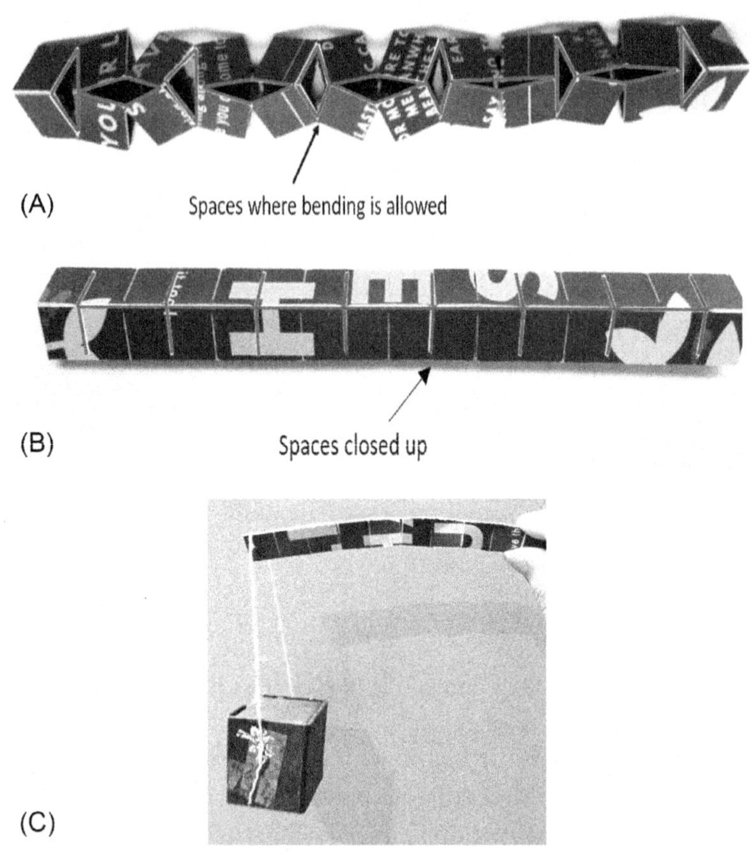

Fig. 22 Changes in the available spaces before and after folding of the structure. (A) Flexible structure because of the availability of open spaces. (B) Rigid structure as the spaces are closed. (C) Stiffened and closed structure can carry loads.

and bending is no longer possible because of the rigid nature of the base material (Fig. 22B).

This interesting observation can be applied to developing a variable stiffness mechanism.

Based on our observations, we reasoned that we could make a tube that would be flexible when the spaces were open but rigid and capable of carrying weight when the spaces were closed. The mechanics of such a load-bearing structure can be assumed to follow that of a hollow triangular beam fixed on one end and free on the other end (Fig. 23). The maximum load that the structure can carry can be estimated as follows.

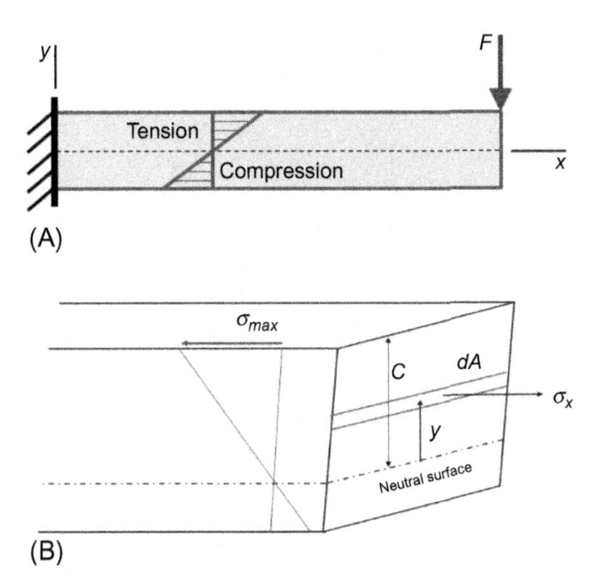

Fig. 23 Beam bending illustrations (σ_{max} is the yield stress of the material; *F* is the load; *x*, *y* are the coordinate axes).

The maximum load that the structure can carry was calculated as follows. For static equilibrium,

$$M = \int y\sigma_x dA = \int y\left(\frac{y}{c}\right)\sigma_{max} dA, \tag{2}$$

$$M_{max} = \frac{\sigma_{max} \times I}{c}, \tag{3}$$

where

$$I = \int y^2 dA, \tag{4}$$

which is also known as the second moment of area.

Here, σ_{max} is the yield stress of the material. Furthermore,

$$M = F \times R \tag{5}$$

The maximum load-bearing capacity of our structure of a hollow triangle (Fig. 24) is given by

$$F_{max} = \frac{3\sigma_{yield} I_{Hollow}}{2d_1 R}, \tag{6}$$

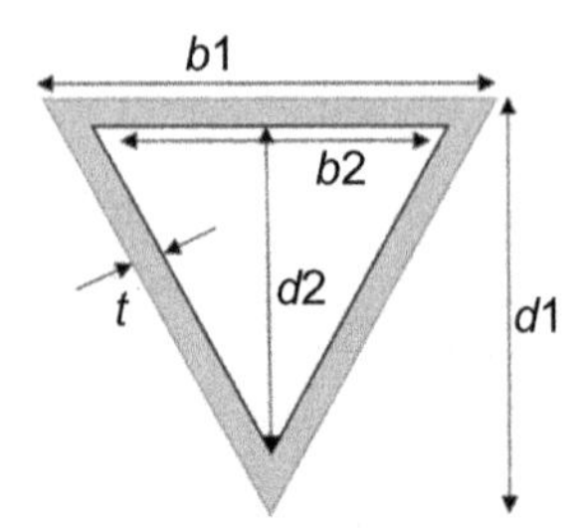

Fig. 24 Maximum load-bearing capacity of a *hollow triangle* (b_1, b_2 defines the edge length of the *outer and inner triangle,* respectively; t is the edge thickness and d_1, d_2 defines the height for the *outer and inner triangle,* respectively).

where

$$I_{Hollow} = \frac{b_1 {d_1}^3}{36} - \frac{b_2 (d_2 - 2t)^3}{36} - \frac{2}{3}(t b_2 d_2). \tag{7}$$

Since this structure is segmented into different sections, it is possible to selectively stiffen certain sections of the tube while keeping the rest of the tube flexible. This translates to a dynamic means of load bearing as it can enable us to achieve complex shapes and maneuvering behavior for our applications.

7.1.2 Silicone rubber

As silicon is a flexible material, we theorized that it will display better characteristics compared to the rigid cardboard material as it can be easily stretched into its auxetic nature. We used 1.5 mm silicon rubber sheets using a similar cutting technique.

Firstly, we observed that even the slightest loads at the ends caused buckling of the tube even under closed conditions. This is because the slits introduce structural problems given that silicon is a very soft and flexible material. We also observed that the rotation of the squares was not as smooth as the friction between the silicon layers, causing many out-of-plane rotations. As a result, the cell-opening mechanism was not satisfactory, and the auxetic behavior was minimal. Additionally, the load-bearing capacity was not as good as that of rigid cardboard paper as there were already initial deformations when a similar load was applied. Furthermore, we noticed that cracks propagated very easily in the silicon layers and the hinges kept tearing away frequently. The hinges of the tube were torn away when a load was applied. Thus, we did not choose silicon sheets as the base material.

7.1.3 High-density foam

As an alternative to silicon, we used another flexible material—high-density foam (Fig. 25). We theorized that the amorphous nature of foam would slow down the crack propagation and it could also be more easily stretched compared to rigid cardboard paper. The same testing process was carried out, and after the tests, we noticed a marked improvement in terms of the tearing of hinges. However, this material failed to stiffen through our theorized means. This is because even after closing the bulk gaps between each of the individual sections, there were still minute natural gaps within each section because of the nature of foam. Hence, the application of a load leads to huge deformities, so a pure foam-like material is not suited for such applications.

From these tests, we concluded that the perfect mix for such applications is a combination of rigid and flexible materials. The rotating squares must be stiff to bear load and exhibit stiffness characteristics when required. However, the hinges connecting the different squares must be made from a flexible material to ease the stretching of the structure into its auxetic form. To meet these requirements, we could use a multimaterial design by using Vero-White as the rigid material for the squares and a mixture of Tango+ with other materials to create an amorphous flexible material for the hinges. Furthermore, we could make design improvements by altering the hinge shapes. Instead of a straight cut hinge, which introduces a lot of stress concentration, we could use a curved cut to reduce the chance of tearing and improve repeatability. However, the use of multimaterial 3D printers would significantly increase the time and cost of fabrication, which goes

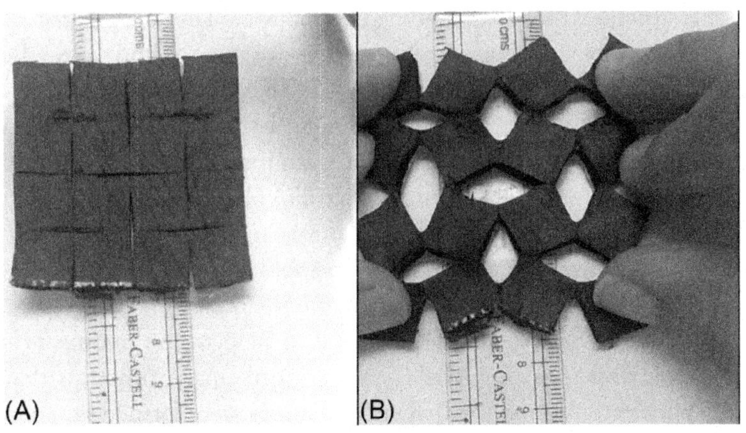

Fig. 25 (A) Nonrigid structure when the spaces are closed; (B) High-density foam auxetic behavior.

against the spirit of our design requirements. Hence, we explored alternative methods for achieving variable stiffness using origami structures.

7.2 Origami methods

While the method of kirigami was promising, it had many issues with respect to reliability and repeatability as the delicate cuts can give way at any time. It also had a low cycle count before the cuts led to tearing. Other designs for auxetic kirigami have similar problems, so we decided our design should minimize or, if possible, eliminate any cutting required.

As an alternative to kirigami, we used origami, which has more robust material characteristics and, as such, is able to withstand higher cycle counts and loads. Through intricate folding patterns using auxetic design, we can integrate multiple units into a single system capable of providing both flexible actuation as well as variable stiffness. The following sections will detail and test such designs. Note that throughout this section, there is not much importance placed on the overall dimensions as origami folding can be highly miniaturized depending on the dexterity of the operator. Instead, care is taken to make the crease pattern clear and visible for first-time users, so the dimensions are generally larger than our design specifications.

7.2.1 Collapsible origami structure

Our starting point of inspiration comes from an on-demand collapsible origami structure that also has load-bearing ability (Zhai et al., 2018). A collapsible structure is a structure that can reconfigure and change shape/size mainly from folding and unfolding. Stents are an excellent example of collapsible structures used in the biomedical industry. The collapsible structure has a stiffness tunable capability, as the method of extension and the method of collapsing follow two different paths. More specifically, extension is easy, whereas after complete extension, collapsing it back is harder. This metamaterial design is inspired by the triangular cylindrical pattern adopted by the authors and it is a relatively common fold in collapsible origami structures (Zhai et al., 2018).

The structure was created using the base fold design as shown in Fig. 26. A simple A4 printing paper was used as the material, and precreasing was done using the Silhouette Curio machine to help in the folding process. Creasing was done by using the score option to make tiny cuts that did not fully penetrate through the paper. The edges were formed into a cylinder and glued using a thin layer of nonwater-based glue. The precrease was then collapsed into the final shape before testing.

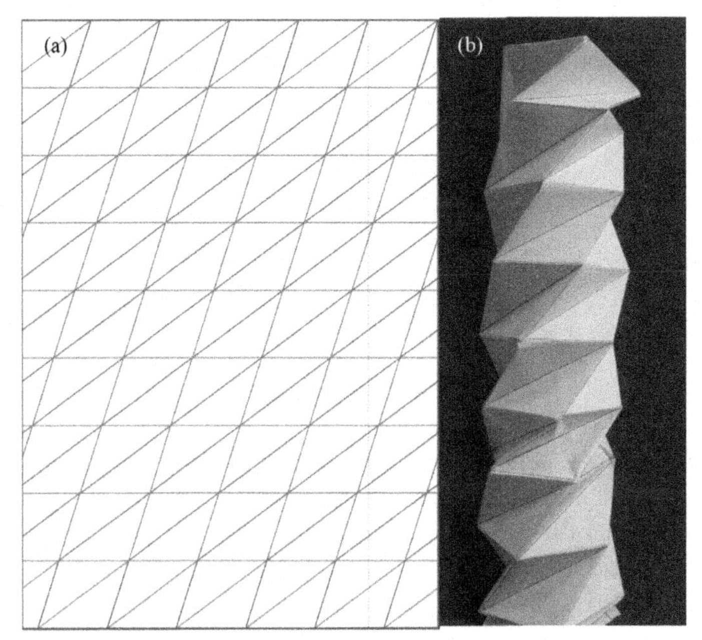

Fig. 26 (A) Folding crease pattern. (B) Completed collapsible tower.

This origami structure can be collapsed into a small unit, but can also show an excellent load-bearing ability when fully extended. This particular structure can carry as much as 200 g before starting to collapse.

The existence of variable stiffness in this structure is not only observed visually but can also be rigorously proven using kinematic models (Zhai et al., 2018). To verify the existence of stiffness variation, the deformation energy of the triangulated cylinder patterns and the extension and collapsing from the energy and strain perspectives is analyzed (Zhai et al., 2018).

A unit cell from the triangulated cylinder and the same triangle in the folded state is shown in Fig. 27.

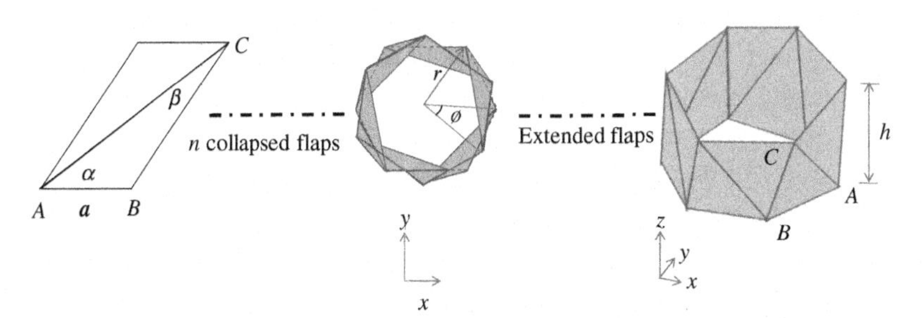

Fig. 27 Unit cell parameters, folded and extended cylinder parameters (Kolken and Zadpoor, 2017).

This unit triangle requires three parameters to be characterized, which include the side a and the angles α and β. It follows that

$$L_{AB} = a, \tag{8}$$

$$L_{BC} = a \times \sin(\alpha)/\sin(\beta), \tag{9}$$

$$L_{AC} = a \times \sin(\alpha + \beta)/\sin(\beta). \tag{10}$$

When in its folded state, the position of the same triangle is characterized by the height h and the angle of twist ϕ as well as the radius r of the cylinder. Again, it follows that

$$l_{AB} = 2r\sin\left(\frac{\pi}{n}\right), \tag{11}$$

where n is the number of repetitions of the unit cell. Since there are different folded states with different heights and twist angles, it is not possible to determine them using the three constants, namely a, α, and β. Thus, there is a need to introduce strains ϵ_{AB}, ϵ_{BC}, ϵ_{AC} to link the variables to the constants, where

$$\epsilon_{AB} = \frac{l_{AB} - L_{AB}}{L_{AB}} \tag{12}$$

and so on. The deformation energy stored in one strip of the paper is then given by

$$U = \frac{nEA}{2}\left(L_{AB}\epsilon^2{}_{AB} + L_{BC}\epsilon^2{}_{BC} + L_{AC}\epsilon^2{}_{AC}\right), \tag{13}$$

where EA is the tensile rigidity. This deformation energy shows that there are apparent bi-stable states where the fully extended as well as the fully collapsed states have minimum energy. There exists an energy barrier between these two states, which is evidence that once fully extended, the material is stiff and requires a higher amount of energy to collapse again. A more detailed derivation of the results can be found in Zhai et al. (2018). By varying the angles α and β, it is possible to select the energy barrier and hence the stiffness change required for bespoke applications.

While this method of using collapsible structures showed promise as a variable stiffness mechanism, it also had a major disadvantage with respect to actuation. As mentioned earlier in the section objectives, we aimed to create a combined actuation and variable stiffness method using a single structure. When tested for actuation capabilities, it was observed that the device was only able to collapse and extend and did not provide useful actuation methods like bending. The structure yielded pure compressions and extensions without any bending. Furthermore, even while compressing and extending, the device undergoes an involuntary twist that cannot be overcome. This is because the complete structure had a certain twist to it as all the mountain and valley folds pointed in one direction. Hence, it was deemed important to eliminate this directional twist and achieve a net zero twist structure while preserving its collapsible nature to add on bending capabilities.

7.2.2 Miura origami structure

Miura origami, or as it is called in English, the herringbone tessellation, is a form of an auxetic structure made from parallelograms. As shown in Fig. 28, this tessellating structure displays excellent auxetic behavior.

For our design, the two-dimensional herringbone tessellation was combined with the collapsible fold from the previous design. Moreover, the parallelograms were alternated in each layer such that the twist from one layer canceled out the opposite twist from its adjacent layer, resulting in a net zero twist. This yielded a hybrid composite structure that was compatible with the current method of actuation and displayed variable stiffness. The folding pattern and the complete folded structure are shown in Figs. 29 and 30. A similar process of precreasing, gluing, and folding was followed to get the complete structure.

After obtaining our complete structure, we then tested its ability for actuation compatibility using a tendon-driven mechanism as well as variable stiffness using the collapsible method. The structure was fixated with tendons routing from the inner channel. Our tests showed that this structure

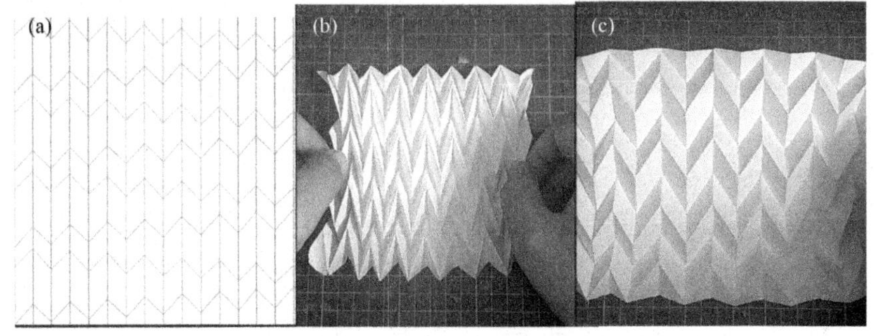

Fig. 28 (A) Herringbone fold crease pattern. (B) Herringbone folded. (C) Herringbone unfolded.

Fig. 29 Hybrid Miura-collapsible crease pattern.

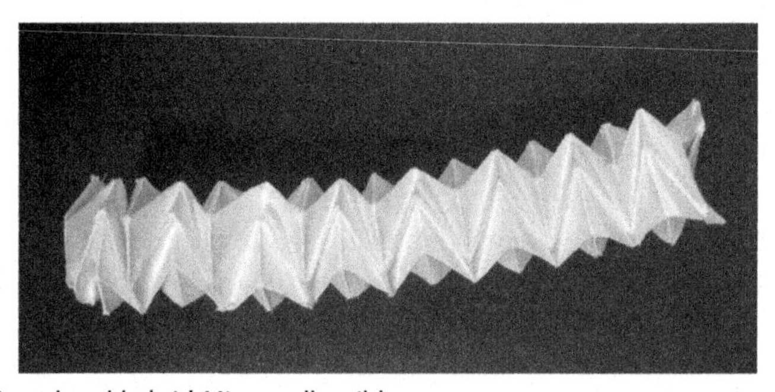

Fig. 30 Completed hybrid Miura-collapsible structure.

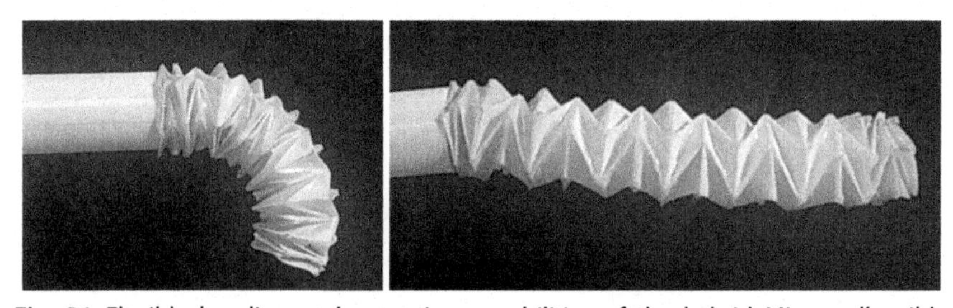

Fig. 31 Flexible bending and extension capabilities of the hybrid Miura-collapsible structure.

exhibited smooth bending as well as compression and extension movements without any involuntary twist (Fig. 31).

Furthermore, the collapsible nature (Fig. 32) derived from the previous design also allowed us to achieve a variable stiffness mechanism.

As shown in Fig. 33, a part of the structure was fully extended while the remaining part was kept flexible. We could suspend a load from the stiff part

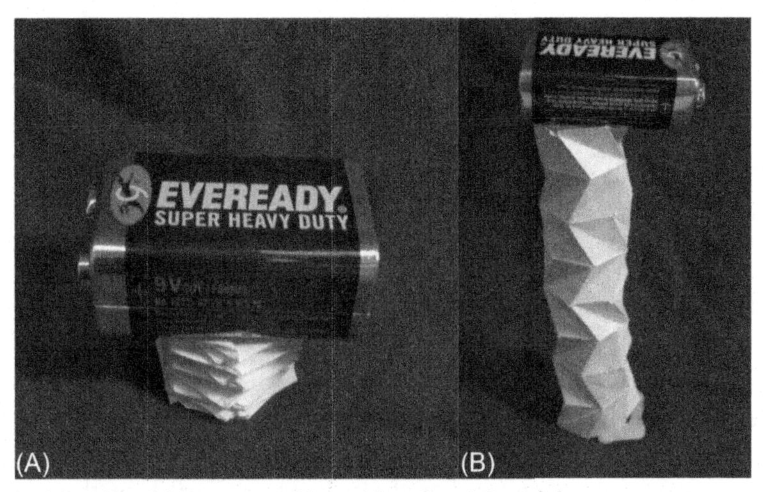

Fig. 32 (A) Collapsible structure; (B) Load-bearing ability of the structure.

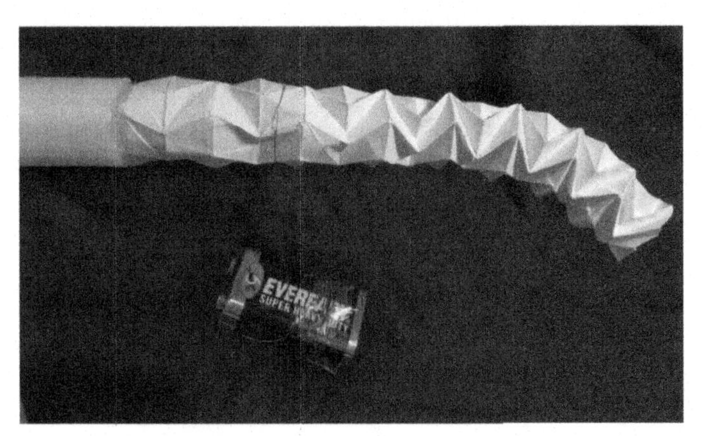

Fig. 33 Variable stiffness along the structure.

without the structure deforming or collapsing. The end of the tube was still flexible and compliant, showing that we can achieve a means of selective variable stiffness.

7.2.3 Waterbomb tube

Another successful design allowed us to combine variable stiffness with actuation in a single structure. The simplest waterbomb base crease pattern and fold design is as shown in Fig. 34.

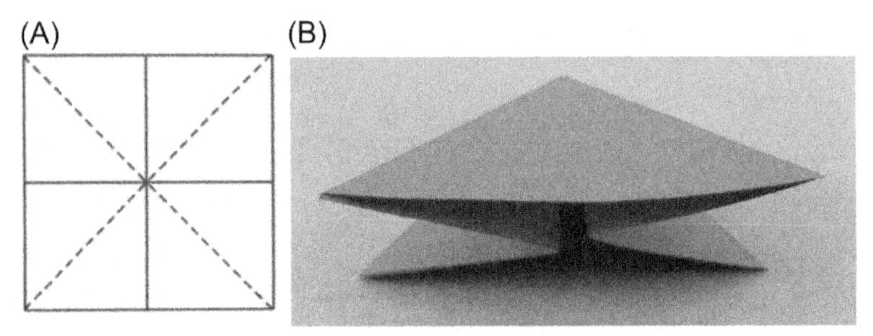

Fig. 34 (A) Waterbomb base crease. (B) Simple waterbomb structure.

It is possible to obtain a more complex geometry that displays much more interesting properties by tessellating the base by $n \times m$ units over a rectangle and with each alternating row (or column) having the waterbomb base phase shifted by 180 degrees. For the two-dimensional base that can be approximated into a circular cylinder with smooth curvature, a minimum of six rows is required, if the cylinder is rolled along its length. The crease pattern and the completed structure is shown in Fig. 35. Furthermore, this origami structure is made up of a variant of the hexagonal re-entrant structure and hence can display interesting auxetic behavior (Fig. 36).

The waterbomb cylinder has recently received much interest recently, especially with respect to practical applications like stent grafts. However,

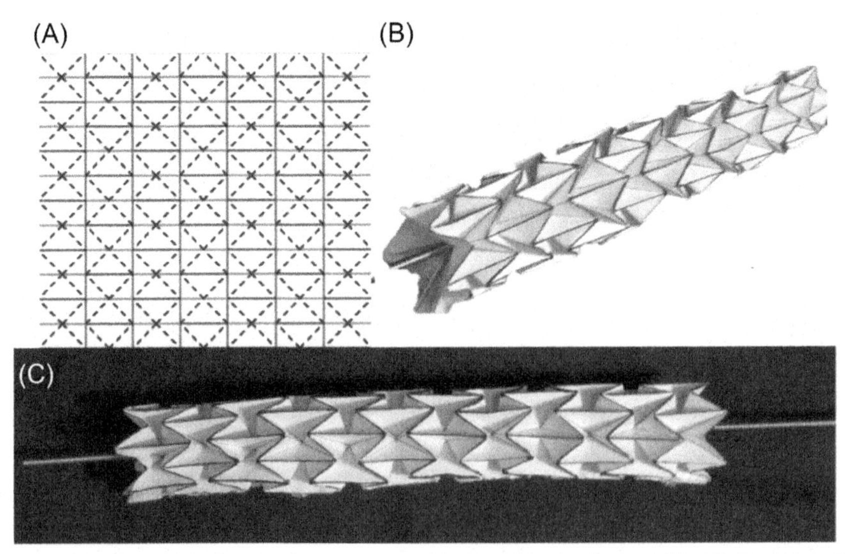

Fig. 35 (A) Waterbomb tessellating crease. (B) Waterbomb phase shifts. (C) Completed waterbomb tube.

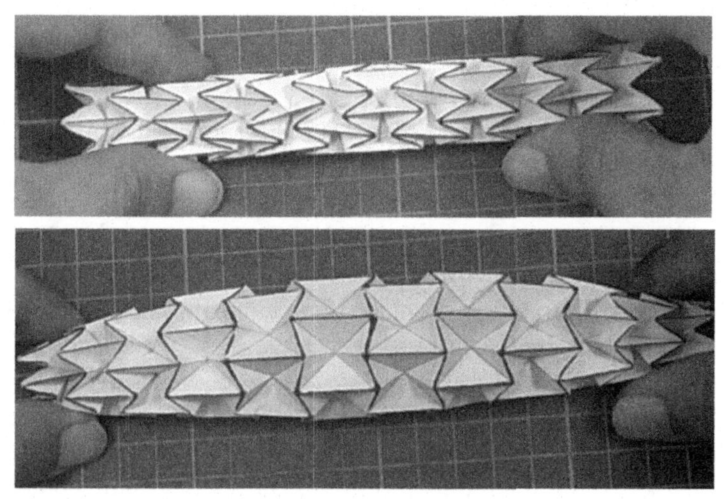

Fig. 36 Waterbomb tube displaying auxetic properties.

most of these applications only make use of its radial expansion characteristics with little to no attention being given to its bending or twisting. By exploring different tendon-routing mechanisms and methods of achieving circular twist along its central axis, we were able to harness the bending characteristics of the tube as well as obtain a method of variable stiffness.

Bending of this tubular structure is tricky as it relies on the careful routing of the tendons across its complex surface structure. We tried three different methods of routing (inner through the main channel, outer crisscross, and outer straight; see Fig. 37), but only one structure, which involves the

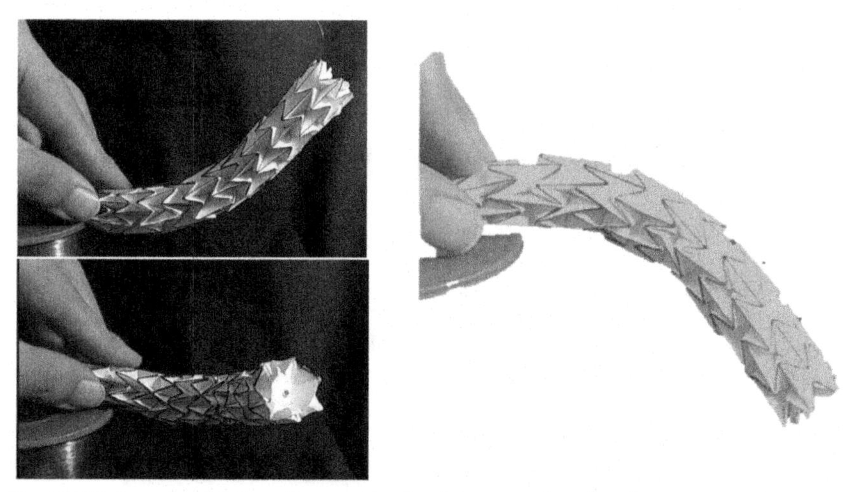

Fig. 37 Various flexible bending methods of the waterbomb tube structure.

crisscrossing of the tendons from the outside, yields bending. As the focus of the chapter is on the means of achieving variable stiffness, we will not go into detail about tendon-routing methods.

Twisting of the tube along its central axis (Fig. 38) yielded another interesting result. This twist motion effectively changed the metamaterial structure in such a way to resist further loading. Hence, by having an on/off state for the twist of the structure, we found that we could control and vary the stiffness of the tubular structure. Furthermore, the twist did not need to be constrained to a binary state. A graded change in stiffness was observed by varying the extent of twist anywhere between no twist and the fully squeezed version. The detailed kinematic analysis of the change in metamaterial structure and the associated change in stiffness was quantified by Feng (Feng et al., 2018).

We harnessed this interesting property by developing a method of controlling the amount of twist from a distal point. A thin tensile spring was run through the central channel of the waterbomb tube and fixated at the top using a layer of silicone rubber. Tensile springs were used since their compliant nature allows them to deform and follow the bending motion of the waterbomb tube. Furthermore, these springs are also able to maintain their curvature and provide a torsional force at their tip when required. This useful property allowed us to deliver a method of twisting the waterbomb tube to modulate the stiffness of the structure. The twist can also be applied when the structure is bent, which would allow the device to maintain its shape throughout manipulation tasks during surgery. Fig. 39 shows the stiffness capability of the waterbomb structure upon the application of twist. Notice the curvature of the structure before and after the application of twist. It is evident that an untwisted structure has a lower bending stiffness. Thus, we

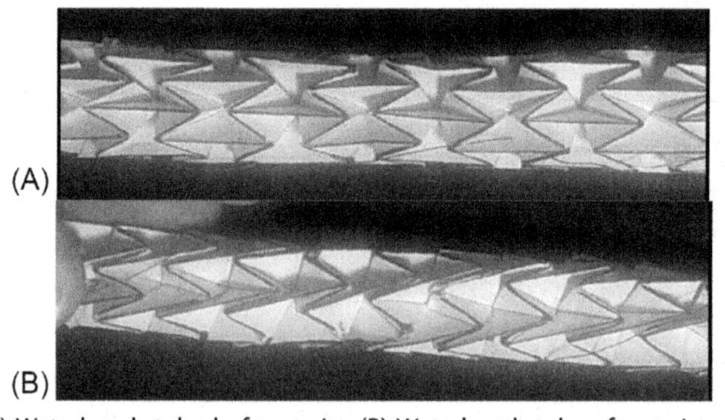

Fig. 38 (A) Waterbomb tube before twist. (B) Waterbomb tube after twist.

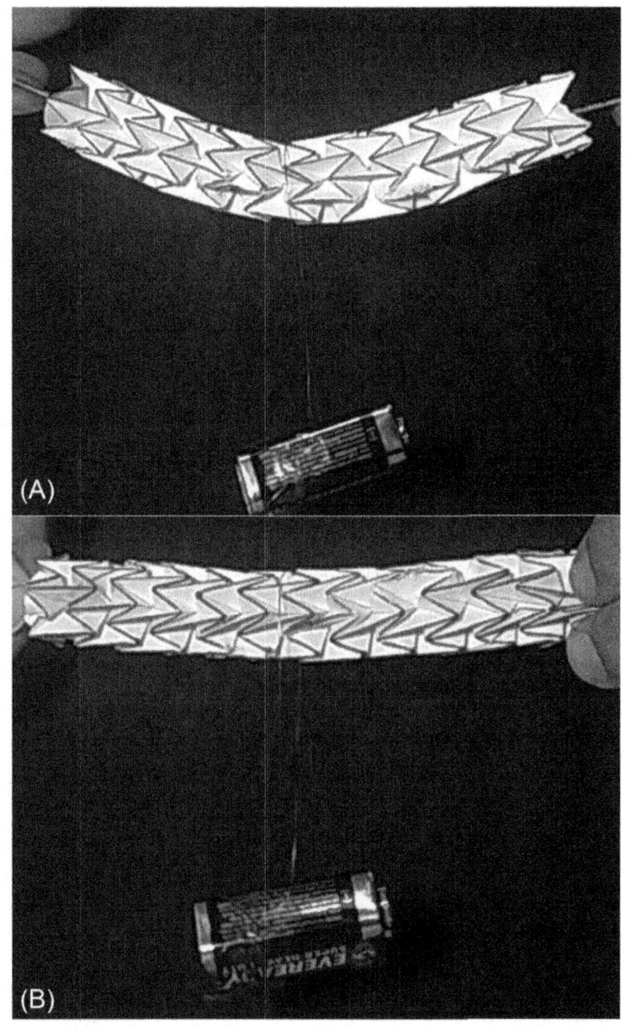

Fig. 39 (A) Waterbomb tube before twist showing flexibility; (B) Waterbomb tube after twist showing rigidity.

have successfully created a method to vary the stiffness of our tubular device that can be well approximated as a surgical stem.

8 Conclusion

MIS procedures have many benefits for patients, including lesser postsurgery pain, shorter hospital stays, and faster recovery time. However, the lack of flexible, miniaturized robotic devices has prevented their full-scale application in the medical industry. One of the biggest challenges for soft and flexible tubular MIS equipment is the load bearing capacity and the ability to transform from a flexible compliant device into a stiff and rigid device when required. Currently available methods of achieving tunable stiffness are either not suited for fast-changing applications or require high temperature and pressure applications, which may pose a safety risk when being used inside human bodies. Furthermore, the change in stiffness happens in bulk across the device, making it difficult to selectively stiffen certain regions independently. Instead, we theorize and design a completely novel means of achieving tunable stiffness using auxetic materials.

We first tested the method of jamming using auxetic materials to achieve a change in stiffness. We also created folding patterns that exhibit stiffness tunability using kirigami and origami methods. These new methods are compatible with our existing tendon-driven mechanism and provide an integrated actuation and stiffness modulation method. These methods are safe and suitable for biomedical applications where the stiffness can be isolated to specific regions.

References

Balasubramanian, A., Standish, M., Bettinger, C.J., 2014. Microfluidic thermally activated materials for rapid control of macroscopic compliance. Adv. Funct. Mater. https://doi.org/10.1002/adfm.201304037.

Behbahani, S.B., Tan, X., 2017. Design and dynamic modeling of electrorheological fluid-based variable-stiffness fin for robotic fish. Smart Mater. Struct. https://doi.org/10.1088/1361-665x/aa7238.

Brown, E., Rodenberg, N., Amend, J., Mozeika, A., Steltz, E., Zakin, M.R., et al., 2010. Universal robotic gripper based on the jamming of granular material. Proc. Natl. Acad. Sci. U. S. A. https://doi.org/10.1073/pnas.1003250107.

Carlson, J.D., Jolly, M.R., 2000. MR fluid, foam and elastomer devices. Mechatronics. https://doi.org/10.1016/S0957-4158(99)00064-1.

Chen, L., Gong, X.L., Li, W.H., 2007. Microstructures and viscoelastic properties of anisotropic magnetorheological elastomers. Smart Mater. Struct. https://doi.org/10.1088/0964-1726/16/6/069.

Chen, J., Lau, H.Y.K., Xu, W., Ren, H., 2016. Towards transferring skills to flexible surgical robots with programming by demonstration and reinforcement learning. In: Proceedings of the 8th International Conference on Advanced Computational Intelligence, ICACI 2016. https://doi.org/10.1109/ICACI.2016.7449855.

Cheng, N.G., Gopinath, A., Wang, L., Iagnemma, K., Hosoi, A.E., 2014. Thermally tunable, self-healing composites for soft robotic applications. Macromol. Mater. Eng. https://doi.org/10.1002/mame.201400017.

Clark, W.W., Brigham, J.C., Mo, C., Joshi, S., 2010. Modeling of a high-deformation shape memory polymer locking link. In: Industrial and Commercial Applications of Smart Structures Technologies 2010. https://doi.org/10.1117/12.847987.

Ditaranto, R.A., 1964. Theory of vibratory bending for elastic and viscoelastic layered finite-length beams. J. Appl. Mech. https://doi.org/10.1115/1.3627330.

Feng, H., Ma, J., Chen, Y., You, Z., 2018. Twist of tubular mechanical metamaterials based on waterbomb origami. Sci. Rep. https://doi.org/10.1038/s41598-018-27877-1.

Haines, C.S., Lima, M.D., Li, N., Spinks, G.M., Foroughi, J., Madden, J.D.W., et al., 2014. Artificial muscles from fishing line and sewing thread. Science. https://doi.org/10.1126/science.1246906.

Hao, T., 2002. Electrorheological fluids. In: Encyclopedia of Smart Materials. https://doi.org/10.1002/0471216275.esm035.

Huan, A.S., Xu, W., Ren, H., 2016. Investigation of a stiffness varying mechanism for flexible robotic system. In: 2016 IEEE International Conference on Mechatronics and Automation, IEEE ICMA 2016. https://doi.org/10.1109/ICMA.2016.7558669.

Jiang, A., Xynogalas, G., Dasgupta, P., Althoefer, K., Nanayakkara, T., 2012. Design of a variable stiffness flexible manipulator with composite granular jamming and membrane coupling. In: IEEE International Conference on Intelligent Robots and Systems. https://doi.org/10.1109/IROS.2012.6385696.

Jiang, A., Dasgupta, P., Althoefer, K., Nanayakkara, T., 2014. Robotic granular jamming: a new variable stiffness mechanism. J. Robot. Soc. Japan. https://doi.org/10.7210/jrsj.32.333.

Kolken, H.M.A., Zadpoor, A.A., 2017. Auxetic mechanical metamaterials. RSC Adv. https://doi.org/10.1039/c6ra27333e.

Li, Z., Du, R., Yu, H., Ren, H., 2014. Statics Modeling of an Underactuated Wire-Driven Flexible Robotic Arm. https://doi.org/10.1109/biorob.2014.6913797.

Li, Z., Feiling, J., Ren, H., Yu, H., 2015a. A novel tele-operated flexible surgical arm with optimal trajectory tracking aiming for minimally invasive neurosurgery. In: Proceedings of the 2015 7th IEEE International Conference on Cybernetics and Intelligent Systems, CIS 2015 and Robotics, Automation and Mechatronics, RAM 2015. https://doi.org/10.1109/ICCIS.2015.7274580.

Li, Z., Zin Oo, M., Nalam, V., Duc Thang, V., Ren, H., Kofidis, T., Yu, H., 2015b. Design of a novel flexible endoscope—cardioscope. J. Mech. Robot. https://doi.org/10.1115/1.4032272.

Li, Z., Ren, H., Chiu, P.W.Y., Du, R., Yu, H., 2016. A novel constrained wire-driven flexible mechanism and its kinematic analysis. Mech. Mach. Theory. https://doi.org/10.1016/j.mechmachtheory.2015.08.019.

Li, Z., Wu, L., Ren, H., Yu, H., 2017. Kinematic comparison of surgical tendon-driven manipulators and concentric tube manipulators. Mech. Mach. Theory. https://doi.org/10.1016/j.mechmachtheory.2016.09.018.

Lindler, J.E., Wereley, N.M., 1999. Double adjustable shock absorbers using electrorheological fluid. J. Intell. Mater. Syst. Struct. https://doi.org/10.1106/468R-DHQM-076W-MAF6.

Liu, B., Boggs, S.A., Shaw, M.T., 2001. Electrorheological properties of anisotropically filled elastomers. IEEE Trans. Dielectr. Electr. Insul. https://doi.org/10.1109/94.919919.

Loeve, A.J., Van De Ven, O.S., Vogel, J.G., Breedveld, P., Dankelman, J., 2010. Vacuum packed particles as flexible endoscope guides with controllable rigidity. Granul. Matter. https://doi.org/10.1007/s10035-010-0193-8.

McEvoy, M.A., Correll, N., 2015. Thermoplastic variable stiffness composites with embedded, networked sensing, actuation, and control. J. Compos. Mater. https://doi.org/10.1177/0021998314525982.

McKnight, G., Doty, R., Keefe, A., Herrera, G., Henry, C., 2010. Segmented reinforcement variable stiffness materials for reconfigurable surfaces. J. Intell. Mater. Syst. Struct. https://doi.org/10.1177/1045389X10386399.

Nakamura, T., Saga, N., Nakazawa, M., 2002. Impedance control of a single shaft-type clutch using homogeneous electrorheological fluid. J. Intell. Mater. Syst. Struct. https://doi.org/10.1106/104538902029068.

Parthasarathy, M., Klingenberg, D., 1996. Electrorheology: mechanisms and models. Mater. Sci. Eng. R. Rep. 17 (2), 57–103. https://doi.org/10.1016/0927-796X(96)00191-X.

Saxena, K.K., Das, R., Calius, E.P., 2016. Three decades of auxetics research—materials with negative Poisson's ratio: a review. Adv. Eng. Mater. https://doi.org/10.1002/adem.201600053.

Schubert, B.E., Floreano, D., 2013. Variable stiffness material based on rigid low-melting-point-alloy microstructures embedded in soft poly(dimethylsiloxane) (PDMS). RSC Adv. https://doi.org/10.1039/c3ra44412k.

See, H., Sakurai, R., Saito, T., Asai, S., Sumita, M., 2001. Relationship between electric current and matrix modulus in electrorheological elastomers. J. Electrostat. https://doi.org/10.1016/S0304-3886(01)00028-6.

Shan, W., Lu, T., Majidi, C., 2013. Soft-matter composites with electrically tunable elastic rigidity. Smart Mater. Struct. https://doi.org/10.1088/0964-1726/22/8/085005.

Shan, W., Diller, S., Tutcuoglu, A., Majidi, C., 2015. Rigidity-tuning conductive elastomer. Smart Mater. Struct. https://doi.org/10.1088/0964-1726/24/6/065001.

Taghavi, M., Helps, T., Huang, B., Rossiter, J., 2018. 3D-printed ready-to-use variable-stiffness structures. IEEE Robot. Autom. Lett. https://doi.org/10.1109/LRA.2018.2812917.

Tan, Z., Ren, H., 2017. Design analysis and bending modeling of a flexible robot for endoscope steering. Int. J. Intell. Robot. Appl. https://doi.org/10.1007/s41315-017-0014-x.

Tangboriboon, N., Sirivat, A., Kunanuruksapong, R., Wongkasemjit, S., 2009. Electrorheological properties of novel piezoelectric lead zirconate titanate Pb(Zr0.5,Ti0.5)O3-acrylic rubber composites. Mater. Sci. Eng. C 29 (6), 1913–1918. https://doi.org/10.1016/j.msec.2009.03.002.

Varga, Z., Filipcsei, G., Zrínyi, M., 2006. Magnetic field sensitive functional elastomers with tuneable elastic modulus. Polymer. https://doi.org/10.1016/j.polymer.2005.10.139.

Wei, K., Meng, G., Zhang, W., Zhou, S., 2007. Vibration characteristics of rotating sandwich beams filled with electrorheological fluids. J. Intell. Mater. Syst. Struct. https://doi.org/10.1177/1045389X06072380.

Wu, J., Gong, X., Fan, Y., Xia, H., 2010. Anisotropic polyurethane magnetorheological elastomer prepared through in situ polycondensation under a magnetic field. Smart Mater. Struct. https://doi.org/10.1088/0964-1726/19/10/105007.

Wu, L., Song, S., Wu, K., Lim, C.M., Ren, H., 2017. Development of a compact continuum tubular robotic system for nasopharyngeal biopsy. Med. Biol. Eng. Comput. https://doi.org/10.1007/s11517-016-1514-9.

Yeh, Z.F., Shih, Y.S., 2005. Critical load, dynamic characteristics and parametric instability of electrorheological material-based adaptive beams. Comput. Struct. https://doi.org/10.1016/j.compstruc.2005.02.028.

Zhai, Z., Wang, Y., Jiang, H., 2018. Origami-inspired, on-demand deployable and collapsible mechanical metamaterials with tunable stiffness. Proc. Natl. Acad. Sci. U. S. A. https://doi.org/10.1073/pnas.1720171115.

Zhou, G.Y., 2003. Shear properties of a magnetorheological elastomer. Smart Mater. Struct. https://doi.org/10.1088/0964-1726/12/1/316.

Appendices for Chapter 2

```
function f=equations(t, m)
% t - time wector
% m - cholesterol mass vector: m(1) - cholesterol mass in I compatment,
% m(2)- cholesterol mass in II compartment
global k
global k12
global k21
global mdiet
global min
global mout
global mtis

f=zeros(2,1);
% f(1) equation that describes I compartment
f(1)=k/m(1)+ k21*m(2)-k12*m(1)+ min-mout;
% f(2) equation that describes II compartment
f(2)=-k21*m(2)+k12*m(1)-mtis+ mdiet;


% Run the two comartment model of chlesterol homeostasis
% this code supports the two-compartment model of cholesterol
homeostasis as a teaching tool
% for more: A. Wrona, J. Balbus, O. Hrydziuszko, K. Kubica,
Two-compartment model as a teaching tool for cholesterol homeostasis,
% Adv Physiol Educ 39: 372-377, 2015; doi:10.1152/advan.00141.2014
clear all
clc
global k
global k12
global k21
global mdiet
global min
global mout
global mtis
%Parameters
k=732;          % the rate constant of de novo cholesterol synthesis
```

```
k12=3.6;      % the rate constant of cholesterol exchange between
compartments (from compartment I to compartment II)
k21=1.0;      % the rate constant of cholesterol exchange between
compartments (from compartment II to compartment I)
mdiet=0;      % dietary cholesterol, 0-3g daily intake, but only up to
0.33g/day is absorbed
min=0.85;     % cholesterol obtained from intestine
mout=1.2;     % cholesterol used as a precursor of bile acid and being
a part of bile
mtis=0.243;   % cholesterol transported to muscles and peripheral
tissue; 350mg/day
v1=6.5;       % volume of blood plasma in the liver compartment in dl
v2=23.5;      % volume of blood plasma in the peripheral blood com-
partment in dl
 % in doi:10.1152/advan.00141.2014 v1 and v2 referred to whole blood
inf = msgbox('Before you run the calculations you have to input: the
number of time intervals, tmax for each interval, values of model
parameters.','Info','warn');
pause(2)
disp('Physiological Parameters:')
disp(strcat('1. k=',num2str(k)))
disp(strcat('2. k12=',num2str(k12)))
disp(strcat('3. k21=',num2str(k21)))
disp(strcat('4. mdiet=',num2str(mdiet)))
disp(strcat('5. min=',num2str(min)))
disp(strcat('6. mout=',num2str(mout)))
disp(strcat('7. mtis=',num2str(mtis)))
prompty = 'Do you want to disturb this system? Y/N: ';
stral = input(prompty, 's');
ada=0;
if stral=='Y'| stral=='y'
  ada=1;
end
  while ada>0
     ada=0;
     str2=input('enter the number of parameter you want to change:');
     switch str2
        case 1
          str3=input('Enter a new value:');
          k=str3;
```

```
        case 2
          str3=input('Enter a new value:');
          k12=str3;
        case 3
          str3=input('Enter a new value:');
          k21=str3;
        case 4
          str3=input('Enter a new value:');
          mdiet=str3;
        case 5
          str3=input('Enter a new value:');
          min=str3;
        case 6
          str3=input('Enter a new value:');
          mout=str3;
        case 7
          str3=input('Enter a new value:');
          mtis=str3;
        otherwise
        disp('bad command')
    end
disp('New parameters:')
disp(strcat('1. k=',num2str(k)))
disp(strcat('2. k12=',num2str(k12)))
disp(strcat('3. k21=',num2str(k21)))
disp(strcat('4. mdiet=',num2str(mdiet)))
disp(strcat('5. min=',num2str(min)))
disp(strcat('6. mout=',num2str(mout)))
disp(strcat('7. mtis=',num2str(mtis)))
        str4=input(prompty,'s');
        if str4=='Y'| str4=='y'
          ada=1;
        end
    end
    m1=k/(mout+mtis-mdiet-min);
    m2=(mdiet°2+mdiet*min-mdiet*mout-k*k12-2*mdiet*mtis-
min*mtis+mout*mtis+mtis°2)/(k21*(mdiet+min-mout-mtis));
    c1=m1/v1;
    c2=m2/v2;
```

```
disp(strcat('m1=',num2str(m1)))
disp(strcat('c1=',num2str(c1)))
disp(strcat('m2=',num2str(m2)))
disp(strcat('c2=',num2str(c2)))

n=input('Enter the number of time intervals= ');
if n<1
      hg = msgbox('Wrong value. Please try again', 'Error', 'warn');
      n=input('Enter the time intervals= ');
end
tmax=[];
parameter=zeros(7,n);
for i=1:n
   format compact
      interval=i
      t=input('tmax[minute]= ');
      if t<0
         hg  =  msgbox('Wrong  value.  Please  try  again',  'Error',
'warn');
         t=input('tmax[minute]= ');
      end
      tmax=[tmax;t];
   parameter(1,i)=k;
   parameter(2,i)=k12;
   parameter(3,i)=k21;
   parameter(4,i)=mdiet;
   parameter(5,i)=min;
   parameter(6,i)=mout;
  parameter(7,i)=mtis;
disp('Parameters:')
disp(strcat('1. k=',num2str(parameter(1,i))))
disp(strcat('2. k12=',num2str(parameter(2,i))))
disp(strcat('3. k21=',num2str(parameter(3,i))))
disp(strcat('4. mdiet=',num2str(parameter(4,i))))
disp(strcat('5. min=',num2str(parameter(5,i))))
disp(strcat('6. mout=',num2str(parameter(6,i))))
disp(strcat('7. mtis=',num2str(parameter(7,i))))
prompt = 'Do you want to change parameters in the current time inter-
val? Y/N: ';
```

```
str1 = input(prompt,'s');
ada=0;
if str1=='Y'| str1=='y'
  ada=1;
end
  while ada>0
    ada=0;
    str2=input('enter the number of parameter you want to change:');
    switch str2
     case 1
       str3=input('Enter a new value:');
       k=str3;
     case 2
       str3=input('Enter a new value:');
       k12=str3;
     case 3
       str3=input('Enter a new value:');
       k21=str3;
     case 4
       str3=input('Enter a new value:');
       Cdiet=str3;
     case 5
       str3=input('Enter a new value:');
       Cin=str3;
     case 6
       str3=input('Enter a new value:');
       Cout=str3;
     case 7
       str3=input('Enter a new value:');
       mtis=str3;
     otherwise
     disp('bad command')
  end
parameter(str2,i)=str3;
disp('New parameters:')
disp(strcat('1. k=',num2str(parameter(1,i))))
disp(strcat('2. k12=',num2str(parameter(2,i))))
disp(strcat('3. k21=',num2str(parameter(3,i))))
```

```
disp(strcat('4. mdiet=',num2str(parameter(4,i))))
disp(strcat('5. min=',num2str(parameter(5,i))))
disp(strcat('6. mout=',num2str(parameter(6,i))))
disp(strcat('7. mtis=',num2str(parameter(7,i))))
      str4=input(prompt,'s');
      if str4=='Y'| str4=='y'
        ada=1;
      end
   end
end
in = msgbox('Calculations, please wait','Info','warn');

% reading parameter values
k=parameter(1,1);
k12=parameter(2,1);
k21=parameter(3,1);
mdiet=parameter(4,1);
min=parameter(5,1);
mout=parameter(6,1);
mtis=parameter(7,1);

% initial values of cholesterol mass in compartment I - m0(1) and in
compartmet II - m0(2)
% intitial m0(1) and m0(2) are calculated on stationary solutions - it
means
% that in the first interval of time mdiet=0 (otherwise it means a
continuous meal)
% mdiet>0 during the time of absorption process

m0(1)=m1;
m0(2)=m2;
t0=0;
[t,m]=ode45(@equations, [t0, tmax(1)], m0);
    plot(t,m(:,2)/v2, 'linewidth',2,'color','r')
    hold on
    m2max=m0(2);
    axis([0, sum(tmax),40,(max(m2max)/v2)+50])
    xlabel('time [minute]','fontSize',20)
    ylabel('concentration [mg/dL]','fontSize',20)
    if n>1
       schift=0;
```

```
        for i=2:n
            m0=[m(end,1) m(end,2)];
            k=parameter(1,i);
            k12=parameter(2,i);
            k21=parameter(3,i);
            mdiet=parameter(4,i);
            min=parameter(5,i);
            mout=parameter(6,i);
            mtis=parameter(7,i);

            [t,m]=ode45(@equations, [t0, tmax(i)], m0);

            schift=schift+tmax(i-1);
            plot(t+schift,m(:,2)/v2, 'linewidth',2,'color','r')
            m2max=[m2max,m(end,2)];
            axis([0, sum(tmax),170,(max(m2max)/v2)+10])
    end
end
```

Index

Note: Page numbers followed by *f* indicate figures, *t* indicate tables and *np* indicate footnotes.

F

G

H

N

R

Y

Z

Printed in the United States
By Bookmasters